THE BIG PICTURE

HISTOLOGY

THE BIG PICTURE

HISTOLOGY

Rick Ash, PhD
Professor of Neurobiology and Anatomy
Director of Histology
University of Utah School of Medicine
Salt Lake City, Utah

David A. Morton, PhD
Associate Professor of Neurobiology and Anatomy
Director of Anatomy
University of Utah School of Medicine
Salt Lake City, Utah

Sheryl A. Scott, PhD
Professor of Neurobiology and Anatomy
Associate Chair Department of Neurobiology and Anatomy
Director of Graduate Studies Department of Neurobiology and Anatomy
University of Utah School of Medicine
Salt Lake City, Utah

New York   Chicago   San Francisco   Lisbon   London   Madrid   Mexico City
Milan   New Delhi   San Juan   Seoul   Singapore   Sydney   Toronto

The Big Picture: Histology

1 2 3 4 5 6 7 8 9 0   CTP/CTP   17 16 15 14 13 12

ISBN 978-0-07-147758-1
MHID 0-07-147758-6

This book was set in Minion by Aptara, Inc.
The editors were Michael Weitz, Brian Kearns, and Susan Kelly.
The production supervisor was Catherine Saggese.
The illustration manager was Armen Ovsepyan and the illustrator was Mathew C. Chansky
Project management was provided by Ruchira Gupta, Aptara, Inc.
The designer was Alan Barnett; the cover designers were Alan Barnett and Libby Pisacreta.
China Translation & Printing Services, Ltd., was printer and binder.

Library of Congress Cataloging-in-Publication Data
Ash, Rick.
 The big picture : histology / Rick Ash, David A. Morton, Sheryl A.
Scott.
      p. ; cm.
 Histology
 Includes index.
 ISBN 978-0-07-147758-1 (softcover : alk. paper) — ISBN 0-07-147758-6
(softcover : alk. paper) 1. Histology—Examinations, questions, etc.
I. Morton, David A. II. Scott, Sheryl A. III. Title. IV. Title:
Histology.
 [DNLM: 1. Histology—Examination Questions.   QS 518.2]
 QM554.A84 2012
 611.0076—dc23
                                        2012013575

McGraw-Hill books are available at special quantity discounts to use as premiums and sales promotions, or for use in corporate training programs. To contact a representative please e-mail us at bulksales@mcgraw-hill.com.

DEDICATION

To my wife, Kaethe, for her endless patience and encouragement, and to our daughter, Tanya, son-in-law, Andy, and grandchildren, Max and Sam: all of you are responsible for the joy and fun. It's time for another river trip!

—*Rick Ash*

To my brothers Gordon, Joe, and Mike, and my sister Daniela, as well as my in-laws Rachel, Caryn, Annette, Jon, Andre, Cali, Alain, Michelle, David, Michelle, Caroline, Aaron, Philippe, Amanda, Michael, Kylie, Katrine, Nathan, and Coryse.

—*David A. Morton*

To my family for their love, support, and inspiration: my husband, Dick; our children Rachel, Chris, and Kate; and our grandchildren, Nick and Ryan.

—*Sheryl A. Scott*

CONTENTS

PREFACE

The Big Picture: Histology was developed from our collective years of experience teaching human histology to first-year medical and dental students at the University of Utah School of Medicine. Our goal in the histology course always has been to try to provide students with an efficient way of learning, and we now have applied that same approach to this book.

Histology is a science that is important because it extends the structural description of the body to the microscopic level, and provides a physical context for other basic sciences such as physiology and biochemistry. The study of human histology helps to prepare students for their pathology course, in which familiarity with normal tissue architecture is required to recognize the structural changes produced by disease and then to understand and predict the functional consequences of these changes. For example, a detailed knowledge of the construction of the walls of the stomach and duodenum is essential to understand the different roles these organs play in the normal process of digestion and absorption of food. This knowledge also helps to explain why bleeding can be a common and serious consequence if ulcers form in either the stomach or the duodenum.

Through years of teaching, we generated a set of notes that provided our students with a "big picture" overview of the course material, and along the way, we continued to supplement these notes with the results of current research. Later, micrographs were added to some of the material that was developed into handouts for our laboratory sessions. Although we told our students that our notes were intended to be only supplemental and that they would be much better off to ignore them and to read the assigned textbooks thoroughly, they rarely did. The students complained that the textbooks faculty assigned generally contained too much information and detail, and most students simply did not have the time to read through large textbooks and cull out what they needed. Therefore, for the histology course, they relied on our notes, which they said were clear and concise. These notes were the starting point for this book.

Our goal in this book was to provide text, illustrations, and micrographs that are complete, yet concise, to present the "big picture" of human histology, and to this point, we feel we have succeeded. To illustrate and emphasize key points, we have included detail, often clinical in nature, to better elaborate the correlation between the structural and functional applications of this science. The format of the book is simple. Each page spread consists of text on the left-hand page with associated illustrations on the right-hand page. In this way, students can grasp the big picture of individual principles in bite-sized pieces, one concept at a time.

- Key structures are highlighted in bold when first mentioned.
- Bullets and numbers are used to simplify important concepts.
- More than 400 full-color illustrations and micrographs depict the essential histology. Except as noted, all histological sections were stained with hematoxylin and eosin (H&E) and were photographed by the authors.
- Icons indicate high-yield, clinically relevant concepts throughout the text.
- Study questions and answers follow each chapter.
- A final examination is provided at the end of the text.

We hope you enjoy reading this text as much as we enjoyed writing it!

—*Rick Ash*

—*David A. Morton*

—*Sheryl A. Scott*

ACKNOWLEDGMENTS

We are deeply grateful to a number of people who played a significant role in the development of this book. We express a warm thank you to Monica Vetter, our Department Chair, for her unwavering support of education. We thank the excellent team assembled by McGraw-Hill Medical Publishing to produce this textbook, including Michael Weitz, for his dedication, assistance, vision, leadership, and friendship through the completion of this title; Susan Kelly, for her eagle eye as an editor and her words of encouragement and wisdom during the past few years on this project; Brian Kearns, Karen Davis, and John Williams for their support; and Matt Chansky for his care and attention in creating the images for this title. We thank our families for their infinite patience and unwavering support, without which this book would never have been possible. We owe special thanks to Dr. Walter Stevens, Jr. (December 6, 1933–July 15, 2004), who created the medical histology course at the University of Utah School of Medicine that brought the three of us, as coauthors, together to form the basis of this textbook. This project was a true collaboration.

Aspen Shadows, Boulder Mountain, Utah

Mouth of Indian Creek, Glen Canyon National Recreation Area, Utah

ABOUT THE AUTHORS

Rick Ash, PhD, received a bachelor's degree in psychology from the University of Illinois, a PhD in biological sciences from Stanford University, and completed postdoctoral training at the University of California at San Diego. He joined the Neurobiology and Anatomy Department at the University of Utah School of Medicine in 1980, and has directed research on membrane biology for the past three decades. Dr. Ash currently serves as Histology Director. He was awarded the University Distinguished Teaching Award.

David A. Morton, PhD, received a bachelor's degree in zoology from Brigham Young University in Provo, Utah, and completed his graduate degrees in human anatomy at the University of Utah School of Medicine. He currently serves as Anatomy Director. He was awarded the Early Career Teaching Award, the Preclinical Teaching Award, and the Certificate of Recognition for distinguished and exemplary service to students with disabilities. Dr. Morton is an adjunct professor in the Physical Therapy Department and the Department of Family and Preventive Medicine. He also serves as a visiting professor at Kwame Nukwame University of Science and Technology in Kumasi, Ghana.

Sheryl A. Scott, PhD, received a bachelor's degree in zoology from Duke University, a PhD in biology (neurobiology) from Yale University, and pursued postdoctoral training at McMaster University Medical Center in Hamilton, Ontario, and at Carnegie Institution of Washington in Baltimore, Maryland. She has directed a research program funded by the National Institutes of Health studying neural development for nearly 30 years, first at Stony Brook University in Stony Brook, New York, and more recently at the University of Utah School of Medicine. Dr. Scott is currently Director of Graduate Studies and Associate Chair of the Department of Neurobiology and Anatomy. She teaches medical histology along with Drs. Ash and Morton.

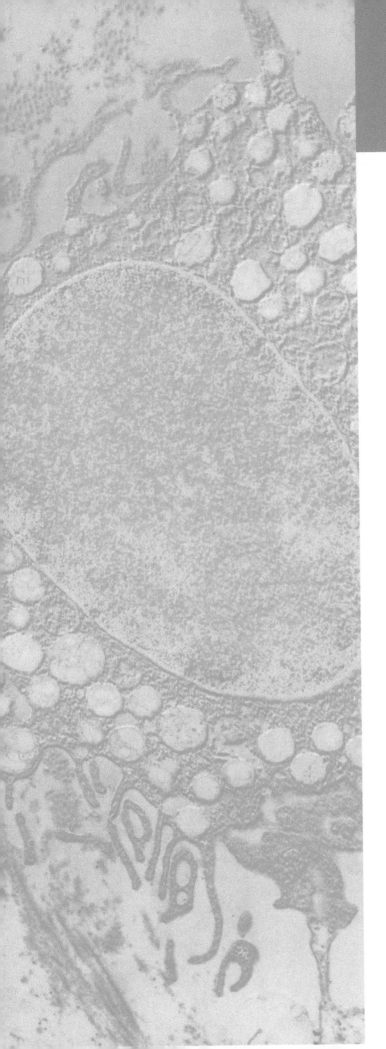

EPITHELIUM

OVERVIEW

Epithelial cells are found in many organs, including skin, intestines, liver, and pancreas, where they perform a wide variety of activities. Learning to identify and classify examples of this diverse tissue can be an interesting challenge because it requires that you consider a set of interrelated features that include the locations, organization, and functions of epithelial cells.

It is useful to begin by recognizing that continuous sheets of epithelial cells cover many of the body's free surfaces (e.g., skin and intestines). Tight connections between epithelial cells at these surfaces are necessary to keep the epithelial sheets physically intact. The continuity is an important feature of the epithelia lining surfaces connected to the external environment because it helps prevent pathogens from entering the body. However, the epithelium lining the intestines must also mediate the transport of dietary nutrients into the body. These dual functions of protection and transport, along with the structural features that provide tight connections between cells, are shared by many epithelia.

Epithelia are also the key elements of many glands, such as the liver and pancreas, which produce and export numerous products onto free surfaces or into the blood. The free surfaces in some of these secretory organs may be insignificant, but the characteristic close connections among the epithelial cells are maintained.

A specialized set of epithelial cells provides additional functions in the body. Epithelial cells serve as sensory receptors for the nervous system in the eye, ear, and mouth. Other epithelial cells serve as small muscles, facilitating the expulsion of glandular secretions or dilating the iris of the eye.

CHARACTERISTICS OF EPITHELIA

The human body contains hundreds of different kinds of epithelial cells, making it difficult to provide a concise and complete characterization of epithelial tissue. It is helpful to note that many epithelia cover a surface. This location requires these cells to exhibit a set of structural features to maintain cohesion and include cell–cell junctions, cell–substrate junctions, and intracellular filament systems. Many of these features are common to all epithelial cells. This chapter begins by considering the common features of epithelial cells, briefly discusses replacement and embryology of these cells, and then outlines the scheme used to name different types of epithelium and provides examples of each type. Finally, some of the major epithelial functions are considered.

LOCATION OF EPITHELIA

Epithelia are found in two primary locations (Figure 1-1A–C):

- Covering the external or internal surface of an open space (e.g., the outer surface of the skin or the inside lining of the digestive system).
- As the parenchyma (functional tissue cells) of internal organs (e.g., the kidney or liver).

ORGANIZATION OF EPITHELIA

There is only a small amount of extracellular material between epithelial cells. The cells may be arranged in one or more layers, and often they may be arranged as continuous sheets when the epithelium covers a surface. Most epithelial cells are held together in close contact by cell–cell junctions called **desmosomes**, and they may communicate with each other via **gap junctions** (Figure 1-1D and E).

- **Desmosomes.** Multimeric structures that provide an intracellular scaffold for transmembrane **cadherin** proteins whose extracellular domains link small portions of adjacent cells together via Ca^{2+}-dependent adhesions. The cytoplasmic faces of desmosomes bind **intermediate filaments** of the **keratin** family.
- **Gap junctions.** Coordinate cellular activities in epithelium. The cytoplasm of adjacent cells is frequently connected by gap junctions, which permit the rapid spread of signaling molecules (e.g., cyclic adenosine monophosphate, or cAMP) among cells.

ASSOCIATION WITH BASAL LAMINA

In any epithelium, some cells are in contact with a specialized extracellular matrix, called a **basement membrane** when seen in light micrographs, or a **basal lamina** when seen in electron micrographs (Figure 1-1D and F). The two terms are often used interchangeably; in this text, however, we will use basal lamina when referring to the layer without specific reference to a microscopic technique. Loose connective tissue is typically found deep to the basal lamina.

Epithelial cells are attached to a basal lamina via adhesive membrane proteins of the **integrin** family.

- **Hemidesmosomes.** Some integrins are associated with hemidesmosomes, which, like desmosomes, are attached to intermediate filaments in the cytoplasm.
- **Focal adhesions.** Other integrins are associated with structures called focal adhesions, which are linked to cytoplasmic actin filaments.

The portion of an epithelial cell closest to the basal lamina is termed **basal**; the **apical** portion is opposite and faces toward the free surface, if there is one.

The basal lamina consists of a complex mixture of macromolecules, including the proteins laminin and type IV collagen, and glycosaminoglycans (GAGs). Strong adhesive interactions between the integrin molecules of the epithelial cells and extracellular proteins (e.g., laminin) link the cells to the basal lamina. Components of the basal lamina also bind a variety of hormones and other signaling molecules that provide information to the epithelium. Normal epithelial cells require contact with a basal lamina to establish basal-apical polarity and to divide.

KERATIN FILAMENTS

Epithelial cells contain many intermediate filaments (10-nm diameter) of the keratin family. Keratin filaments link to desmosomes and hemidesmosomes, a network that provides physical support to epithelial cells. Humans express more than 50 keratin genes in a cell-specific fashion.

▽ **Epidermolysis bullosa simplex** is a genetic disease that causes the skin to blister in response to pressures such as those produced by mild scratching or shifting position in a chair. The disease is associated with mutations in genes coding for keratin proteins K5 and K14. Defects in filaments assembled from these mutant proteins render basal epidermal cells hypersensitive to mechanical damage, illustrating the important structural role that keratins play. ▽

OTHER COMPONENTS AND NUTRITION

Epithelia may contain sensory nerve endings and immune system cells; however, epithelia rarely contain blood vessels, which are found in connective tissue deep to the basal lamina. Thus, exchange between epithelial cells and blood occurs across a basal lamina.

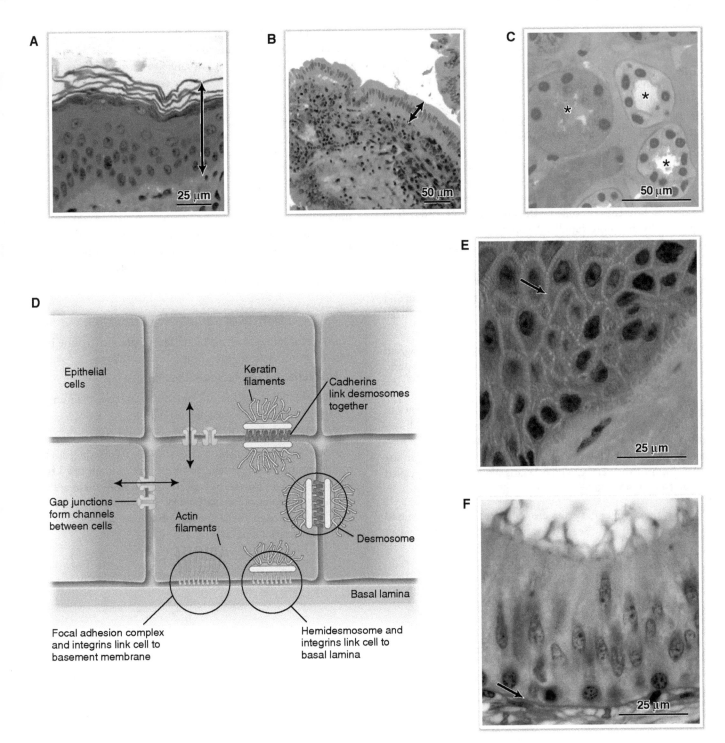

Figure 1-1: Characteristic locations of epithelia within the body. **A.** Section of the surface of the skin, which consists of layers of epithelial cells that provide covering and protection (*arrow* indicates the height of epithelium). **B.** Section of the small intestine (jejunum), lined with a single layer (*arrow*) of epithelium specialized for absorption and secretion. **C.** Section of a kidney showing the parenchyma, which contains several types of epithelial-lined tubules (*) involved in urine formation. **D.** Junctions involved in organizing epithelia. Desmosomes contain adhesive cadherin proteins that make strong connections between adjacent epithelial cells. Hemidesmosomes and focal adhesion complexes contain integrins that link epithelia to the basal lamina. Keratin filaments bind to the cytoplasmic face of desmosomes and hemidesmosomes, whereas focal adhesions are associated with the actin cytoskeleton. Gap junctions directly connect the cytoplasms of neighboring cells, coordinating the physiology of an epithelium. **E.** Desmosomes link skin epithelial cells together. Slight cell shrinkage causes these attachments to appear as spines between the cells (*arrow*). **F.** Section of the epididymis. In most tissues, the basement membrane is too thin to be distinguished with the light microscope; the periodic acid–Schiff (PAS) reaction shown has stained the carbohydrate components of the basement membrane red (*arrow*). (PAS stain)

REPLACEMENT OF EPITHELIAL CELLS

Epithelial cells have a finite life span, from a few days to months. Dead cells are replaced by the mitotic division of **progenitor cells**, which are most commonly adult **stem cells** found in the epithelium. For example, stem cells are responsible for epithelial replacement in the skin and intestines. In the liver, however, parenchymal cells can serve as progenitors and true stem cells are either rare or absent.

Epithelial stem cells are undifferentiated cells that maintain contact with the basal lamina and generally divide slowly. Stem cells divide to produce more stem cells, but some mitoses produce a different cell, called a **transit amplifying cell**, which typically divides more rapidly. A cohort of transit-amplifying cells will eventually differentiate into parenchymal cells (Figure 1-2A).

- The rate at which stem cells divide, the frequency at which they produce transit-amplifying cells, and the expansion and differentiation of transit-amplifying cells are all influenced by extracellular signaling factors.

- This allows an epithelium to adjust its replacement program from that required for replacing normal cell loss to respond to an increased rate after damage by injury or disease.

▽ A variety of stimuli can produce observable changes in tissues, which usually result from a change in the differentiation of progenitor cells. If the changes are benign, the process is called **metaplasia**; for example, the normal change in the epithelium of the cervix, which begins after puberty, is called **squamous metaplasia** (Figure 1-2B). ▽

 If the changes in tissues are associated with an increased probability of disease development, the process is called **dysplasia**. For example, if epithelial cells in the cervix become infected with the human papillomavirus, the cells often exhibit dysplastic changes that can progress to cervical cancer. In general, dysplasia and metaplasia are reversible. ▽

▽ Epithelial tissues give rise to almost all tumors in adults. **Cancer** development is a multistep process whereby cells initially gain the ability to proliferate abnormally. In the case of epithelial tumors, called **carcinomas**, malignancy is usually defined at the point that the tumor cells gain the ability to destroy the basal lamina and spread to other organs. Many epithelial tumors derive from glands and are then called **adenocarcinomas**.

Carcinoma cells often maintain the expression of the specific keratin genes that were active in the normal epithelium from which they arose. For this reason, keratin-specific antibodies are key reagents for identifying and typing carcinomas. ▽

EMBRYONIC ORIGINS OF EPITHELIUM

Many embryology and histology texts state that epithelia are derived from all three embryonic layers, as follows:

- **Ectoderm** (e.g., epidermis of the skin).
- **Mesoderm** (e.g., cells lining blood vessels [endothelium] and serosal membranes [mesothelium]).
- **Endoderm** (e.g., epithelium of the digestive tract).

However, pathologists do not consider endothelium and mesothelium to be epithelium. These cells contain few keratin filaments and, importantly, cancers arising from them are classified by pathologists as **sarcomas**, not carcinomas, because their appearance and behavior resemble those of tumors derived from connective tissue cells.

A

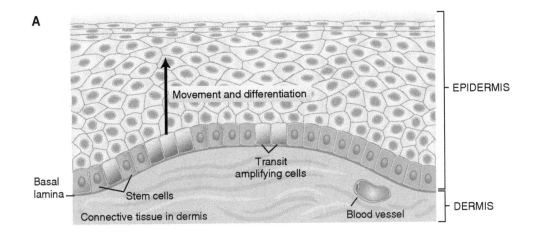

EPIDERMIS

Movement and differentiation

Transit amplifying cells

Basal lamina

Stem cells

Connective tissue in dermis

Blood vessel

DERMIS

B

Mature stratified squamous epithelium

Metaplastic epithelium

100 μm

Figure 1-2: Epithelial cell replacement. **A.** Stem cells divide to replace epithelial cells. The schematic shows a stratified epithelium in which the differentiated cells exit the basal lamina to occupy higher layers, as occurs in skin or in the cervix before puberty. The rate of the various cell divisions and the fates of daughter cells are controlled by signaling factors. This allows the tissue to respond to changing conditions and adjust the number and types of cells produced. **B.** Section of the surface of the cervix showing metaplasia. The stratified epithelium to the left of the arrow eventually replaces the larger epithelial cells on the right (which appear somewhat disorganized in the area where the conversion is underway) and becomes a single layer of columnar cells. This is a normal process that begins at puberty and continues for years, transforming the entire lining of the cervix.

CLASSIFICATION OF EPITHELIA

Epithelia are classified by three major categories, according to cell shape and layering, with important surface features sometimes added as part of the nomenclature. The epidermis of the skin, for example, consists of a multilayer of epithelial cells with thin cells at the surface coated by an extracellular layer of keratin. This is called a stratified squamous keratinized epithelium. The criteria used for classification are precise and easy to diagram; however, they were developed over many years by analysis of thousands of sections and are not always simple to apply. Without the experience of having examined many examples, it is difficult to identify an epithelium on a given section.

CELL SHAPE

There are three epithelial cell shapes: squamous (or low), cuboidal, and columnar. Because the cell boundaries are difficult to distinguish with the light microscope, the key to detecting the cell shape is often the shape and orientation of the nuclei (Figure 1-3A).

- **Squamous (low) cells.** Flat or plate-like cells whose long axis is parallel to the basal lamina. The nuclei are often thin.
- **Cuboidal cells.** Taller than squamous cells and are about as wide as they are high. The nuclei are often round and central.
- **Columnar cells.** Possess a long axis that is perpendicular to the basal lamina. The nuclei may be round or oval.

CELL LAYERS

There are three types of cell layers: simple, stratified, and pseudostratified. The shape of the cells at the free surface becomes part of the nomenclature (Figure 1-3B).

- **Simple epithelia.** Composed of a single layer of cells, all of which contact the basal lamina. If the cells are cuboidal, the epithelium is called simple cuboidal.
- **Stratified epithelia.** Composed of multiple layers of cells. Cells in the upper layers do not have contact with the basal lamina. If the cells at the free surface are squamous, the epithelium is called stratified squamous, regardless of the cell shape in the lower layers.

- **Pseudostratified epithelia.** Appears stratified because the cells' nuclei are at several levels. However, it is a simple epithelium because all cells contact the basal lamina. Learning to recognize this type of epithelium is aided by using clues provided from adjacent tissues because it is usually not possible to determine if all cells extend contacts to the basal lamina (Figure 1-3C).

APICAL SPECIALIZATIONS

The cells at the apical surface may have significant features, including microvilli, stereocilia, cilia, or extracellular keratin. When present, such features may become part of the nomenclature, including "stratified squamous keratinized epithelium" for skin epithelium.

- **Microvilli.** Short, finger-like membrane projections from the apical epithelial surface, which are supported internally by microfilaments. Microvilli increase the surface area and thus the absorptive capacity of epithelial cells increases. They are especially prominent on cells lining the intestinal tract, where they are called the brush border (Figure 1-3D).
- **Stereocilia.** Long projections from the apical cell membrane, which are structurally similar to microvilli. They are found on sensory cells in the ear and lining the epididymis and ductus deferens, where they function in absorption and maturation of sperm. Stereocilia are characterized by their extreme length, which distinguishes them from microvilli, and their lack of motility, which distinguishes them from cilia.
- **Cilia.** Long, motile, apical membrane projections originating from basal bodies in the cytoplasm. Cilia contain microtubules and an associated protein, called dynein, which has adenosine triphosphatase (ATPase) activity. Interactions between microtubules and dynein produce movements that bend cilia. Ciliary movements propel the ovum in the uterine tube from the ovary to the uterus and move mucus and dirt out of the upper respiratory tract (Figure 1-3C).
- **Keratin.** An intermediate filament found in the cytoplasm of all epithelial cells. In skin, epithelial cells near the surface disintegrate and deposit a layer of keratin.

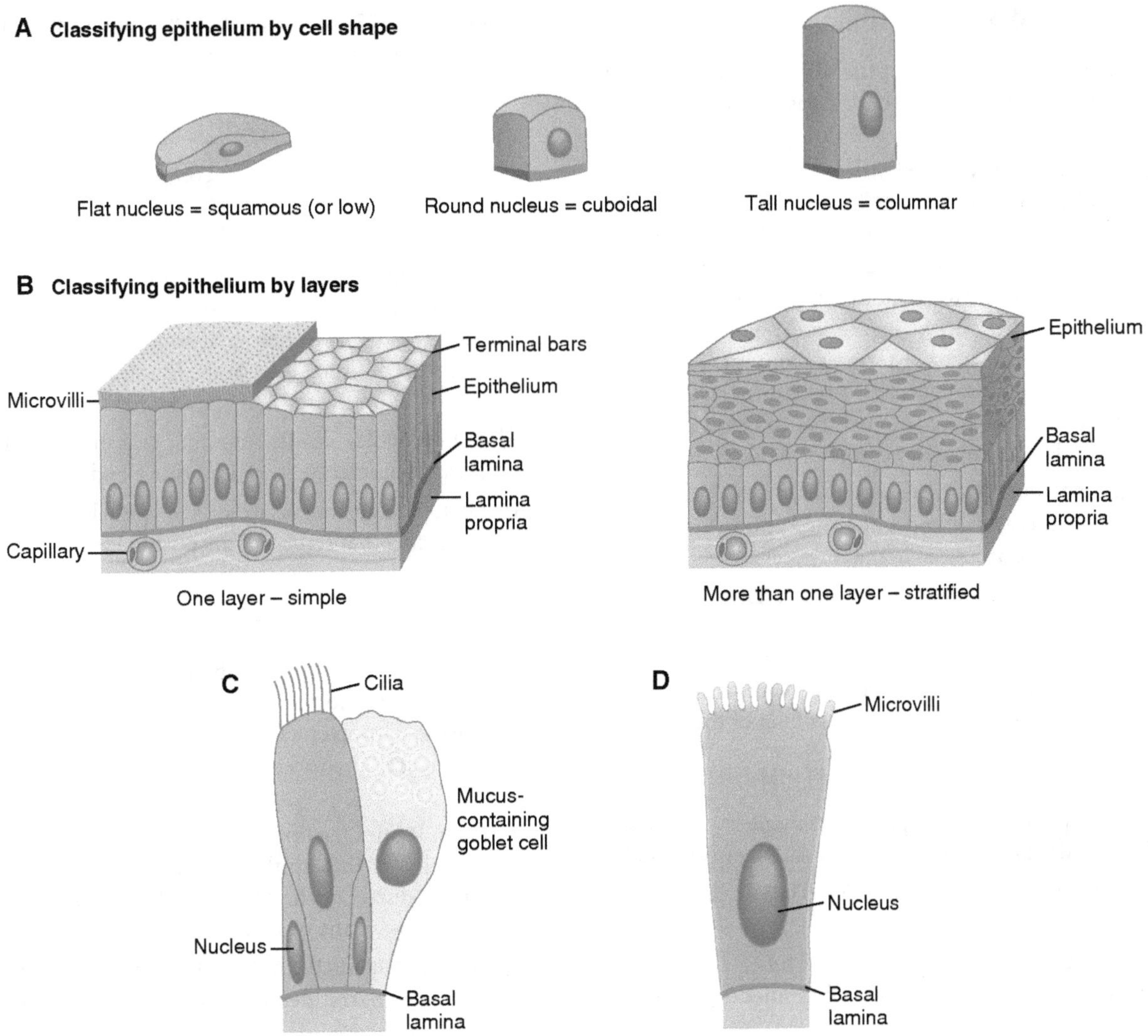

A Classifying epithelium by cell shape

Flat nucleus = squamous (or low)   Round nucleus = cuboidal   Tall nucleus = columnar

B Classifying epithelium by layers

Microvilli

Terminal bars

Epithelium

Basal lamina

Lamina propria

Capillary

One layer – simple

Epithelium

Basal lamina

Lamina propria

More than one layer – stratified

C

Cilia

Mucus-containing goblet cell

Nucleus

Basal lamina

D

Microvilli

Nucleus

Basal lamina

Figure 1-3: Important features of epithelia: cell shape, layering, and surface specializations. **A.** Three epithelial cell shapes: squamous (or low), cuboidal, and columnar. Nuclei often provide the best indication of cell structure. Nuclei of the thin squamous cells tend to be elongated and are parallel to the surface when seen in cross-section. The nuclei of cuboidal cells are often round and centrally located. The nuclei of columnar cells are usually elongated and aligned with the long axis of these cells. **B.** Cell layering is determined by cell association with the basal lamina. Simple epithelia consist of a single layer of cells in contact with the basal lamina, whereas stratified epithelia have at least one layer lacking such contact. **C.** All cells in pseudostratified epithelial, such as those lining large airways, are in contact with the basal lamina, but the nuclei are often found at different levels, producing a stratified appearance. Cilia on cells lining airways move mucus produced by the epithelium along the apical surface. **D.** Microvilli on intestinal cells help to increase the surface area for absorption.

SIMPLE EPITHELIA

Both the progenitor cells and all of the differentiated cells in a simple epithelium are in contact with the basal lamina.

SIMPLE LOW EPITHELIUM

Simple low epithelium is composed of a single layer of flat cells, all of which contact the basal lamina. Many histologists cite endothelial and mesothelial cells as examples of simple low, or simple *squamous*, epithelia; recall, however, that pathologists do not consider these tissues to be epithelia. In addition, many pathologists prefer to limit the term "squamous," meaning scale-like, to describe stratified tissues where the cells at the surface layers are thin; we have adopted that convention in this book. Two examples of simple low epithelia include:

- **Glomerular capsule** (parietal cell layer of Bowman's capsule). These cells create a fluid-impermeable barrier to contain the initial urine produced in the corpuscle (Figure 1-4A).
- **Alveoli** (type 1 pneumocytes in the lung). These thin cells promote the exchange of gases between inspired air and blood and are difficult to see because they are so thin. They can be easily confused with the endothelial cells of capillaries, which lie immediately across the basal lamina of the alveoli.

SIMPLE CUBOIDAL EPITHELIUM

Simple cuboidal epithelium is composed of a single layer of cube-shaped cells, all of which contact the basal lamina. Examples include:

- **Hepatocytes** (liver parenchymal cells). These cells produce many blood proteins.
- **Glandular ducts.** The early portions of tubes that deliver secretory products to a surface are often lined by a simple cuboidal epithelium.

- **Kidney tubules.** The cells in these tubules are involved in urine formation (Figure 1-4B).

SIMPLE COLUMNAR EPITHELIUM

Simple columnar epithelium is composed of a single layer of tall, thin cells, all of which contact the basal lamina. Examples include:

- **Mucosal epithelium.** The cells lining the mucosa of the stomach and intestines are involved both in the absorption of nutrients and the secretion of mucus and other products. Absorptive columnar cells in the intestine have a dense array of microvilli on their apical surface (Figure 1-4C).
- **Glandular ducts.** Simple columnar cells often line the terminal, large diameter ducts of exocrine glands.
- **Oviduct.** The apical surface of the epithelium of this organ contains cilia that serve to transport the ovum to the uterus.

PSEUDOSTRATIFIED COLUMNAR EPITHELIUM

Pseudostratified columnar epithelium is composed of a single layer of tall and short cells, giving the false appearance of stratification. However, all cells of pseudostratified columnar epithelium are anchored to the basal lamina. Examples include:

- **Trachea.** It lines the upper respiratory track and the trachea. Many cells in this epithelium are ciliated and transport mucus produced by the **goblet cells** (also present in the epithelium) toward the oral cavity (Figure 1-4D).
- **Epididymis and ductus deferens.** These are all simple epithelia that appear stratified because of the uneven distribution of their nuclei.

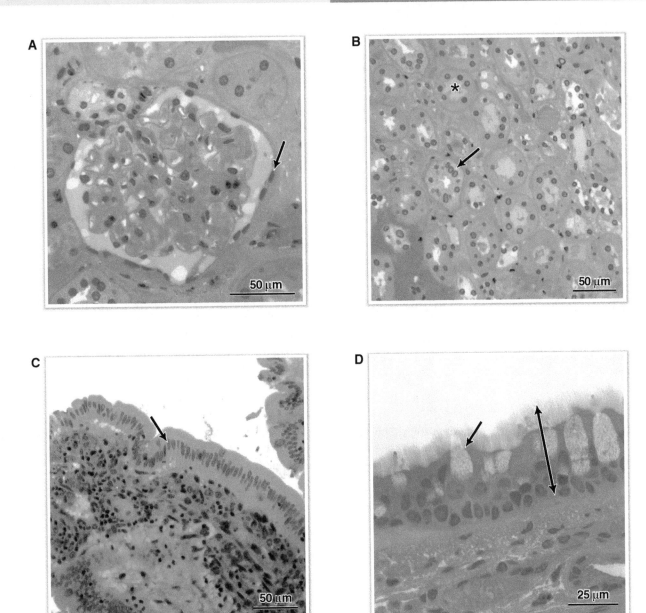

Figure 1-4: Examples of simple epithelia. **A.** Section of the kidney containing simple low parietal cells (*arrows*) that line renal corpuscles. These cells form a barrier that prevents urine from escaping into the surrounding tissue. **B.** Section of the kidney showing simple cuboidal cells (*) that line some of the tubules in the kidney medulla, which conduct and process urine. Other tubules are lined by columnar cells (*arrow*). **C.** Section of the jejunum containing simple columnar epithelial cells that line the surface of this region of the small intestine; the arrow indicates the row of nuclei in the columnar cells. The apical portion of these cells is covered with microvilli. **D.** Section of the trachea showing the pseudostratified ciliated columnar epithelium that lines this airway. The vertical arrow indicates the height of the epithelium. This epithelium appears to be stratified, instead of simple, because of the staggered array of nuclei. The small arrow indicates a mucus-secreting cell. Many of the cells bear long, thin cilia on their surface.

STRATIFIED EPITHELIA

The progenitor cells in a stratified epithelium remain in contact with the basal lamina. Some cells lose this contact and move toward the apical surface as they differentiate. Stratified epithelia rely on desmosomes to keep these multilayers intact.

STRATIFIED SQUAMOUS EPITHELIUM

Stratified squamous epithelium is composed of multiple layers of cells; the apical cell layer is squamous shaped. The basal layer of cells may be cuboidal or columnar epithelium. Examples include:

- **Skin** (epidermis). Exposed to the external environment for protection.
- **Esophagus** (mucosa). Protects esophagus from mechanical stress (Figure 1-5A).

In skin and parts of the oral cavity, the surface layers consist only of keratin, which remains after the squamous apical cells die; the epithelium then is called stratified squamous keratinized epithelium. The surface keratin is absent in the esophagus and vagina.

STRATIFIED CUBOIDAL EPITHELIUM

Stratified cuboidal epithelium is an uncommon epithelium that lines portions of the ducts of some glands, including the sweat and salivary glands (Figure 1-5B).

STRATIFIED COLUMNAR EPITHELIUM

Stratified columnar epithelium also is an uncommon epithelium and is found on the conjunctiva of the eye. It lines portions of the large excretory ducts of the salivary glands and the pancreas (Figure 1-5C). There are times when distinguishing the difference between stratified columnar epithelium and stratified cuboidal epithelium requires an artistic rather than a scientific judgment.

UROTHELIUM

Urothelium consists of a multilayer of cells that are specialized to prevent leakage of urine back into tissues while being able to stretch as the organs that are covered by the epithelium expand.

- **Relaxed state.** The cells at the free surface are often dome-shaped and normally may be binucleated.
- **Stretched state.** When the epithelium is stretched, these apical cells can become low, although this epithelium is never referred to as squamous.

Urothelium was originally called **transitional epithelium** because it was considered a transition between the stratified squamous epithelium at the end of the urethra and the columnar epithelium found in the kidney; however, it now is recognized as a specific type of epithelia. Urothelium is found only in the urinary system, where it lines portions of the kidney, ureter, and urinary bladder (Figure 1-5D), and also is found in the initial portion of the urethra.

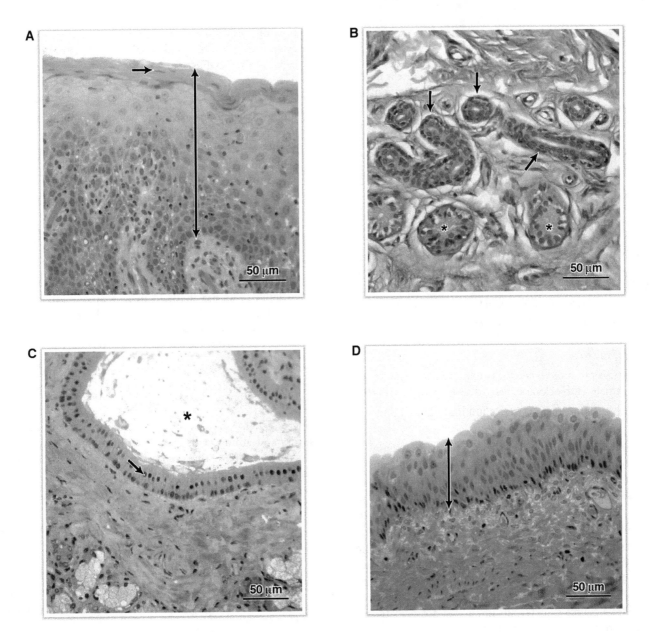

Figure 1-5: Examples of stratified epithelia. **A.** Section of the vagina showing the stratified squamous nonkeratinized epithelium that covers the surface. The vertical arrow indicates the height of the epithelium. The squamous nature of the cells near the surface can be deduced by their elongated nuclei (*small arrow*). **B.** Section of skin showing stratified cuboidal epithelia that lines the sweat gland ducts. Several duct profiles (*arrows*) are cut in different orientations and show two distinct cell layers. The sweat glands (*) are seen below the ducts. **C.** Section of a sublingual salivary gland showing stratified columnar epithelium that lines an interlobular duct. Tall columnar cells (*arrow*) are seen at the lumen (*) of the duct. **D.** Section of the urinary bladder showing urothelium that covers the luminal surface of the organ. The arrow indicates the height of the epithelium.

FUNCTIONS OF EPITHELIA

Epithelia commonly cover surfaces such as skin and the lining of the intestines and respiratory system. Epithelial cells in these locations must provide the body with several functions, including protection, transport, and secretion. Of course, cells in other tissues can perform these functions, but they are prominent enough in epithelia to warrant description in this chapter. Also mentioned briefly in this chapter are the specialized epithelial functions of sensory transduction and contraction.

PROTECTION

Barrier epithelia are found in several locations, including the skin, gut, urinary tract, and respiratory system. They cover the free body surfaces that are directly exposed to the environment and serve as the **first line of defense against invading pathogens**. The basic form of protection that barrier epithelia provide is to block access to underlying tissue. For instance,

- The tight connections between epithelial cells in the skin allow it to serve as a physical impediment that prevents the entry of microbes.

- The mucus produced by epithelia that cover the airways traps inhaled material, blocking access to deeper portions of the lung.

Epithelial cells also provide **chemical protection**. For instance:

- **Antibody transport.** Some epithelia, such as those lining the gut and the secretory cells of salivary glands, can transport antibodies from the extracellular space to the lumen by transcytosis, which links the processes of endocytosis to exocytosis to move large molecules across a cell sheet. The antibodies can then react with pathogens in the lumen (Figure 1-6A).

- **Defensins.** Many kinds of cells can synthesize **defensins**, which are small cationic peptide antibiotics that can kill bacteria. Barrier epithelia are rich sources of defensins. A specialized set of intestinal columnar cells, the Paneth cells, provides a source of defensins in an area normally teeming with bacteria (Figure 1-6A and B).

TRANSCELLULAR TRANSPORT

The active uptake of solutes (e.g., sugars and amino acids) is a critical activity of the simple columnar absorptive cells of the intestines. The structures and processes involved in this activity are used in other epithelial and nonepithelial cells for various purposes.

JUNCTIONAL COMPLEXES The efficient transport of solutes from the lumen of the gut toward the bloodstream requires that the epithelial sheet prevents diffusion between cells back to the gut. This is accomplished by the formation of membrane complexes between cells near their apical surfaces (Figure 1-6C).

- **Tight junctions**, or **zonula occludens**, are regions of close apposition of the plasma membranes of neighboring cells, which prevent all diffusion between cells or limit diffusion to specific molecules. This junction completely encircles the tops of epithelial cells involved in transcellular transport, establishing luminal and basal compartments outside the cells. It also establishes two membrane domains: an apical domain above and a basolateral domain below the zona occludens. Membrane proteins differ in the two domains.

- **Adherens junctions** are found below the tight junctions and in transporting epithelial cells. These junctions are called **zonula adherens** because they also completely encircle the cell. Although zonula adherens are not the same as desmosomes, they function in a similar manner and support tight junctions by holding adjacent cells together. Zonula adherens serve as attachment sites for actin-based microfilament bundles, called the **terminal web**, which extend up into the apical microvilli. **Cadherin** proteins are extracellular components of these adherens junctions and provide the linkage between cells.

- **Terminal bars** are the light microscopic manifestations of junctional complexes (zonula occludens plus zonula adherens) that can be seen with the electron microscope (Figure 1-6D).

MICROVILLI The surface area of the apical membrane above the junctional complexes is vastly expanded by a dense array of microvilli, called the **brush border**.

PUMPS AND CARRIERS Energy is required to actively accumulate sugars and amino acids from the gut lumen. This is accomplished by coupling three types of membrane transporters: ion pumps, cotransporters, and simple carriers.

- The **sodium pump, or Na,K-ATPase**, is a membrane enzyme that uses energy generated from the hydrolysis of one molecule of ATP to move 3 Na^+ out of the cell and 2 K^+ into the cell. Pump activity directly makes the inside of cells electrically negative, which is increased by the additional leakage of K^+ out through ion channels. The result is a steep, inwardly directed Na^+ electrochemical gradient. On absorptive cells of the gut, the sodium pump is found below the tight junctions (Figure 1-6C).

- **Sodium-dependent carriers, or cotransporters, in the apical membranes of the microvilli** use the Na^+ gradient to energize the uptake of desired solutes (Figure 1-6C). For example, Na^+-dependent amino acid cotransporters couple Na^+ entry to amino acid uptake. This allows cells to accumulate high concentrations of amino acids from the gut lumen. The Na^+ that enters along with amino acids and other cotransported solutes is pumped out by the sodium pump.

- **Sodium-independent carriers in the basolateral membranes** facilitate the diffusion of the high concentration of solutes out of the cells (Figure 1-6C). These solutes can diffuse across the basal lamina, enter capillaries in the connective tissue below the basal lamina, and be distributed to other areas of the body via the bloodstream. Tight junctions between the epithelial cells prevent leakage of the transported solutes back to the gut lumen.

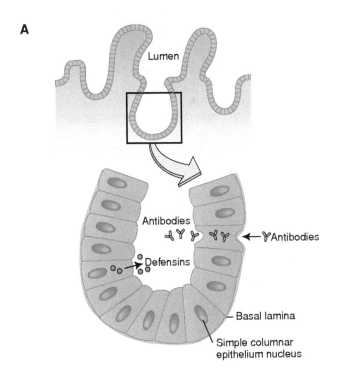

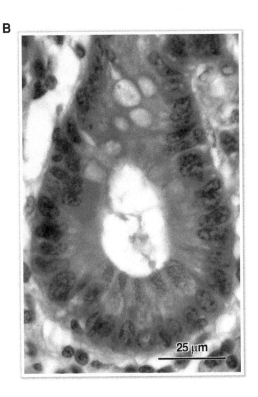

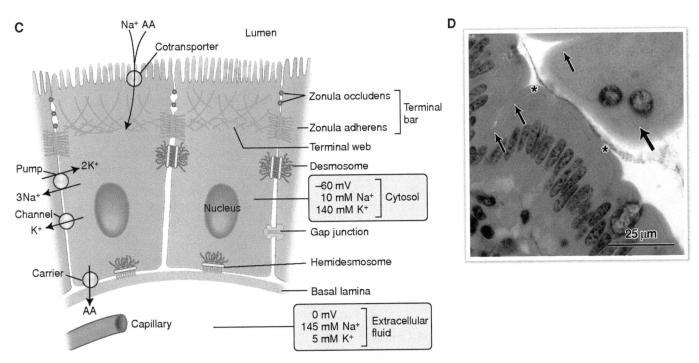

Figure 1-6: Epithelial functions include protection and transport. **A.** Protective functions provided by intestinal epithelium. Membrane junctions linking cells provide a physical barrier to invasion by microbes. Many cells of the intestinal tract are able to transport protective antibodies into the lumen of the gut by transcytosis. Paneth cells produce and export antibiotic peptides (defensins). **B.** Paneth cells in the jejunum. These cells contain vesicles that stain bright orange and are filled with antibiotic peptides. **C.** Features of transporting epithelia. Microvilli of the brush border contain Na^+-dependent cotransporters. Below the brush border, tight junctions (zona occludens) prevent diffusion between cells and zona adherens link cells. The Na,K-ATPase in membranes below these junctions creates an internally directed Na^+ gradient that energizes the Na^+-dependent import of solutes, such as amino acids (AA), through the apical cotransporters. Solutes exit cells through carriers on basolateral membranes. The junctional complexes prevent leakage back to the lumen and the solutes diffuse into the bloodstream. **D.** Simple columnar cells of the small intestine (jejunum). The brush border (*) covers the apical portion of the cells. Terminal bars, which lie in the area occupied by junctional complexes, appear as dark lines encircling the tops of cells caught in a glancing cross-section (*large arrow*), or as a line of small dots below the brush border where cells meet (*small arrows*).

EXOCRINE GLANDS

Many organs, including the gut, contain epithelial cells that are extensively involved in producing and exporting water and ions as well as peptides, proteins, mucus, and other organic molecules. If the exported material is delivered to a free surface, the process is called **exocrine secretion**. For example, sweat glands are exocrine glands that secrete sweat to the external surface of the skin and serve to cool the body. **Endocrine secretion**, or secretion into the blood, is performed by epithelial and nonepithelial cells and will be discussed later in the text for those organs where it occurs.

- **Unicellular exocrine glands.** Some surface-covering epithelia contain individual secretory cells that secrete products directly onto the epithelium they inhabit. These cells can then be classified as unicellular exocrine glands. The goblet cells seen in the pseudostratified epithelium of the trachea are an example of a mucus-secreting unicellular gland (Figure 1-4D).

- **Multicellular exocrine glands.** Arise early in development by extension of portions of a surface epithelium into the deeper connective tissue supporting the epithelium. These ingrowths differentiate into **secretory portions**, which produce the secretory product, and **ducts**, which remain physically connected to the surface and conduct the secretory product. Exocrine glands may be large, with a complex three-dimensional structure.

The nomenclature used to describe the structure of ducts and secretory portions in exocrine glands is based on four features: branching of the ducts, the shape of the secretory units, the secretory product, and the mode of secretion. When looking at a section, it is often impossible to see all of the organization clearly, and therefore these classification criteria must be imposed mentally.

CLASSIFICATION OF EXOCRINE GLANDS

DUCT BRANCHING Exocrine glands contain one of two types of duct branching. In **simple** exocrine glands (e.g., sweat glands), there is a single unbranched duct connecting the secretory unit to the surface (Figure 1-7A). In **compound** glands (e.g., the submandibular salivary gland), the duct system branches and connects several secretory units to the surface (Figure 1-7B).

SHAPE OF SECRETORY UNIT The secretory portions of exocrine glands adopt one of three shapes. In **tubular** glands (e.g., sweat glands), the secretory cells are arranged in tubes. In **acinar** glands (e.g., sebaceous glands), the secretory cells are arranged as hollow spheres. This arrangement may also be referred to as alveolar. In **tubuloacinar** glands (e.g., the submandibular salivary gland), the secretory cells are arranged in both tubular and acinar structures (Figure 1-7A and B).

TYPE OF SECRETION The fluids produced by exocrine glands have one of three viscosities. **Serous** secretions, produced by sweat glands and the pancreas, are thin and watery and may contain proteins and enzymes. **Mucous** secretions, produced by goblet cells in the trachea, are thick and contain a variety of glycoproteins. **Seromucous** secretions, produced by the submandibular salivary gland, are a mixture of products released by both serous and mucous cells.

MODE OF SECRETION The products produced by exocrine glands are released from the secretory cells by one of three mechanisms. **Merocrine** secretion, performed by sweat and salivary glands, involves the release of product without the general loss of cell components. This is the most common mode of exocrine secretion and usually is accomplished by the exocytosis of small vesicles from the apical portion of cells, although water and salts may be exported directly from the cytoplasm via carriers present on the plasma membrane. **Holocrine** secretion occurs in sebaceous glands and involves the rupture of an entire cell, which then becomes the secretory product (Figure 1-7C). **Apocrine** secretion occurs in the mammary gland and results from the release of the apical portion of a cell, leaving the basal portion intact.

EXAMPLES OF EXOCRINE GLANDS

Exocrine glands can be described using all four criteria listed above. For example,

- **Sweat glands** are simple tubular, serous, and merocrine glands. The secretory portions of these glands are highly coiled, which obscures their simple tubular form, and they are often called simple coiled tubular glands.

- **Sebaceous glands** are compound, acinar, and holocrine glands. Sebaceous glands secrete sebum, containing a mixture of lipids, onto hair shafts. In many cases, two or more sebaceous gland acinar secretory units appear to empty into a single duct; some texts call this arrangement simple branched rather than compound.

- **Salivary glands** are compound, tubuloacinar, seromucous, and merocrine glands. The serous secretory cells are arranged in acini, and the mucous cells are mostly in tubules. The ratio of serous to mucous secretory cells and their product varies among different glands.

- **Goblet cells** of the trachea are classified as unicellular, mucous, and merocrine glands.

SPECIALIZED EPITHELIAL CELLS

Two other types of cells, myoepithelial and neuroepithelial, resemble muscle or nerve cells but are classified as epithelia.

MYOEPITHELIAL CELLS

Myoepithelial cells are found in some exocrine glands surrounding the secretory cells, including those in salivary glands and in the breast (Figure 1-7D). Myoepithelial cells contain smooth muscle-like actomyosin and help expel secretions into the duct system for export. The dilator pupillae muscle that dilates the iris in the eye is also composed of myoepithelial cells.

NEUROEPITHELIAL CELLS

Neuroepithelial cells are specialized to serve as **sensory receptors** and include rods and cones in the eye, hair cells in the inner ear, and taste bud receptors in the oral cavity. These sensory transducers are critical elements of the nervous system.

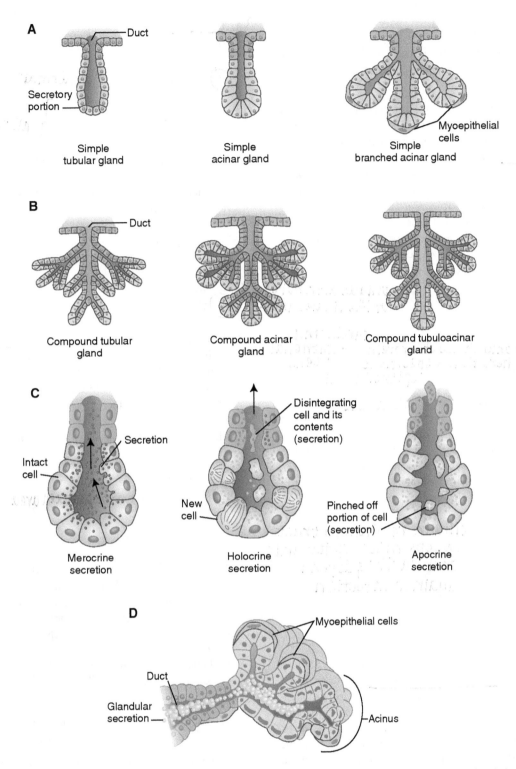

Figure 1-7: Exocrine gland structures, secretory modes, and myoepithelial cells. **A.** Simple and simple branched glands have only one excretory duct, which is connected to one (simple) or more (simple branched) secretory units. Secretory units that connect to the ducts are either acinar (spherical) or tubular shaped. **B.** Compound glands have a branched duct system. The secretory portions can be entirely acinar, entirely tubular, or a mixture of the two. **C.** Merocrine secretion involving the release of material stored in vesicles from the apical cell membrane by exocytosis. This is the most common mode of secretion. In holocrine secretion, which occurs in sebaceous glands, entire secretory cells rupture and the contents form sebum on hair shafts. In apocrine secretion, which occurs in the mammary gland, some of the apical portion of the cell is released to produce milk. **D.** Myoepithelial cells at the basal surfaces help move secretory material along the duct system. These cells may contract, forcing material from the secretory portions down the ducts, or they may act to prevent expansion of the glands and initial portions of the ducts, forcing the expulsion of material due to the pressure created by the addition of secretions into a fixed volume.

STUDY QUESTIONS

Directions: Each of the numbered items or incomplete statements is followed by lettered options. Select the **one** lettered option that is **best** in each case.

1. A researcher was using reverse transcription polymerase chain reaction (RT-PCR) to screen blood samples for carcinoma cells that had metastasized into the bloodstream of a patient. RT-PCR is a technique used for amplifying a defined sequence of RNA, and can provide a sensitive method of detecting a small number of cells that contain unique mRNAs in a tissue sample. Carcinomas are malignant tumors derived from epithelial cells. Blood can be considered as a specialized connective tissue. The researcher's RT-PCR primers should most likely be designed to amplify the mRNA encoding of which of the following proteins?

   A. Actin
   B. Collagen
   C. Elastin
   D. Keratin
   E. Vimentin

   [handwritten: • unique attributes of the tissue sample; carcinomas – epithelial cells; Keratin – only choice of protein here that's specific to epithelial cells]

2. Inherited defects in epithelial cell cilia function can result in human infertility; that is, sperm do not swim properly in males and eggs are not transported efficiently in females. A young couple arrives at an infertility clinic because they have not been able to conceive. Because these defects affect many epithelial cells throughout the body, the physician might be able to quickly identify the affected partner simply by asking each of them, "Which of the following symptoms do you experience?"

   A. Dry mouth
   B. Excessive bruising
   C. Fragile skin
   D. Frequent diarrhea
   E. Numerous colds

   [handwritten: cilia in respiratory tract sweeps mucus up towards oral cavity & defends against infection]

3. A researcher is studying an epithelial cell line that forms a sheet of cells and resembles gut epithelium in culture. These cells contain a heat-sensitive mutation. In cultures grown at 34°C, a small amount of fluorescent dye injected into one cell quickly diffuses to neighboring cells. At 39°C, however, the dye remains only in the injected cell. Careful microscopic examination shows no differences between the sheets at either temperature. It would be reasonable to expect to find the mutation in genes coding for which of the following structures?

   A. Basal lamina *[handwritten: X]*
   B. Desmosomes
   C. Gap junctions *[handwritten: → provide connections b/t cells, facilitate mvmt. of small molecules]*
   D. Keratin filaments
   E. Microvilli

4. Dysplasia in a stratified epithelium would be indicated by observing which feature of some of the cells above the basal layer?

   A. They are dividing *[handwritten: — "abnormal"]*
   B. They are squamous
   C. They contain desmosomes
   D. They contain keratin filaments
   E. They contain nuclei

   [handwritten: all "normal"]

5. Epithelial cells that serve as integral elements of neural networks function as

   A. Ependymal cells
   B. Glial cells
   C. Interneurons
   D. Motor neurons
   E. Sensory receptors

6. A patient is seen in clinic complaining of blisters that appear at the slightest touch of his skin. Histologic examination of a skin biopsy shows that at the sites of these blisters, the epithelium has completely lifted off its basement membrane but is otherwise intact. If an autoimmune condition is responsible for this condition, antibodies in the patient's blood might be found in which of the following structures?

   A. Gap junctions
   B. Desmosomes
   C. Hemidesmosomes *[handwritten: — Basal cell area]*
   D. Keratin filaments
   E. Zonula occludens

7. Mutations in human keratin genes K8 and K18 are associated with an increased susceptibility to several liver diseases. Intermediate filaments in liver epithelial cells expressing these mutant proteins show abnormal reorganization after the cells are stressed. Epithelial cells from the skin of these patients behave normally under all conditions. The most likely reason that skin epithelial cells behave normally is that the keratins in the epidermis are

   A. Not associated with desmosomes
   B. Not involved with mechanical stability
   C. Only expressed in the basal cell layer
   D. Only found in dead cells at the apical surface
   E. The product of genes other than K8 and K18

ANSWERS

1—D: The only protein in the choices given that is specific for epithelial cells is keratin, making it the most likely target of the researcher's test. Specific keratins are expressed by carcinomas, and identifying these keratins is a useful tool for determining properties of cancer, such as the likely site of origin of metastatic tumors.

2—E: Cilia in the large airways of the lungs and the trachea sweep mucus up toward the oral cavity. The mucus traps pathogens that enter the respiratory system and provides an important defense against infection. Therefore, the partner who experiences numerous colds will most likely be the affected partner.

3—C: Gap junctions provide connections between cells that facilitate the movement of small molecules. These junctions allow the spread of signaling molecules across an epithelium and help coordinate cellular activities.

4—A: Epithelial cells normally require contact with the basal lamina in order to divide. The other features listed can all be found in normal epithelia.

5—E: Neuroepithelial cells, such as rods and cones, hair cells in the ear, and taste bud receptors in the mouth serve as sensory receptors for the nervous system.

6—C: Hemidesmosomes attach the basal layer of skin epithelial cells to the basal lamina; therefore, if they were the target of the patient's autoantibodies, the entire epithelium would lift off, as described in the question. If the antibodies reacted against desmosomes, cells above the basal layer would separate. Keratin filaments are inside cells and would not be available for antibody binding.

7—E: Humans express more than 50 keratin genes, so it is likely that epidermal cells express different keratins than those found in the liver; K8 and K18 are not highly expressed in skin.

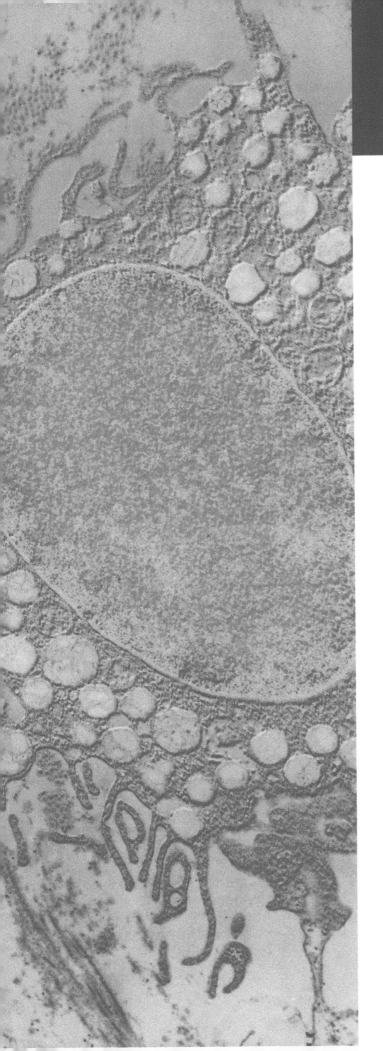

CONNECTIVE TISSUE PROPER

OVERVIEW

Connective tissues are aptly named—they connect epithelia to underlying body structures, link muscles to bones, and hold joints together. However, connective tissues provide the body with more than structural connections and form—they play vital roles in defense, repair, storage, and nutrition. The extent and composition of **extracellular matrix (ECM)** is a key feature of connective tissues. Unlike epithelia, muscle, and nerve tissue, which consist mostly of cells, connective tissues contain significant amounts of extracellular material, which may be synthesized by only a small number of resident cells. This chapter focuses on what are termed **connective tissues proper**; that is, loose and dense connective tissues and adipose tissue. A few other rare types of connective tissue are mentioned briefly in this chapter; however, several specialized connective tissues (i.e., cartilage, bone, blood, and hematopoietic tissues) require detailed discussion and will be considered in Chapters 3, 4, and 8.

CHARACTERISTICS AND FUNCTIONS OF CONNECTIVE TISSUES

LOCATION OF CONNECTIVE TISSUES

Connective tissues are found throughout the body, characteristically in the following locations:

- **Beneath basal lamina of epithelia.** For example, in skin, the connective tissue of the **dermis** lies below the stratified squamous keratinized epithelium (see Figure 1-1, Chapter 1).

- **Stroma.** Connective tissues within organs often serve as the **stroma**, a support tissue, rather than as the parenchyma, a functional tissue. For example, connective tissues form organ capsules and surround blood vessels, nerves, and accessory structures within organs. The portal areas in the liver, as seen in Figure 2-1A, are an example of a stromal connective tissue area that provides support to the parenchymal epithelial liver cells.

- **Between organs.** Loose connective tissue fills spaces between organs.

ORGANIZATION OF CONNECTIVE TISSUES

Connective tissues are constructed from cells surrounded by ECM, which consists of **fibrous proteins** and **ground substance**. The cells and ECM will be described more fully later in the chapter, but it is instructive at this point to compare the general organization of connective tissues to epithelia, as follows:

- Cells in an epithelium are held closely together with desmosomes, and there is little extracellular material.

- Connective tissue cells are usually solitary and can move about through the extracellular material.

Loose and dense connective tissues are defined by the ratio of their components, as follows:

- Loose connective tissues are composed largely of **cells and ground substance**.

- Dense connective tissues contain numerous **collagen fibers** and many fewer cells.

This distinction is easy to visualize when the two tissues are found close together, as seen in the duodenum (Figure 2-1B).

FUNCTION OF CONNECTIVE TISSUES

- **Structural support.** The connective tissues described in this chapter, along with cartilage and bone discussed in Chapters 3 and 4, respectively, form the structural framework on which the rest of the body is assembled.

- **Nutrition.** Epithelia are not directly vascularized, whereas arteries and veins supplying organs are surrounded by connective tissue (Figure 2-1A). The exchange of nutrients and waste products between the blood in capillaries and venules and cells of an organ occurs by diffusion through connective tissues, except in restricted locations such as the brain and thymus.

- **Storage.** ECM of connective tissues is an important site for storing water, electrolytes (e.g., NaCl), and plasma proteins. Lipids are stored by adipocytes.

- **Defense.** ECM provides a physical barrier against microorganisms that manage to penetrate epithelia and the basal lamina, which helps confine infections by these invaders. Thus, connective tissue is often the site of inflammatory and immune responses to pathogens. These responses are first mediated by connective tissue residents, principally macrophages and mast cells, and by transient cells, which may be recruited in large numbers from the blood in response to signals generated by an acute infection.

- **Repair.** Tissue loss due to infection or injury usually stimulates the growth of local connective tissues in an organ. The deposition of new ECM can help close wounds and provides a support for organized tissue replacement. In some cases, total repair is not possible and connective tissue **scars** fill in gaps created by lost parenchymal tissue. Persistent damage to an organ may cause excess production of connective tissue, resulting in diminished function of the remaining parenchymal tissue. This condition is known as **fibrosis**.

- **Information.** Many molecules secreted by cells that are involved in cell–cell signaling (e.g., growth factors and inflammatory mediators) do not diffuse freely through connective tissues but are bound by ECM components. This binding can result in several effects, including limiting the spread or producing locally high concentrations of the signal. The informational role of the basal lamina in regulating epithelial cells, which was discussed in Chapter 1, is also true for interactions among connective tissue cells themselves or other cells that come into contact with the ECM.

REPLACEMENT OF CONNECTIVE TISSUES

Connective tissue cells are derived from **mesenchymal stem cells** of mesodermal origin and have both **permanent** and **transient cell** residents (Figure 2-1C). Many of the transient cells arise from hematopoietic stem cells in bone marrow or the thymus, enter a connective tissue from the bloodstream, and may exit the tissue or die after a short time.

Permanent resident cells may be born locally by the division of stem cells or differentiated resident cells, or if they are born elsewhere, they enter from the blood. Adult mesenchymal stem cells may also produce smooth muscle, endothelial, and mesothelial cells. Fibroblast cells have the ability to divide and differentiate into other types of connective tissue cells when placed in culture, although the extent to which this happens in tissues is uncertain.

▼ Most cancers of solid connective tissue origin are called **sarcomas**; this term, however, also includes malignant growths arising from muscle and blood vessels. Sarcomas are much less common in adults than are carcinomas. ▼

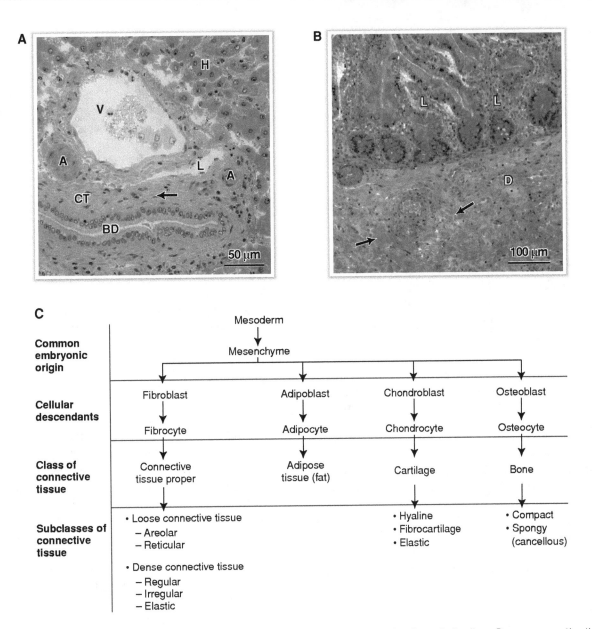

Figure 2-1: Connective tissue; locations, organization, and production. **A.** Stromal connective tissue in the liver. Dense connective tissue (CT) surrounds the veins (V), arteries (A), lymphatics (L), and bile ducts (BD) in a portal triad in the liver. The arrow indicates the nucleus of a fibroblast cell of the connective tissue. H, hepatocytes. **B.** Loose and dense connective tissues in the duodenum. The loose connective tissue (L) of the lamina propria underlies the simple columnar epithelium and consists of many closely packed cells. Below that is the layer of dense connective tissue (D) of the submucosa, which is composed of numerous large extracellular fibers (*arrows*) and few cells. The lumen of the duodenum is above the view shown here. Mucus-secreting cells of the epithelium are stained red in this preparation. (H&E stain with periodic acid–Schiff reaction). **C.** Production of connective tissue cells. Stem cells derived from embryonic mesoderm give rise to all the differentiated cells listed in the table. Stem cells with similar capabilities persist in adults.

CELLS OF CONNECTIVE TISSUE—PERMANENT RESIDENTS

As mentioned previously, some cells in connective tissue are permanent residents (e.g., fibroblasts, adipocytes, macrophages, and mast cells) (Figure 2-2A). The permanent resident cells are discussed in this section; in the next section, the transient cells of connective tissue that usually wander in from blood, often in response to specific signals, are discussed.

FIBROBLASTS

Fibroblasts are typically the most abundant cells found in connective tissues and have the following properties:

- They synthesize and secrete fibers and ground substance, the major components of ECM.

- They are normally nonmitotic; however, during tissue repair, fibroblasts can proliferate rapidly while also increasing their synthetic activity.

- They are difficult to distinguish from other cells when using the light microscope because usually only the nuclei of many of the connective tissue cells can be seen. The light micrograph rarely provides enough information to identify particular cells although there is a close association to collagen bundles, which would suggest that the cell is a fibroblast (Figure 2-2B). When observed with the transmission electron microscope, fibroblasts appear as spindle-shaped cells associated with extracellular collagen fibers.

- They contain abundant rough endoplasmic reticulum (RER) and an expanded Golgi apparatus when actively synthesizing and secreting matrix components. Fibroblasts can also produce and export factors that influence the growth and differentiation of neighboring cells.

ADIPOCYTES

Adipose tissue is a special connective tissue, and is discussed later in this chapter. Adipocytes are also found scattered in other organs, and have the following properties:

- In adults, adipocytes store lipid in a single droplet that occupies most of the cell's volume. The lipid is extracted during the processing required to prepare a slide, leaving a large, clear center surrounded by a thin rim of cytoplasm containing a flattened nucleus (Figure 2-2C).

- The number of adipocytes is determined early in life; these cells rarely divide in adults.

- Adipocytes produce a variety of endocrine hormones and cytokines involved in regulating metabolism and the immune system (see Chapter 14).

MACROPHAGES

Macrophages are cells derived from **monocytes** found in the blood, which in turn are derived from stem cells in bone marrow. Macrophages have the following characteristics:

- Monocytes migrate from small blood vessels into connective tissues and differentiate into macrophages. These cells take up long-term residence and are sometimes referred to as **histiocytes** when they are present in connective tissue. The process of cell passage from blood across a vessel wall into a tissue is called **diapedesis**.

- Macrophages live for months and can divide to produce more macrophages.

- Macrophages are difficult to distinguish from fibroblasts when viewed through a light microscope without using special stains, but their presence may be inferred when examining certain pathologic conditions (Figure 2-2D). When observed by transmission electron microscopy, macrophages contain abundant lysosomes and residual bodies (foreign matter in various stages of digestion).

Macrophages have several functions, as follows:

- They **phagocytose** and attempt to destroy foreign particles, microorganisms, and damaged tissues.

- They are involved in the normal turnover of ECM by **secreting hydrolytic enzymes** (e.g., collagenase). This activity is increased during inflammation and repair.

- They are key players in **immune reactions**. Macrophages can initiate immune responses by presenting antigens to lymphocytes and by secreting cytokines, the signaling molecules that are important for immune system activities. These roles will be discussed further in Chapter 9.

MAST CELLS

Mast cells resemble basophils found in the blood; they also are derived from bone-marrow precursors, and function as follows:

- Mast cells store chemical mediators of the inflammatory response in large basophilic granules (Figure 2-2E).

- They are responsible for allergic reactions, referred to as immediate hypersensitivity reactions, which occur when these cells rapidly release granules when exposed to an antigen. Chemical mediators stored in granules include histamine (promotes vascular leakiness and smooth muscle contraction), heparin (blocks local coagulation), and factors such as eosinophil chemotactic factor, which attract other immune system cells.

- They are abundant in connective tissue that underlies exposed body surfaces covered by barrier epithelia. Although all mast cells appear similar, the granular content of connective tissue mast cells in the skin and peritoneal cavity differs from that of mucosal mast cells in the intestinal mucosa and lungs. There are no mast cells in the spleen or brain.

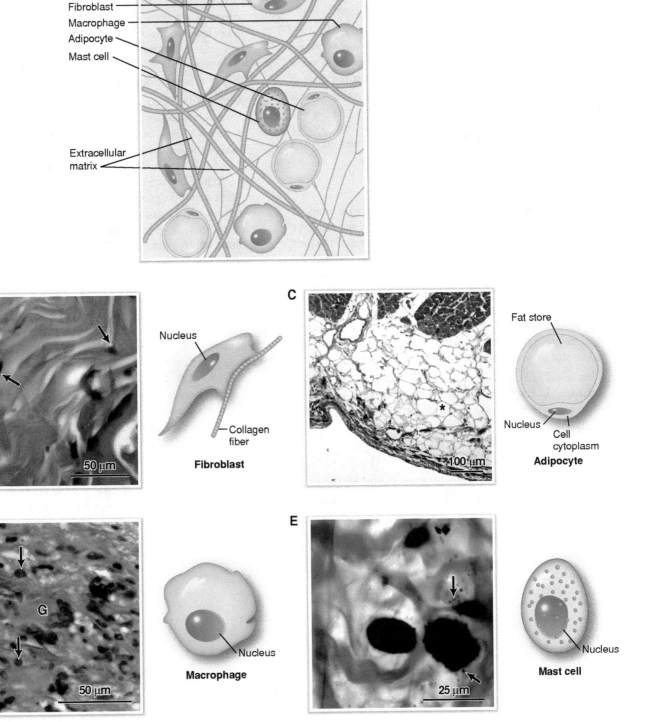

Figure 2-2: Permanent cellular residents of connective tissue. **A.** Principal cells of connective tissue. Fibroblasts, macrophages, adipocytes, and mast cells are situated between components of the extracellular matrix, and are the permanent residents in connective tissue. **B.** Fibroblasts between bundles of collagen in the dermis of skin. Nuclei (*arrows*) are the only cellular feature clearly visible in this section. **C.** Adipocytes in the epicardium of the heart. This outer layer of the heart is commonly the site of fat deposition, which serves as a mechanical cushion for the organ. The center of an adipocyte (*) is clear, leaving only a thin rim of cytoplasm. **D.** Macrophages and giant cells in the dermis of skin undergoing an inflammatory reaction. Macrophages (*arrows*) accumulate in late stages of erythema nodosum, a skin inflammation that causes tender red bumps. Macrophages often fuse, forming multinucleated giant cells (G). Section courtesy Scott Florell, MD, University of Utah School of Medicine, Salt Lake City, Utah. **E.** Mast cells in a loose connective tissue spread. The intensely stained granules of two mast cells on the bottom half of this micrograph obscure their nuclei. A few individual granules have been released from the cells and some are visible at the cell edges (*arrows*).

CELLS OF CONNECTIVE TISSUE—TRANSIENT CELLS

As mentioned previously, some cells in connective tissue are permanent residents (e.g., fibroblasts). However, some connective tissue cells are transients and enter the tissue from small blood vessels via diapedesis. This entry is often in response to local signals such as those generated by microbial infection. Transient cells include plasma cells, lymphocytes, neutrophils, and eosinophils. These transient cells will be discussed briefly here as connective tissue cells, and their origins and activities will be described in Chapters 8 and 9.

PLASMA CELLS

Plasma cells are derived from mitotic populations of B lymphocytes that have been selected and mitotically activated by exposure to a foreign antigen (Figure 2-3A). Some of the proliferating B lymphocytes eventually differentiate into plasma cells, which take up residence in connective tissue and then actively secrete antibodies reactive against the antigen.

- Plasma cells are characterized by an eccentric nucleus, a "wheel-and-spoke" or a "clock face" arrangement of nuclear heterochromatin, stacks of RER, and a prominent Golgi apparatus involved in antibody secretion.

- Plasma cells may be found in any connective tissue and are common in the lymph nodes, spleen, and mucosa of the digestive system. The average lifespan of a plasma cell is 10–20 days.

LEUKOCYTES

Beside macrophages and plasma cells, four other white blood cells, or leukocytes, frequently migrate into connective tissues.

- **Lymphocytes.** These cells are the smallest free cells in connective tissue. Lymphocytes are characterized by a deeply stained, usually featureless nucleus surrounded by a thin rim of cytoplasm. Lymphocytes accumulate at any site of acute infection and tissue inflammation, and are commonly found in the loose connective tissue (lamina propria) of the respiratory and gastrointestinal tracts (Figure 2-3B). In these areas, many lymphocytes may be long-term residents; in other sites, they may pass through quickly.

- **Neutrophils.** These cells are common in blood but are rare in normal connective tissue. Neutrophils are attracted in large numbers to sites of bacterial infection. Neutrophils contain membrane-bound, cytoplasmic granules that contain a variety of substances, including defensins, which can kill bacteria. When attracted to the site of a bacterial infection by inflammatory signals, they avidly phagocytose bacteria and destroy them. Neutrophils have a distinctive four- or five-lobed nucleus and therefore are often called **polymorphonuclear neutrophils,** or **PMNs.** When active in a tissue, they often lyse, releasing nuclear fragments and enzymes that degrade ECM (Figure 2-3C).

- **Eosinophils.** These cells also are rarely found in normal connective tissues but may be recruited from the blood in large numbers by signals generated from allergic reactions and parasitic infections. Eosinophils can kill parasites by releasing cytotoxins from cytoplasmic granules. They also release enzymes that destroy histamine and heparin, and thereby have the potential to modulate allergic reactions. In humans, eosinophils have a characteristic bilobed nucleus. The granules in these cells stain bright orange, making them visible in typical tissue sections (Figure 2-3C). Eosinophils are often attracted to the connective tissue surrounding small airways in the lung and may contribute to diseases such as asthma by inappropriate release of their granule contents. Neutrophils and eosinophils, and the rare basophils, are called **granulocytes** because these cells contain numerous cytoplasmic granules.

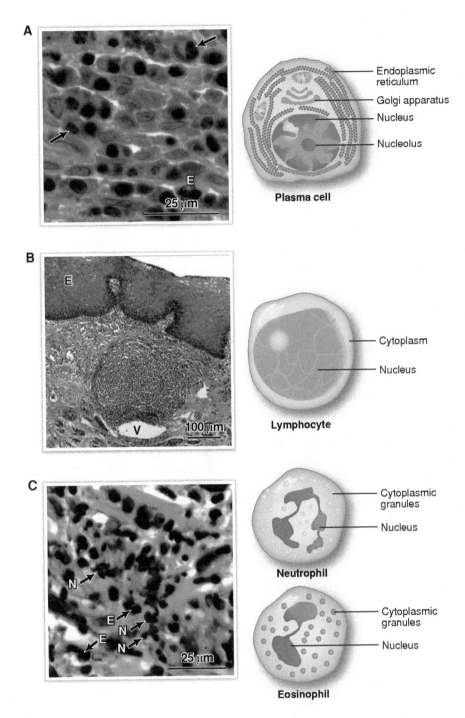

Figure 2-3: Transient cells of connective tissue. **A.** Plasma cells in the cervix. Plasma cells are characterized by an eccentric nuclei, containing large clumps of heterochromatin (*arrow*), dark cytoplasm, and prominent Golgi apparatus, which appears as a lighter pink area near the nucleus. The diagram of a plasma cell illustrates the expanded rough endoplasmic reticulum and the Golgi apparatus required for the production and export of antibodies by these cells. **B.** Lymphocyte aggregate in the esophagus. The mass of lymphocytes appears as a dense collection of nuclei at low magnification because the cells contain little cytoplasm. Individual lymphocytes can also be seen immediately below the epithelium (E). The diagram of a lymphocyte emphasizes that these cells contain little cytoplasm surrounding a dense nucleus. V, vein. **C.** Neutrophils (N) and eosinophils (E) in the dermis of skin affected by a rare skin disease called acute neutrophilic dermatosis. Neutrophils accumulate for unknown reasons and may damage the dermis. The multilobed nuclei of three neutrophils can be seen among a mixture of cells. Section courtesy Scott Florell, MD, University of Utah School of Medicine, Salt Lake City, Utah.

EXTRACELLULAR MATRIX

The components of ECM are **fibers** (collagen and elastic) and **ground substance** (Figure 2-4A).

FIBERS

The ECM fibers consist of proteins and polysaccharides, which are secreted locally and assembled into an organized meshwork in close association with the cells that produced them, typically fibroblasts.

Both collagen and reticular fibers are constructed from collagen proteins. Elastic fibers are formed from elastin proteins surrounded by microfibrils.

- **Collagens.** More than 15 members of the collagen family of proteins have been identified. Collectively, collagen is the most abundant protein in the body (about 30% of the total dry weight); some type of collagen is found in all connective tissue. Both **collagen fibers** and **reticular fibers** are constructed from collagens.

- **Collagen structures.** Collagen molecules assemble into several different, mostly fibrillar structures. The basic assembly unit is usually a trimer of one or more collagen alpha chains (Figure 2-4B). Chemical differences among the alpha chains are responsible for the structural and mechanical characteristics of the many types of collagen. Many collagens form long fibrils themselves or associate with fibrils formed from other collagens. For example, type I collagen forms fibrils that self-associate into extremely large fibers. Type III collagen fibrils form diffuse, thin reticular fibers. Type IV collagen assembles into a diffuse meshwork in basal laminae, and type VII collagen forms short fibrils that link basal laminae to connective tissues.

- **Collagen fibers (type I).** Cells (primarily fibroblasts) synthesize and insert the glycine-rich, alpha chain proteins $\alpha 1$ and $\alpha 2$ into the endoplasmic reticulum. Many proline and lysine residues are posttranslationally modified to hydroxyproline and hydroxylysine (which requires vitamin C), and many hydroxylysines are subsequently glycosylated. Three alpha chains ($2\text{-}\alpha 1 + 1\text{-}\alpha 2$) assemble into a triple helical **procollagen** molecule in the RER. Procollagen is secreted from the cell and undergoes proteolytic cleavage extracellularly to form **tropocollagen**. Tropocollagens self-assemble end-to-end and side-to-side into fibrils that can further associate to form large collagen fibers 1–20 μm in diameter. The nature of these assemblies depends in part on interactions with other collagens and adhesive glycoproteins such as fibronectin. Hydrogen bonding provided by hydroxyproline residues and covalent cross-linking of hydroxylysines stabilize the assemblies. Collagen fibers are the most numerous fibers in connective tissue and are abundant in the dermis, organ capsules, and tendons (Figure 2-4C).

- **Reticular fibers (type III).** These thin collagen fibers (0.5–2 μm) are also called **reticulin**. They are constructed from type III collagen and are more highly glycosylated than the fibers assembled from type I collagen. Reticular fibers are not visible with a light microscope unless the fibers are specifically stained by silver salts. Reticular fibers form a flexible meshwork surrounding smooth muscle cells in the walls of arteries and the digestive system. They form a supportive network in lymphatic and hematopoietic organs and provide a loose framework for parenchymal epithelial cells in the endocrine glands and liver (Figure 2-4D).

- **Elastic fibers.** Elastic fibers (0.2–2 μm) consist of a core of the protein **elastin**, rich in proline and glycine, which is deposited onto a thin sheath of microfibrils formed from a number of proteins, including **fibrillin**. Monomeric elastin proteins adopt a random coil configuration. Mature elastin contains two unique amino acids, desmosine and isodesmosine, which covalently link elastin monomers together, forming a large, randomly organized, cross-linked network. Stretching the elastin networks orders the elastin monomers, which is entropically unfavorable, causing the network to spontaneously recoil to its original state, similar to vulcanized rubber. Elastic fibers are found in connective tissues in varying proportions and are most abundant in organs that experience stretching (e.g., vocal cords, trachea, and skin); they are present to some extent in many locations throughout the body (Figure 2-4C). Elastic fibers in loose connective tissue are more dispersed and, therefore, easier to visualize. Elastin is abundant in a more sheet-like or membrane form in the media of the aorta and other major arteries.

- There are other fibers in the body that might be considered to be specialized precursors of typical elastic fibers. The **zonule fibers** that suspend the lens in the eye consist largely of microfibrillar aggregates without elastin. Other fibers that consist only of microfibrils, called oxytalan fibers, are found in the periodontal ligaments that hold teeth to the jawbone. Elaunin fibers consist of microfibrils and a small amount of elastin and are found in the dermis of the skin, along with true elastic fibers.

▽ **Marfan syndrome** is a common genetic disease caused by a set of mutations in the fibrillin gene. These mutations result in various symptoms, ranging from misalignment of the lens to fatal rupture of the aorta. ▽

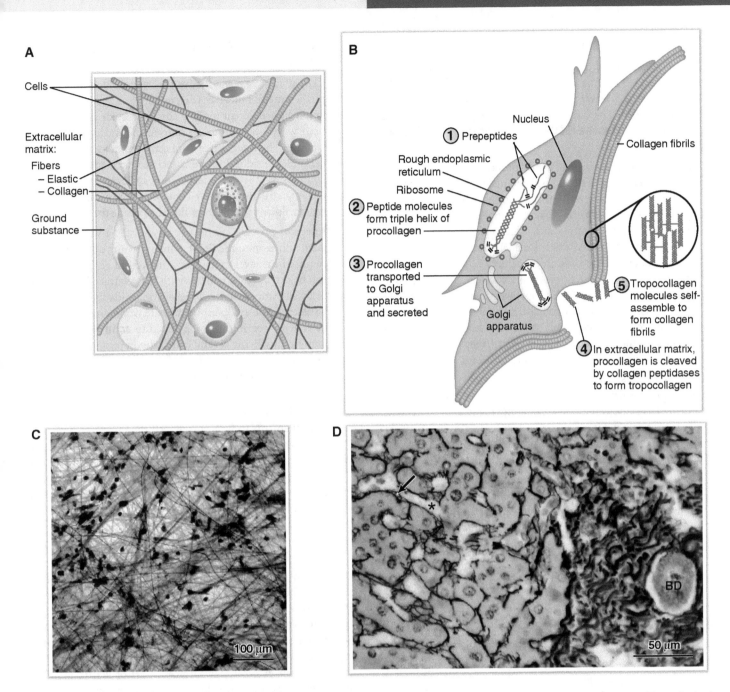

Figure 2-4: Fibers in the extracellular matrix. **A.** The main components of extracellular matrix: fibers and ground substance. The principal fibers are collagens and elastic fibers. **B.** Steps in collagen synthesis. (1) Collagen peptides are synthesized on membrane-bound ribosomes and simultaneously inserted into the lumen of the rough endoplasmic reticulum (RER). The peptides contain extensions to their N- and C-termini, called propeptides. (2) The propeptides play two important roles: they help guide self-association of triple-stranded, helical procollagen molecules in the RER, and they also help to prevent formation of larger aggregations. (3) Proline and lysine residues on the peptides are hydroxylated and the proteins are sent on the secretory pathway via the Golgi apparatus. (4) When procollagen is secreted from cells, extracellular proteases remove the propeptides to form tropocollagen. (5) Tropocollagens can now self-assemble into long collagen fibrils, initially in close association to the cell's plasma membrane and in the presence of other secreted matrix proteins. **C.** Collagen and elastic fibers in a loose connective tissue spread. The large collagen type I fibers are stained pink, and the thin, branched elastic fibers are stained dark blue. Mast cells are also darkly stained in this preparation. (Verhoeff stain). **D.** Reticular fibers in the liver. Reticular fibers appear as thin dark segments (*arrow*) on the surfaces of hepatocytes (light pink), primarily on surfaces facing large capillaries (*). A portion of a portal area is visible on the right; note the much larger bundles of type I collagen and the bile duct (BD). Section courtesy Frederick Clayton, MD, University of Utah School of Medicine, Salt Lake City, Utah. (Reticulin, or silver, stain)

GROUND SUBSTANCE

Ground substance is a viscous, gel-like material with a high water content. It fills the spaces between cells and fibers in connective tissue and serves as a lubricant and barrier to penetration by invaders. The main components of ground substance are **glycosaminoglycans (GAGs)**, **proteoglycans**, and **adhesive glycoproteins** (Figure 2-5A).

GAGs. These are large, unbranched polysaccharide chains composed of repeating disaccharide subunits. At least one member of the pair is an N-acetylated hexosamine, which explains the name of this class of molecules. Examples of GAGs include hyaluronic acid (D-glucuronic acid, D-acetylglucosamine)$_n$, chondroitin sulfate (D-glucuronic acid, D-acetylgalactosamine)$_n$, and dermatan sulfate (L-iduronic acid, D-acetylgalactosamine sulfate)$_n$ (Figure 2-5B). Some hyaluronic acid chains are large ($n = 25,000$), and other GAGs are shorter ($n < 300$). The sugars of many GAGs are usually sulfated, and the resulting polymers are stiff, extended, and highly negatively charged. As a consequence, on the basis of molecular weight, they occupy a much larger volume than do most other biomolecules.

Proteoglycans. With the exception of hyaluronic acid, most GAGs are covalently attached to a core protein backbone, which forms large molecules called **proteoglycans** (Figure 2-5C). There are many different core proteins, and GAGs constitute 80–90% of the weight of most proteoglycans.

- Proteoglycans and individual GAGs (e.g., hyaluronic acid) form large aggregates. Because of their chemistry, especially the large number of fixed negative charges they carry, these molecules bind a great deal of water and cations (e.g., sodium). This provides the ground substance with its viscous, gel-like structure, which helps to impede invasion by pathogens.

- Successful pathogens often secrete hydrolytic enzymes reactive against GAGs and proteoglycans, which reduce the viscosity of ECM, facilitating spread of the invaders.

Adhesive glycoproteins. These are globular proteins to which shorter, branched carbohydrates are covalently attached. They usually contain more protein than carbohydrate by weight, unlike proteoglycans.

- Examples of adhesive glycoproteins include **fibronectin**, which is found in many connective tissues, and **laminin**, which is concentrated in basal laminae. The fibronectin protein has binding sites for itself and for collagen and heparin, which allows it to participate in a complex set of molecular interactions involved in organizing the ground substance.

- Cells attach to these adhesive proteins via **integrin** receptors on their membranes (see Chapter 1). Members of the integrin family have different binding specificities; for example, some bind with high affinity to fibronectin and others bind to laminin. The integrins of connective tissue cells involved in extracellular adhesion are associated with **focal adhesion** complexes, not hemidesmosomes. Focal adhesions are dynamic structures involved both in linking a cell to ECM and in transducing signals from the ECM inside the cell.

- Integrins provide connective tissue cells a variable, regulated adhesion to the ECM, which permits cell movements required in development and in wound repair. Integrins can be activated by contact with their extracellular ligands or by alterations of the cytoskeleton inside the cell. In focal adhesions, the cytoplasmic domain of one member of the activated integrin dimers is linked to **actin** filaments via a complex involving several cytoskeleton proteins, including **talin**, **vinculin**, and **filamin**. The extracellular domains of the dimers attach tightly to specific sites on adhesive glycoproteins (Figure 2-5D).

- Alterations in the cytoskeleton can uncouple the linkage of actin filaments to integrins, which then undergo a conformational change that greatly decreases their extracellular binding affinity to the adhesive glycoproteins (Figure 2-5E). Cycles of attachment and release of integrins, coupled to changes of cell shape, allow connective tissue cells to move through the ECM or divide.

EXTRACELLULAR MATRIX TURNOVER

Destruction of the ECM was mentioned previously as a strategy used by pathogens to facilitate their spread. The degradation of ECM also occurs at a low rate as part of normal turnover of ECM components and is increased after tissue damage or infection. Components are digested by specific hydrolytic enzymes (e.g., glycosidases or proteases), which are secreted from cells or bound to their extracellular surfaces. A set of **matrix metalloproteases** is important for the controlled turnover of collagens and adhesive glycoproteins of ECM. Under normal circumstances, these proteases are used to loosen the matrix, allowing cells to move through the ECM or divide. Tumor cells employ these enzymes to break down basal laminae and ECM to facilitate their metastatic spread.

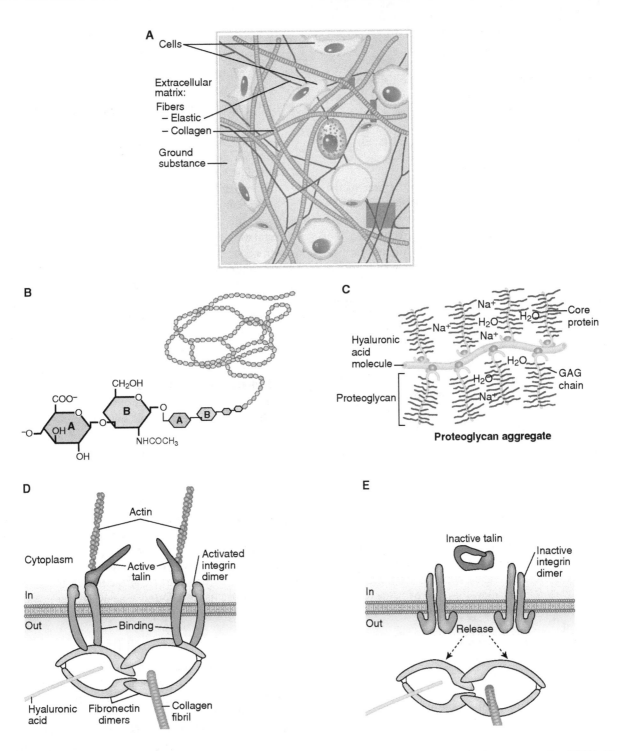

Figure 2-5: Ground substance in the extracellular matrix (ECM) and cell-matrix interactions. **A.** The main components of ECM: fibers and ground substance. The principal components of ground substance are glycosaminoglycans (GAGs), proteoglycans, and adhesive glycoproteins. **B.** GAG structure. GAGs are heteropolysaccharides produced from repeating disaccharide units, one of which is N-acetylated. Shown here is the most abundant GAG, hyaluronic acid, a single molecule that can consist of thousands of disaccharides (e.g., D-glucuronic acid, N-acetyl-D-glucosamine) and occupy a large volume in relation to its molecular weight. **C.** Proteoglycan structure. Most GAGs are shorter than hyaluronic acid; they are not found free but are covalently attached to one member of a set of core proteins, forming a proteoglycan. The GAGs in proteoglycans are usually sulfated and highly negatively charged, attracting cations (e.g., sodium) and binding water. **D.** Activated integrin dimers attach cells to the ECM. Shown is a schematic model of how the extracellular portions of activated integrin complexes in a focal adhesion complex on a fibroblast cell are attached to the adhesive glycoprotein, fibronectin. Maintenance of the activated state requires interactions between integrins and the actin cytoskeleton. **E.** Inactive integrins detach from the ECM. When cytoskeleton binding is lost, the integrins undergo a conformational change that greatly decreases the extracellular binding affinity, and they detach from adhesive glycoproteins. Also indicated in this figure are additional interactions of fibronectin with hyaluronic acid and collagen, suggesting the complex networks that can be formed in ECM.

CLASSIFICATION OF CONNECTIVE TISSUE PROPER

Several schemes exist for classifying connective tissue proper, and most include the following categories discussed in this section. The term connective tissue proper is limited to loose and dense connective tissues.

LOOSE CONNECTIVE TISSUE

Loose (areolar) connective tissue is the most widespread of the connective tissues. Ground substance and cells are more abundant than fibers. Collagen, reticular, and elastic fibers are arranged in a loose, irregular meshwork. Loose connective tissue is flexible, well vascularized, and only slightly resistant to stress. All cell types described in this chapter may be found in loose connective tissue.

Loose connective tissue is found beneath some epithelia, including the stomach and intestines, where it is called **lamina propria.** Movement of solutes absorbed by epithelial cells of the intestines into the circulatory system is a critical activity of the lamina propria. Areas of loose connective tissue wrap and cushion organs (Figure 2-6A). This tissue plays a role in inflammation as a storehouse of immune cells in many organs.

DENSE IRREGULAR CONNECTIVE TISSUE

In dense irregular (collagenous) connective tissue, the collagen fibers are abundant and arranged in a dense three-dimensional network. There are relatively fewer cells than in loose connective tissue, and they consist mostly of fibroblasts. It is formally correct to include the term **collagenous** in the name of this tissue because there are a few types of connective tissues containing abundant dense, irregular arrays of elastic fibers that will be discussed below. However, in practice, this distinction is not made and the term dense irregular connective tissue is used. The extensive collagen fibers allow this type of tissue to withstand tension exerted in many directions, which provides structural strength and flexibility to organs.

Dense irregular connective tissue is found in the dermis of the skin, the submucosa of the gut, below the loose connective tissue of the lamina propria, and in the capsule surrounding organs (Figure 2-6B). In some of these locations, particularly the dermis, elastic fibers are also present to provide flexibility.

DENSE REGULAR CONNECTIVE TISSUE

Collagen fibers also are abundant in dense regular (collagenous) connective tissue, where they are arranged in parallel bundles. There are very few cells present and, again, these are predominately fibroblasts. The arrangement of collagen bundles in dense connective tissue allows it to resist stretching in one direction. This tissue is found in tendons and ligaments (Figure 2-6C).

ADIPOSE TISSUE

There are two types of adipose tissue: unilocular, or white fat, which provides lipid storage, and multilocular, or brown fat, which generates heat early in life.

- **Unilocular (white fat) tissue.** This tissue consists principally of large collections of adipocytes (Figure 2-6D). If a diet is adequate, cells contain a single mass of lipid. This tissue normally accounts for about 20% of body weight in adults. White fat serves to store and release lipid and provide thermal insulation. Lipid storage and release from these cells is under complex control by hormones and the nervous system. Adipocytes are part of a network of cells involved in regulating the energy balance in the body. Adipocytes produce several hormones, including **leptin**, which stimulates a general increase in metabolism and acts in the hypothalamus to signal satiety, and **adiponectin**, which affects blood glucose levels and lipid catabolism. Perturbation of the endocrine activity of adipocytes is likely to play an important role in the progression of metabolic alterations that produce type 2 diabetes mellitus (non–insulin-dependent) (see Chapter 14).

- **Multilocular (brown fat) tissue.** This tissue is found only in the fetus and the newborn, primarily in the neck, axilla, and on the kidneys. These fat cells have a centrally located nucleus and contain many fat droplets. Cells contain numerous mitochondria that oxidize the lipid to generate heat in response to cold stress.

ELASTIC TISSUES

Elastic tissues are less common tissues in which elastin-based polymers predominate over collagen fibers. **Dense irregular elastic tissue** is found in the middle layer of large arteries, including the aorta, where elastic recoil smoothes the pressure wave created by the heart's contraction (Figure 2-6E). **Dense regular elastic tissue** is found in a few ligaments, including the suspensory ligament of the penis.

RETICULAR TISSUE

Some organs contain such an abundance of loosely arrayed reticular fibers (type III collagen) that reticular tissue is often considered a separate class of loose connective tissue. In these cases, the fibroblasts that specialize in the synthesis of reticular fibers are called **reticular cells.** Reticular tissue is found in hematopoietic bone marrow, lymph nodes, and the spleen, where it provides an important supportive stroma.

MUCOUS CONNECTIVE TISSUE

Mucous connective tissue consists mostly of ground substance, predominately hyaluronic acid and a few fibroblasts. It is found inside forming teeth and in the umbilical cord, where it is called Wharton's jelly, which serves as an experimental source of fetal mesenchymal stem cells.

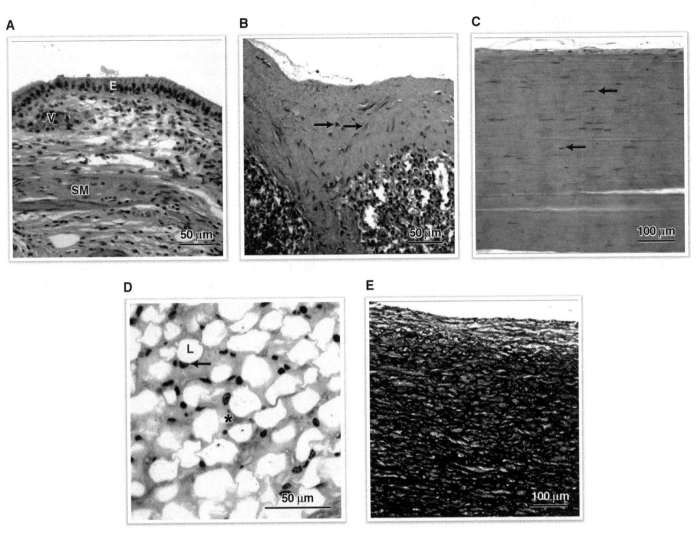

Figure 2-6: Examples of different types of connective tissues. **A.** Loose connective tissue below the simple columnar epithelium (E) of the gallbladder. Bundles of smooth muscle (SM) run through the connective tissue. V, vein. **B.** Dense irregular connective tissue in the capsule of the spleen. Fibroblast nuclei (*arrows*) are visible among the thick bundles of collagen. An extension of the capsule penetrates down into the parenchyma of the spleen. **C.** Dense regular connective tissue of a tendon. The elongated fibroblast nuclei (*arrows*) in widely separated, regular rows show that the large type I collagen bundles are organized unidirectionally in this tissue. **D.** Unilocular (white fat) adipose tissue. A single lipid droplet (L) occupies the center of the adipocyte cells, leaving a rim of cytoplasm and the nucleus (*arrow*) at the periphery. Capillaries (*) are found between the adipocytes. **E.** Elastic fibers in the wall of the aorta. The section has been treated with Verhoeff stain, which renders elastin-containing elements black. The middle layer of the aorta contains a large amount of elastic fibers arranged in sheet-like structures, called laminae or membranes. Here, smooth muscle cells, not fibroblasts, are responsible for the synthesis and organization of elastic fibers. (Verhoeff stain)

STUDY QUESTIONS

Directions: Each of the numbered items or incomplete statements is followed by lettered options. Select the **one** lettered option that is **best** in each case.

1. A 7-year-old girl repeatedly sprains her ankles badly. The injuries appear to result during normal play activity. The joints are extremely extensible, and the tendons and skin are unusually flexible. These findings can best be explained by an inherited disorder that affects which of the following protein families?

   A. Actin

   B. Collagen

   C. Elastin

   D. Fibrillin

   E. Keratin

   F. Myosin

2. What are the three basic components from which all types of connective tissue are constructed?

   A. Arteries, veins, and capillaries

   B. Cells, fibrous proteins, and ground substance

   C. Collagen, hyaluronic acid, and fibronectin

   D. Fibroblasts, fibroproteins, and proteoglycans

   E. Mast cells, lymphocytes, and adipocytes

3. Which constituent of connective tissue is most responsible for impeding the movement of pathogens through the lamina propria that underlies epithelia?

   A. Collagen fibers

   B. Ground substance

   C. Mast cells

   D. Plasma cells

   E. Reticular fibers

4. Marfan syndrome results from a genetic mutation in the gene that encodes the protein fibrillin. Patients with Marfan syndrome experience various symptoms, ranging from misalignment of the lens to fatal rupture of the aorta. Which component of connective tissue is most directly responsible for these wide-ranging symptoms?

   A. Collagen fibers

   B. Elastic fibers

   C. Ground substance

   D. Proteoglycans

   E. Reticular fibers

5. One warm spring day, a patient sees her physicians complaining that she has been sneezing and has a runny nose and that her eyes are watery and itchy. A physical examination indicates that her nasal passages are inflamed and swollen but she has no other signs of illness. She reports that she spent the previous day gardening and pruning her fruit trees. The physician determines that the patient's symptoms result from "hay fever," more properly known as allergic rhinitis, a mild allergic reaction to an allergen such as pollen or dust. Which type of cell is most responsible for triggering this patient's symptoms?

   A. Adipocytes

   B. Fibroblasts

   C. Macrophages

   D. Mast cells

   E. Plasma cells

ANSWERS

1—B: Collagen is a structural protein that ensures strength and minimizes flexibility. If the inherited disease affected collagen, the joints then would be more flexible and, therefore, more prone to injury. Actin, myosin, and fibrillin are proteins found in muscle. Elastin is a connective tissue protein; if it were damaged, the skin would lose flexibility. Keratin is a structural protein of the skin, fingernails, and hair.

2—B: Connective tissues are constructed from cells that are surrounded by extracellular matrix, which consists of fibrous proteins and ground substance.

3—B: Ground substance has a viscous, gel-like consistency, which helps impede invasion by pathogens. Successful pathogens often secrete hydrolytic enzymes, which can reduce the viscosity of extracellular matrix and facilitate spread of the invaders.

4—B: Elastic fibers consist of a core of the protein elastin deposited onto a thin sheath of microfibrils, including fibrillin, which is essential for the formation of elastic fibers.

5—D: Mast cells are responsible for allergic reactions that occur when the cells release granules upon exposure to an allergen. Chemical mediators stored in the granules, such as histamine, trigger the symptoms.

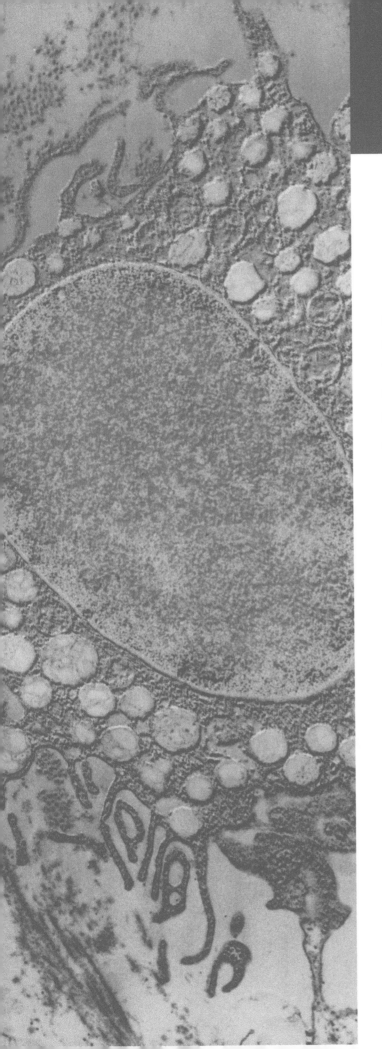

CHAPTER 3

CARTILAGE

OVERVIEW

The three types of cartilage—hyaline, elastic, and fibrocartilage—are connective tissues that provide specialized support to both soft and hard tissues, allowing them to resist external pressure.

- **Hyaline cartilage** is the most common. In adults, it is found arranged in rings and plates that help keep the walls of the large airways open, and it also lines the ends of bones at movable joints, producing resilient gliding surfaces.

- **Elastic cartilage** facilitates speech by providing flexible support to the larynx.

- **Fibrocartilage** links the vertebral bones together with shock-absorbing spacers.

As is the case with all connective tissues, the composition of the **extracellular matrix (ECM)** is the major determinant of the properties of cartilage, but cartilage differs from connective tissue proper in two significant ways: it is avascular and it contains only one cell type, the **chondrocyte**.

GENERAL FEATURES OF CARTILAGE

To understand the three types of cartilage, it is useful to consider the following four features (Figure 3-1A):

Chondrocytes. The only cells found in normal cartilage are chondrocytes, which are produced from progenitor cells called **chondroblasts**. When chondrocytes divide, the resulting daughter cells remain closely associated in clonal groups, called **isogenous groups** or **cell nests**, and are surrounded by the ECM they produce.

ECM. This material contains both fibrous proteins and ground substance. Fibrils formed from **type II collagen** are abundant in hyaline cartilage; elastic cartilage contains many elastic fibers. The chondrocytes in fibrocartilage produce large amounts of type I collagen and a lesser amount of type II collagen. Extremely high molecular weight proteoglycans, which bind large amounts of water, are produced by chondrocytes in many cartilages. The resulting matrix endows these tissues with high tensile strength combined with resiliency.

Perichondrium. This layer consists of a dense connective tissue capsule that covers most cartilages and contains blood vessels, nerves, and a resident population of chondroblasts. The fibroblasts in this layer secrete type I collagen.

Nutritional support. Because all cartilage is **avascular**, chondrocytes must receive nutrition by diffusion from adjacent structures, such as from blood vessels in the perichondrium. The high water content typical of the matrix facilitates this exchange.

TYPES OF CARTILAGE

HYALINE CARTILAGE

Hyaline cartilage is the most common type of cartilage. It supports the airways in the respiratory system from the nose to the bronchi in the lungs, covers the articular surfaces of bones at joints, and lines the ends of ribs where they meet the sternum. The development of long bones in the embryo begins with the formation of hyaline cartilage models that are soon replaced by bone, except for the **epiphyseal plates**, where the persistence of hyaline cartilage is responsible for the lengthening of bones during growth through adolescence (see Chapter 4). The ends of these plates remain as **articular surfaces** at **synovial joints**,

the joints with a space that allows significant movement, and is surrounded with a capsule and synovial membrane.

Cells. The large chondrocytes are located in gaps, or **lacunae**, in the matrix. The matrix immediately surrounding the chondrocytes is called **territorial**, and differs in composition and staining properties from **interterritorial** matrix. The cells in an isogenous group are separated by a thin band, or **septum**, of territorial matrix (Figure 3-1B).

Fibers. Collagens constitute about half of the mass of the extracellular material produced by the chondrocytes in hyaline cartilage. **Type II collagen** is the major form of fiber and assembles into small-diameter fibrils rather than the larger fibers and bundles produced by type I collagen. The lack of texture produced by large fibers contributes to the smooth, glassy appearance of this cartilage, both grossly and microscopically; fresh specimens are translucent. In fact, the name hyaline means glassy. The type II fibrils are associated with many other components, including type IX collagen and **chondronectin**, a structural glycoprotein that links the fibrils to proteoglycans in the matrix.

Ground substance. The proteoglycans in this cartilage form extremely large aggregates. For example, individual molecules of **aggrecan**, an abundant proteoglycan in hyaline cartilage, have molecular weights in the millions and are linked to chains of hyaluronic acid to form structures that are 100 times more massive (Figure 3-1C). The complex interactions among the proteoglycans and the collagen networks, along with the large amount of water bound by these components, are responsible for this tissue's mechanical properties. Territorial matrix contains a higher proportion of proteoglycans to collagen than interterritorial matrix; thus, territorial matrix usually stains more intensely.

Perichondrium. The inner surface of the perichondrium contains chondroblasts that occasionally differentiate into chondrocytes, move deeper, and become surrounded by matrix. These peripheral chondrocytes are usually smaller and form smaller isogenous groups than those found deeper in the tissue. Unlike hyaline cartilage in other locations, articular cartilage lacks a perichondrium (Figure 3-1D). The chondrocytes in these important structures receive nourishment from the fluid produced by cells of the synovial membrane. The chondrocytes in articular cartilage must maintain a low rate of mitosis throughout life to maintain a steady population of cells in this tissue, which lacks a source of chondroblasts.

A

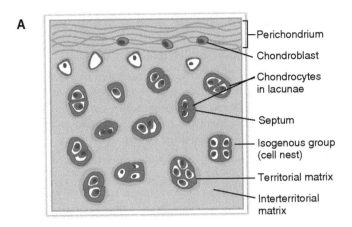

- Perichondrium
- Chondroblast
- Chondrocytes in lacunae
- Septum
- Isogenous group (cell nest)
- Territorial matrix
- Interterritorial matrix

C

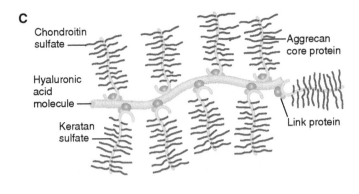

- Chondroitin sulfate
- Hyaluronic acid molecule
- Keratan sulfate
- Aggrecan core protein
- Link protein

B

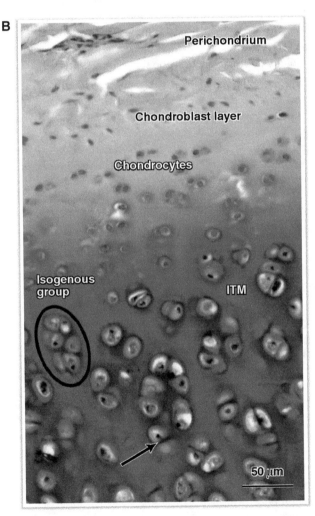

Perichondrium

Chondroblast layer

Chondrocytes

Isogenous group

ITM

50 μm

D

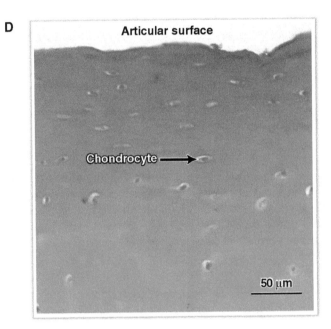

Articular surface

Chondrocyte →

50 μm

Figure 3-1: Features of cartilage; hyaline cartilage. **A.** General features of cartilage in an adult. All cartilage, except fibrocartilage and articular (hyaline) cartilage, is covered with a dense connective tissue perichondrium, which contains blood vessels, progenitor cells, and chondroblasts. Chondrocytes reside in lacunae in the matrix, and their cell division forms isogenous (clonal) cell groups. The matrix immediately surrounding cells (territorial matrix) differs in composition from the interterritorial matrix between isogenous groups. **B.** Hyaline cartilage from the larynx. The perichondrium is composed of thick bundles of collagen and a deeper, more cellular layer that contains chondroblasts. Below the cellular layer of the perichondrium are individual chondrocytes separated by matrix, which quickly adopts the dark staining characteristics of mature matrix. The interterritorial matrix (ITM) between the isogenous groups of chondrocytes stains less darkly than the territorial matrix. Chondrocytes in a group are separated by septae (*arrow*). **C.** An aggrecan–hyaluronic acid complex in the matrix of hyaline cartilage. Each aggrecan core protein can carry about 100 glycosaminoglycans (GAGs), typically chondroitin sulfate and keratan sulfate, yielding a molecular weight of two to three million. Dozens of aggrecans may be linked to hyaluronic acid, forming a single polymer several microns long. **D.** Hyaline cartilage at an articular surface. Chondrocytes are found as individual cells in lacunae that extend up to the surface of this tissue, which covers the ends of bones at synovial joints.

ELASTIC CARTILAGE

Elastic cartilage is resilient and flexible; it easily returns to its original shape after bending, and it possesses more flexibility than hyaline or fibrocartilage. This type of cartilage is an important component of the epiglottis and the larynx. It is also present in structures associated with the auditory system, both in the external portions (auricles) as well as in the external auditory canal and eustachian tubes.

■ **Cells.** Elastic cartilage typically contains a higher ratio of chondrocytes to matrix than found in hyaline cartilage, but usually has fewer cells per isogenous group.

■ **ECM.** In addition to collagen type II, the matrix contains many elastic fibers, which are revealed by stains specific for elastin (Figure 3-2A).

■ **Perichondrium.** Similar to hyaline cartilage, elastic cartilage is enveloped by a perichondrium.

▽ Unlike hyaline and fibrocartilage, elastic cartilage does not calcify with age. However, the elastic cartilage of the epiglottis often shrinks as people age and is partially replaced with adipose tissue (Figure 3-2B). ▼

FIBROCARTILAGE

Fibrocartilage provides high tensile strength, durability, and firm support (much like dense regular connective tissue), as well as resistance to compression (much like hyaline cartilage). It is found in the intervertebral discs, the symphysis pubis, and in the attachments of some tendons and ligaments to bone surfaces lined by hyaline cartilage (Figure 3-2C and D).

■ **Cells.** There are fewer chondrocytes in fibrocartilage than in hyaline or elastic cartilage, and they are often aligned along collagen bundles, either singly or in small isogenous groups.

■ **ECM.** Unlike hyaline and elastic cartilage, the chondroblasts of fibrocartilage secrete large amounts of type I collagen along with a small quantity of type II collagen. The type I molecules assemble into large fibers and bundles. The mass of type I collagen exceeds that of ground substance in fibrocartilage, another difference between this tissue and the other cartilages.

■ **No perichondrium.** In contrast to the other types of cartilage, fibrocartilage lacks a perichondrium. However, fibrocartilage is always found adjacent to a dense connective tissue, and the two tissues blend together where they meet.

▽ The fibrocartilage found in intervertebral discs weakens over time and can tear, allowing the gel of the **nucleus pulposus** (normally kept in the center of the disc) to extrude (**herniated disc**). Inflammatory signals arising from the torn cartilage along with the pressure placed on nearby spinal nerves by the extruded material can cause intense pain. ▼

GROWTH, REPAIR, AND DISEASE

CARTILAGE GROWTH AND REPAIR

Embryonic cartilage initially develops from a mass of progenitor cells that produce matrix, separate themselves into isogenous groups, and later may be covered by a connective tissue perichondrium. Subsequently, cartilage grows by two processes: appositional growth and interstitial growth.

■ **Appositional growth** occurs when chondroblasts in the perichondrium differentiate into chondrocytes, adding cells and matrix to the periphery of the existing cartilage body. This is the principal way that cartilage expands during development.

■ **Interstitial growth** is the expansion of cartilage produced by the division of resident chondrocytes and their production of additional matrix. Interstitial growth in the epiphyseal plates lengthens long bones and is the mode by which the articular cartilage of joints is maintained throughout life (see Chapter 4).

The matrix components in cartilage are continuously turned over and replaced. Chondrocytes release both the hydrolytic enzymes to degrade collagens and proteoglycans, as well as producing and secreting new components.

Cartilage is often slow to recover from damage, which could be due in part to its avascularity, where the perichondrium is ablated and the chondrocytes lose trophic support. Repair will depend first on the replacement of the perichondrium, but fibroblasts from this layer may proliferate and form a scar, preventing complete regeneration of the cartilage.

DISEASES OF CARTILAGE

■ **Osteoarthritis.** Osteoarthritis is a degenerative disease of the joints that results in the loss of hyaline articular cartilage. This is a significant health problem associated with aging that is not well understood. As the disease progresses, levels of both proteoglycans and type II collagen decrease in the matrix and chondrocytes die. The remaining chondrocytes initially divide more rapidly but are unable to produce matrix with the proper mechanical properties. The softened cartilage may disappear entirely, exposing the underlying bone, which becomes the joint surface. This results in changes in the bone, reduction in movement, and pain.

■ **Cartilage tumors.** Several tumors arise from bony structures that produce cartilage-like tissue, mostly with features of hyaline cartilage. Most of these growths, such as **chondromas**, are benign but can cause pain by interfering with nerves. **Chondrosarcomas** are tumors that can produce large nodules of disorganized hyaline cartilage. Most chondrosarcomas are slow growing and remain local; however, some do metastasize.

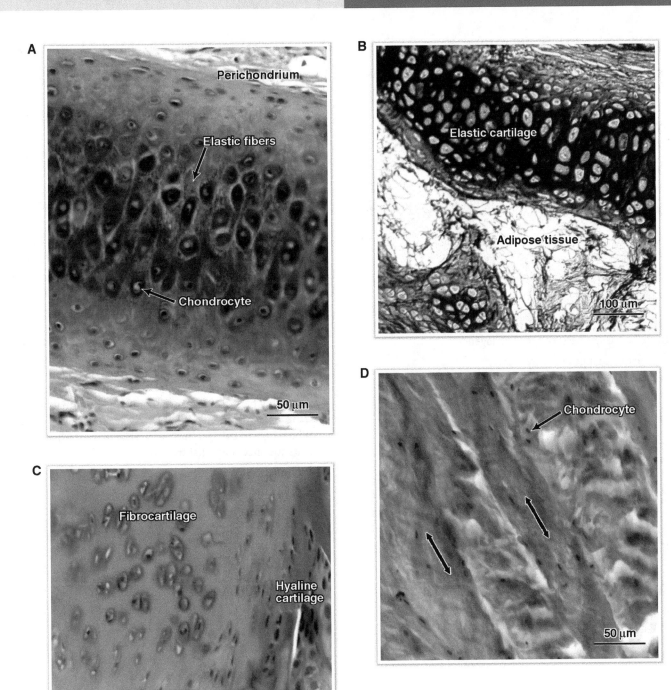

Figure 3-2: Elastic cartilage and fibrocartilage. **A.** Elastic cartilage from the external ear. The dense connective tissue of the perichondrium is evident at upper and lower surfaces of this cartilage. The chondrocytes are large and present mostly as single cells. The abundant elastic fibers appear as a wispy texture between the chondrocytes. (Verhoeff stain) **B.** Elastic cartilage from the epiglottis of an older person. Note the age-related changes in this tissue: the cartilage has shrunk and in places has been replaced with a disorganized tissue containing adipocytes. (Verhoeff stain). **C.** Fibrocartilage and hyaline cartilage in the developing pubic symphysis. The pubic symphysis is a joint composed of fibrocartilage that holds the two hip bones together at the front of the pelvis. Chondrocytes of the fibrocartilage are visible as single and double cells separated by bundles of type I collagen. The hyaline cartilage visible on the right side of this image will be replaced by bone as development proceeds. Fibrocartilage lacks a perichondrium; in this case, it merges smoothly with the presumptive perichondrium of the hyaline cartilage. **D.** Fibrocartilage of the annulus fibrosus of an intervertebral disc. Intervertebral discs consist of layers of fibrocartilage surrounding a central fluid-filled area, called the nucleus pulposus (not shown). The cartilage layers are assembled with bundles of collagen in one layer oriented at right angles to bundles in adjacent layers. The double-headed arrows indicate the orientation of collagen bundles in two layers; between them, the bundles are oriented perpendicular to the plane of the micrograph.

STUDY QUESTIONS

Directions: Each of the numbered items or incomplete statements is followed by lettered options. Select the **one** lettered option that is **best** in each case.

1. What is the molecular basis for shock absorption within articular cartilage?
   A. Attachment of chondrocytes to type I collage fibers
   B. Binding of anions by glycosaminoglycans
   C. Cross-linking of type I collagen fibers in the perichondrium
   D. Electrostatic interactions of proteoglycans with type IV collagen
   E. Hydration of glycosaminoglycans in proteoglycan aggregates

2. Which cartilaginous structure will be most severely affected by defective synthesis of type I collagen?
   A. Cartilage rings in the trachea
   B. Epiglottis
   C. Epiphyseal growth plates
   D. Intervertebral discs
   E. Larynx
   F. Nasal septum

3. Which of the following constituents of cartilage is most responsible for interstitial growth?
   A. Chondroblasts
   B. Chondrocytes
   C. Interritorial matrix
   D. Perichondrium
   E. Proteoglycan aggregates
   F. Territorial matrix

4. Cartilage is a specialized type of connective tissue, composed of cells, fibers, and ground substance. What is one functionally important difference between the composition of cartilage and connective tissue proper?
   A. Cartilage is avascular, whereas connective tissue proper is vascularized
   B. Cartilage cells are not mitotically active, whereas cells in connective tissue proper divide frequently
   C. Cartilage matrix contains no glycosaminoglycans (GAGs), whereas GAGs are abundant in connective tissue proper
   D. Cartilage matrix never contains elastic fibers, whereas elastic fibers are often found in connective tissue proper
   E. Cartilage matrix never contains type I collagen fibers, whereas type I collagen fibers are abundant in connective tissue proper

5. A 25-year-old woman is seen by her physician because of complaints of frequent nosebleeds, a chronic runny nose, and loss of sense of smell. She admits that she has been snorting cocaine on a regular basis. Physical examination shows that her nasal passages are inflamed and that her nasal septum has been degraded and has nearly disappeared. Which type of collagen has been released due to her dangerous substance-abuse habit?
   A. Type I collagen
   B. Type II collagen
   C. Type III collagen
   D. Type IV collagen
   E. Type X collagen

ANSWERS

1—E: Articular cartilage is composed of hyaline cartilage. The mechanical properties of hyaline cartilage, such as its ability to function as a shock absorber, result from complex interactions among the proteoglycan and collagen networks, as well as the large amount of water bound by these components. The major fiber in hyaline cartilage is type II, not type I, collagen. Articular cartilage lacks a perichondrium.

2—D: The matrix of all types of cartilage contains type II collagen. In addition, the matrix of fibrocartilage, which is found in intervertebral discs and the pubic symphysis contains type I collagen fibers.

3—B: Interstitial growth of cartilage occurs by the division of resident chondrocytes and their production of additional matrix. Interstitial growth in the epiphyseal plates lengthens long bones and is the mode by which articular cartilage is maintained throughout life.

4—A: Cartilage is avascular, which has important consequences. Chondrocytes must receive nutrition by diffusion from adjacent structures, such as from blood vessels in the perichondrium.

5—B: The nasal septum is constructed from hyaline cartilage, which contains only type II collagen in the extracellular matrix. Chronic intranasal use of cocaine can degrade the cartilage in the nasal septum.

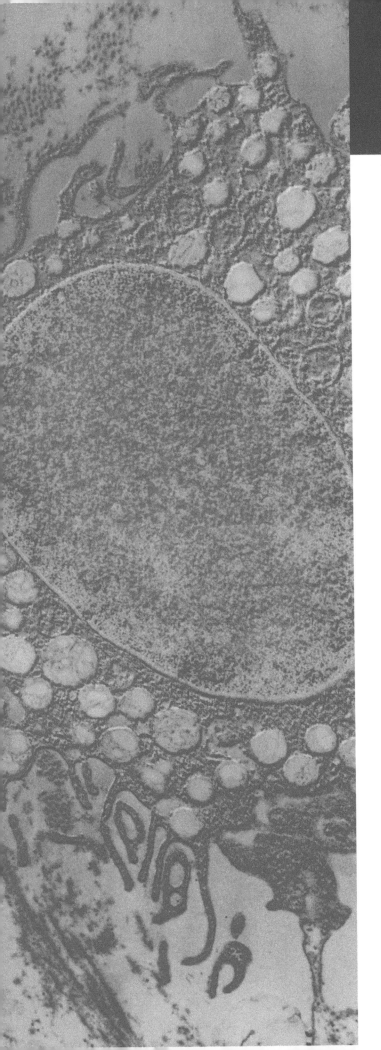

CHAPTER 4

BONE

OVERVIEW

Bone is a specialized connective tissue that comprises most of the skeleton, which supports the entire body. Unlike cartilage, bone tissue is extremely hard and inelastic. This allows the long bones to serve as levers on which muscles act to hold the body upright and move it through space. Bones also provide protective sites for housing the central nervous system (the brain and the spinal cord) as well as the hematopoietic tissue, which is responsible for forming blood cells. The extracellular matrix (ECM) of bone is hardened by the deposition of calcium salts. Bone is thus a reservoir of Ca^{2+}, allowing it to play an important role in regulating the Ca^{2+} concentration in the blood, a vital activity. Bones begin forming early in embryogenesis and continue growing for about 20 years after birth. Although bone is hard, it does not become a static tissue after growth stops. Bone tissue is turned over by a mechanism that maintains the shape and strength of bone even while portions are being removed and replaced every day. Bone fractures are initially repaired by processes that recapitulate those used in early development, and the restored tissue then is returned to the normal adult form by the continuous activity of turnover. Features of the structure, development, and physiology of bone are intertwined, which complicate a simple understanding of this interesting tissue.

GENERAL FEATURES OF BONE

Bone is a connective tissue; therefore, the various resident bone cells and the ECM are its key elements. However, bone is surprisingly complicated and to understand it fully, one must also consider interrelated features of its gross and microscopic organization, development, and physiology.

BONE CELLS

Bone is produced, maintained, and turned over by the following four distinct resident cells:

Osteoprogenitor cells. Fibroblast-like stem cells that divide to become osteoblasts. In adult bone, osteoprogenitor cells are quiescent, except in the repair of bone fractures or during the normal, localized turnover of bone, when they produce osteoblasts.

Osteoblasts. Responsible for the synthesis and secretion of the organic components of bone matrix, called **osteoid**, and its initial mineralization. Osteoblasts become osteocytes when they encase themselves in this matrix (Figure 4-1A). Osteoblasts are nonmitotic and can live for a few months and then either die, become quiescent bone-lining cells, or remain as long-lived osteocytes when trapped in bone matrix.

Osteocytes. The main resident cells of bone. Osteocytes sit in lacunae in the matrix that they maintain. These cells project long, thin processes that link osteocytes together via gap junctions by coursing through small channels, or **canaliculi** (Figure 4-1B). Osteocytes can live 20 to 30 years.

Osteoclasts. Large, multinucleated cells responsible for the removal and turnover of bone matrix. Osteoclasts are derived from the same stem cells that produce macrophages; they are formed by the fusion of many cells and live less than 1 month. They dissolve bone matrix by secreting matrix-degrading enzymes (e.g., an acid-stimulated collagenase) and hydrogen ions to dissolve bone salts. Osteoclasts sit in pits, which they form in bone, called **Howship's lacunae** (Figure 4-1C). The opposing activities of osteoblasts and osteoclasts are carefully coordinated to increase and sculpt bone during embryonic development and postnatal growth through adolescence and to maintain bone mass in adults.

EXTRACELLULAR MATRIX

As with all connective tissues, the cells are surrounded by ECM, which consists by dry weight of equal amounts of organic components, fibrous proteins plus ground substance, and inorganic salts.

Organic components. Type I collagen constitutes 90% of the organic material of ECM and is associated with glycosaminoglycans, proteoglycans, and several structural glycoproteins involved in matrix assembly.

Inorganic components. Calcium phosphate is the main inorganic component in ECM and is present as hydroxyapatite crystals, $Ca_{10}(PO_4)_6(OH)_2$, as well as in other forms. The crystals are arrayed with collagen fibrils and ground substance, forming an inelastic composite material of considerable strength for its weight.

BONE SURFACES

Bone surfaces are lined with connective tissues that contain osteoprogenitor cells (Figure 4-1D).

Periosteum. This tissue covers the outer surface and contains blood vessels and nerves as well as fibroblasts that elaborate collagen fibers, called **Sharpey's fibers**, which penetrate the bone matrix and link this layer to the bone. **Osteoprogenitor cells** are found near the bone surface.

Endosteum. This tissue covers the inner surface, and generally consists of a simple layer of osteoprogenitor cells and quiescent osteoblasts.

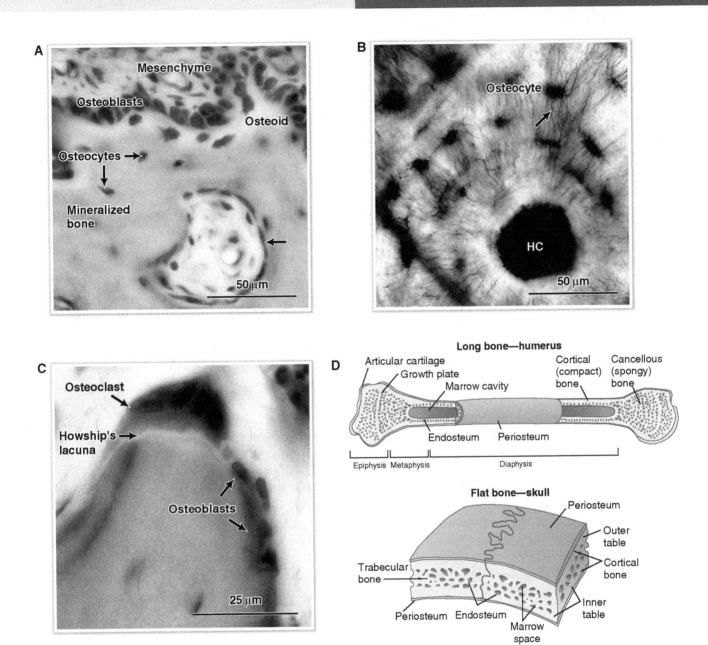

Figure 4-1: Resident cells and general structure of bone. **A.** Osteoblasts and osteocytes in a forming bone. Cuboidal active osteoblasts line the surface of an embryonic bone. Some osteoblasts become osteocytes as they are trapped in the osteoid and mineralized bone. The arrow indicates thin, inactive osteoblasts and osteoprogenitor cells in the endosteum. The bone salts were extracted by chemical treatment before preparing this section. **B.** Osteocytes and canaliculi in adult bone. This is a thin section of mineralized bone that was ground, polished, and treated with ink to show the lacunae that house the osteocyte cell bodies and the thin canaliculi (*arrow*) occupied by the thin processes that link the cells together. HC, Haversian canal. (Ground bone, India ink) **C.** Osteoclast. A large, multinucleated osteoclast has created a pit, or Howship's lacuna, on the surface of a developing bone (*arrow*). **D.** Surface coverings and gross features of long and flat bones. Bones are constructed with a dense outer layer of compact bone and supported internally with thin, porous spongy bone. The outer bone surface is covered with the periosteum, a dense connective tissue, and the inner surfaces are covered with endosteum, a thin layer of simple osteoprogenitor and inactive osteoblasts called the endosteum.

ORGANIZATION OF BONE

GROSS ORGANIZATION OF BONE

Bones are constructed from two macroscopic types of bone tissue: compact bone and spongy bone.

- **Compact bone.** The outer portion of bone consists of a solid mass of **cortical** or **compact bone.**

- **Spongy bone.** The interior region, or marrow cavity, of bone is supported by a very porous structure, variously named **cancellous, trabecular,** or **spongy bone,** that consists of thin struts called **trabeculae.**

Both compact and spongy bones have the same histologic structure. The open space in the marrow cavity is filled with either adipose tissue (yellow bone marrow) or hematopoietic tissue (red bone marrow) and contains blood vessels that supply the inner bone surface. Figure 4-2A illustrates the organization of compact and spongy bone in long and flat bones.

MICROSCOPIC ORGANIZATION OF BONE

COLLAGEN ASSEMBLIES IN BONE Osteoblasts produce two arrangements of collagen fibers in bone matrix, resulting in two microscopic structural organizations.

- **Primary bone.** The collagen fibers first deposited in osteoid by osteoblasts (e.g., during embryogenesis or fracture repair in an adult) are **randomly** arrayed. If this material is mineralized, the result is called **primary** or **woven bone.**

- **Secondary bone.** Primary bone is eventually removed entirely or replaced with **secondary** or **lamellar bone,** in which the collagen fibers are arranged in layers with all fibers in a **parallel** layer.

- Woven bone is more quickly assembled, but lamellar bone with its plywood-like structure is stronger.

OSTEONS AND VASCULARIZATION Much of the volume of **compact bone** requires an internal blood supply that is organized in structures called **osteons** or **Haversian systems,** the main organizational units of this bone (Figure 4-2A).

- Osteocytes near bone surfaces receive nutrition from nearby blood vessels and can transfer nutrients to neighboring osteocytes by diffusion through gap junctions at the ends of their slender processes in **canaliculi.** However, the thickness of compact bone renders diffusion insufficient to supply distant cells.

- Osteons, found deeper in compact bone, are organized around central **Haversian canals** that run parallel to the long axis of the bone and contain blood vessels supported by loose connective tissue. Haversian canals are lined by endosteum and are surrounded by concentric arrays of linked osteocytes, as many as 20 cell layers, forming an osteon. **Volkmann's canals** run radially in bone, interconnecting the vessels and nerves in Haversian canals and linking them to sources in the periosteum and marrow cavity.

LAMELLAE All secondary bone, whether found in areas of compact or spongy bone, is lamellar bone. A section of compact bone is shown in Figure 4-2B and one of spongy bone in Figure 4-2C.

- Collagen fibers in the **cylindrical lamellae** of osteons run in helices with alternating windings so the fibers in adjacent lamellae are oriented at 90 degrees, adding great strength to these structures.

- The surfaces of compact bone are organized in **inner** and **outer circumferential lamellae** that run circularly around the shaft of a long bone.

- Some **interstitial lamellar** bone is found bridging the spaces between osteons, which is largely the result of **bone remodeling** (see Maintenance of Bone, below).

- **Trabecular lamellae** in spongy bone tend to run along the length of the trabeculae (Figure 4-2C).

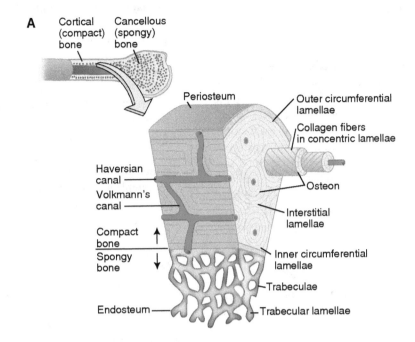

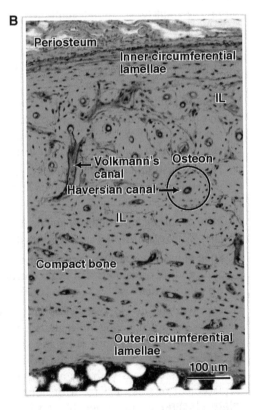

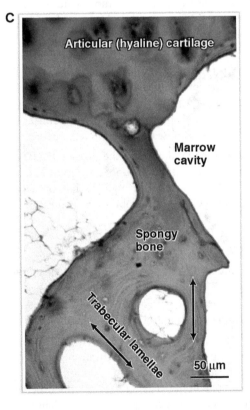

Figure 4-2: Microanatomy of adult bone. **A.** Organization of lamellar bone. The central portion of adult compact bone is composed of cylindrical osteons oriented with the long axis of the bone. The Haversian canal at the center of each osteon contains a blood vessel to nourish internal osteocytes. Collagen fibers in one cylindrical lamella are aligned at right angles to adjacent lamella. Interstitial lamellae are found between osteons, and the inner and outer surfaces of compact bone are organized into circumferential lamellae. Volkmann's canals run radially to the osteons, linking them to blood vessels in the periosteum and marrow cavity. Trabeculae supporting the marrow cavity are also composed of lamellar bone. **B.** Compact bone in a cross-section of the shaft of a long bone, showing the structural elements diagrammed in part A. Interstitial lamellae (IL) appear as irregular areas between the osteons. **C.** Trabecular bone in the epiphysis of a long bone. The lamellae are apparent as thin lines and are generally oriented parallel to large surfaces in spongy bone (*arrows*). This section was taken at the base of the articular cartilage of the long bone; note the hyaline cartilage at the top of the image.

DEVELOPMENT OF BONE

Bone formation begins early in embryogenesis as mesenchymal cells coalesce under the control of homeobox genes in regions from which specific bones arise. Some cells differentiate into osteoprogenitor cells that produce osteoblasts and, in the case of long bones, chondroblasts and cartilage also are formed. Bone formation then commences by one of two processes, either intramembranous ossification or endochondral ossification. Because bone matrix soon becomes inflexible, bone expands by appositional growth as osteoblasts add material to an outer surface. To increase the diameter of a bone while maintaining a relatively hollow marrow cavity, osteoclasts must remove bone from the inside. The combination of differential addition and subtraction of material to create the shape of a growing bone is called **modeling**.

INTRAMEMBRANOUS OSSIFICATION

In intramembranous ossification, **osteoblasts** begin secreting **osteoid** directly into portions of the condensed **mesenchymal connective tissue**.

- **Primary bone** is formed in discontinuous patches separated by areas of connective tissue, and the bony patches are able to merge as they expand by appositional growth.

- Modeling proceeds and as growth slows, discontinuous areas on the outer surface merge, forming compact bone. Some remaining islands of connective tissue are converted into Haversian canals as osteons form. The bony patches in the marrow cavity become spongy bone, and the connective tissue is replaced with adipose or hematopoietic tissue.

Most of the skull, the mandible, and the maxilla of the jaw, as well as the diaphyseal shafts of long bones, are produced in this way. Figure 4-3A shows a section through the shaft of a developing limb bone undergoing intramembranous ossification.

ENDOCHONDRAL OSSIFICATION

In endochondral ossification, cartilage models of bones are first elaborated by **chondroblasts** in the condensed **mesenchyme** and, beginning at 7 weeks' gestation, portions of the cartilage matrix begin being replaced with bone. The development of long bones utilizes both endochondral and intramembranous ossification to modify the initial cartilage model (Figure 4-3B and C).

- **Bone collar.** Intramembranous ossification begins in the perichondrium along the middle of the presumptive diaphysis, forming a **bone collar**. The underlying chondrocytes are then programmed to hypertrophy, deposit calcium phosphate into the matrix, and die.

- **Primary ossification center.** Blood vessels penetrate the bone collar through channels created by **osteoclasts**. **Osteoprogenitor cells** enter and produce **osteoblasts**, which begin depositing primary bone on the calcified cartilage matrix, creating the **primary ossification center** in the shaft of the bone. As bone modeling proceeds, the diaphysis expands in width by intramembranous ossification and the marrow cavity is enlarged by osteoclasts.

- **Secondary ossification center.** After birth, blood vessels invade the epiphyses, creating a **secondary ossification center** near the ends of the bone. A band of hyaline cartilage, the **epiphyseal growth plate**, remains between the primary and secondary centers and continues to expand and supply matrix for ossification. The bone increases in length in this fashion until this plate of cartilage disappears, sometime in the second decade of life. A rim of hyaline cartilage, the **articular cartilage**, remains on the ends of the epiphyses to become part of a synovial joint.

FUNCTIONAL ZONES OF THE EPIPHYSEAL PLATE The cartilage in the epiphyseal growth plate displays a set of recognizable zones related to its interstitial growth and development (Figure 4-3B). From the **epiphyseal** to the **metaphyseal side**, these zones are as follows:

- **Zone of resting** (or reserve) **cartilage.** Chondrocytes in this zone serve as a reservoir of cells to supply the rest of the zones.

- **Zone of proliferation.** Chondrocytes are actively dividing in an orderly fashion, which creates columns of cells parallel to the long axis of the bone. The cells secrete type II collagen and other matrix components typical of hyaline cartilage.

- **Zone of maturation and hypertrophy.** The cells cease dividing, swell markedly, and secrete type X collagen and proteins that promote calcification.

- **Zone of calcification and cell death.** As the matrix accumulates hydroxyapatite, the chondrocytes die.

- **Zone of ossification.** Osteoprogenitor cells invade the matrix and produce osteoblasts that begin creating woven bone on the calcified matrix.

The growth and development of epiphyseal cartilage is influenced by various signaling molecules, including somatotropin, thyroid hormone, sex hormones, and several growth factors produced by the chondrocytes and nearby cells. The mix of these factors changes with age and results in the complete replacement of this cartilage with bone, termed **epiphyseal plate closure**, at about age 20, at which point elongation ceases.

A Intramembranous ossification

Mesenchyme · Osteoblasts · Mesenchyme · Osteocytes · Primary bone · Osteoclasts · Marrow cavity · 50 μm

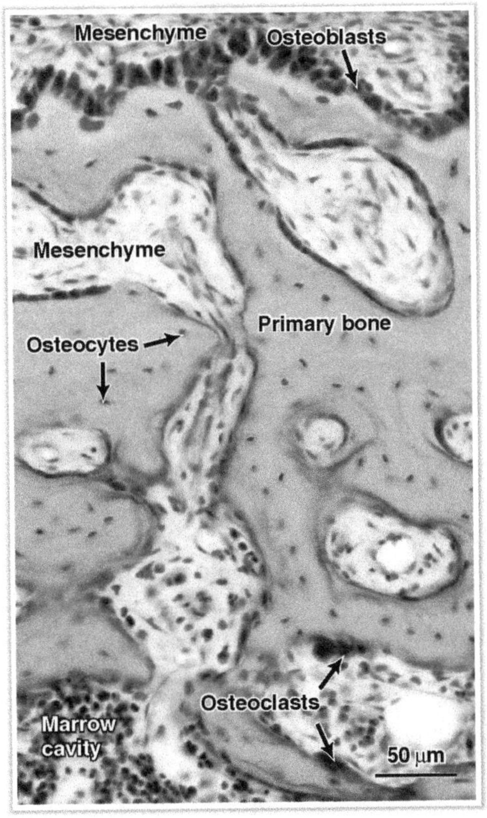

B Endochondrial ossification

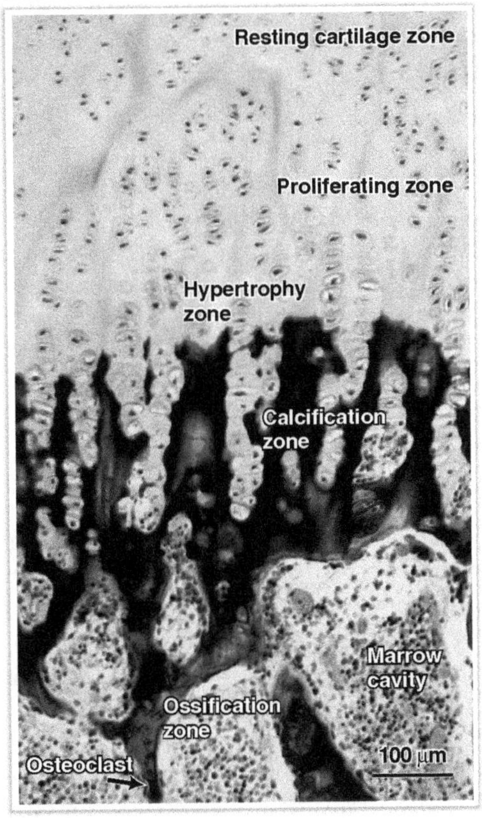

Resting cartilage zone · Proliferating zone · Hypertrophy zone · Calcification zone · Marrow cavity · Ossification zone · Osteoclast · 100 μm

C

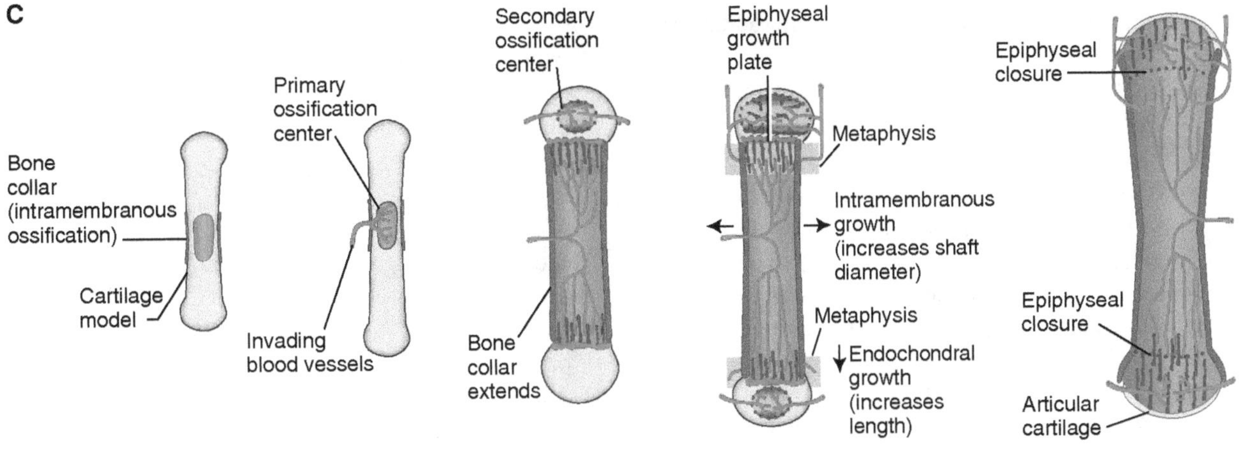

Bone collar (intramembranous ossification) · Cartilage model · Primary ossification center · Invading blood vessels · Secondary ossification center · Bone collar extends · Epiphyseal growth plate · Metaphysis · Intramembranous growth (increases shaft diameter) · Metaphysis · Endochondral growth (increases length) · Epiphyseal closure · Epiphyseal closure · Articular cartilage

Figure 4-3: Bone development. **A.** Intramembranous ossification in a cross-section through a canine embryonic limb bone. At the top of the image a row of osteoblasts, which arose from osteoprogenitor cells in the mesenchyme, is actively adding material to the outer surface (appositional growth). The patches of primary bone are interspersed between islands of mesenchymal connective tissue; some may remain to become Haversian canals of forming osteons. The marrow cavity, filled with small, dark hematopoietic cells, is seen at the bottom of the image. Osteoclasts are frequently seen at the marrow side of the bone. **B.** Endochondral ossification at the epiphyseal growth plate in a human embryonic long bone. Five zones of cartilage are labeled from top to bottom: resting cartilage zone, proliferating zone, hypertrophy zone, calcification zone, and ossification zone. The organized columns of proliferating and hypertrophic chondrocytes produce interstitial expansion that moves the epiphyseal plate toward the top, away from the midpoint of the diaphysis. Hypertrophic chondrocytes are intermixed with those undergoing calcification (dark blue areas). In the pink regions, the calcified cartilage has been converted to bone, which is then pared away by osteoclasts on the marrow surfaces. **C.** Long bone development by endochondral and intramembranous ossification. The diagram illustrates how a hyaline cartilage model of a long bone is converted into a growing bone, beginning with the formation of a collar around the middle of the diaphysis. The complete ossification (closure) of the epiphyseal growth plate occurs sometime in the second decade of life and ends the lengthening process.

MAINTENANCE OF BONE

Bone is a dynamic connective tissue that undergoes constant renewal and change.

REMODELING

During the first year of life, all of the **primary bone** produced to that point is replaced with **secondary bone**, except for the areas involved in active expansion. Secondary bone, along with any newly formed primary bone, is then continuously turned over and replaced with new secondary bone throughout life. This process is called **remodeling**. Remodeling rates are high in children. Once growth stops, remodeling slows, although about 10% of the bone mass in adults is in the process of being actively remodeled at all times.

Remodeling involves the organized excavation of a small volume bone by osteoclasts, followed by tissue replacement and resulting in the creation of a new lamellar bone. Remodeling in the middle of compact bones results in the formation of new osteons (Figure 4-4A). A brief description of this process begins with osteoclast formation from monocytes.

1. Osteoblasts can display, on their surface, the RANK-ligand (RANKL) protein, a ligand for a growth factor receptor present on monocytes. Osteoblasts can also secrete the cytokine macrophage colony-stimulating factor (M-CSF), which interacts with monocytes. These two signals from osteoblasts induce the fusion of monocytes to form osteoclasts.

2. Osteoblasts, and perhaps osteocytes, can regulate both the production and activity of osteoclasts by secreting additional positive and negative signaling molecules.

3. A set of osteoclasts bores a cylindrical tunnel in bone; osteoblasts line the tunnel and begin filling in the gap with new matrix. The factors responsible for initiating the large-scale organized osteoclast activity are not well understood.

4. A small blood vessel grows down the center of the tunnel to supply nutrition and becomes the central element of a Haversian canal, which is forming. Imprecise overlapping of old and new osteons is responsible for the creation of the interstitial lamellae seen in compact bone.

The continuous processes of modeling and then remodeling allow bone to respond to physical stresses, helping to shape the skull to accommodate the growing brain or, for example, to allow the orthodontist's braces to realign teeth in the jaw.

▽ **Bone fractures** result in the rupture of blood vessels in the bone, clot formation, and the death of osteocytes and loss of bone matrix (Figure 4-4B). The periosteum and endosteum proliferate, and osteoblasts secrete a matrix that temporarily bridges the gap that forms. A **callus** of connective tissue and hyaline cartilage then forms in the area and is eventually replaced with primary bone by endochondral ossification. Remodeling replaces the primary bone with secondary bone, and physical stresses guide the process to restore a normal structure. ▼

▽ Remodeling maintains a constant bone mass in individuals from the age of 20 to 40 years, at which point osteoclast activity begins to predominate. The resulting loss of mass can eventually result in porous bones, which are highly susceptible to fracture. This **senescent osteoporosis** is a significant problem of the aging population and tends to affect postmenopausal women more severely. ▼

CALCIUM HOMEOSTASIS

Ca^{2+} levels in blood are precisely maintained because this element is critical for many types of intracellular signaling, such as initiating muscle contraction. Bones contain more than 90% of the body's Ca^{2+} and play a key role in this process, which involves both simple chemistry and hormonal controls (Figure 4-4C). Ca^{2+} can be quickly released from bone by direct dissociation of calcium salts, which is thought to occur mainly from the surfaces of spongy bone and newly formed lamellar bone and near osteocyte membranes. In addition, there are two hormones involved with the regulation of blood Ca^{2+} that affect bone physiology: parathyroid hormone and calcitonin.

Parathyroid hormone (PTH). PTH binds to receptors on osteoblasts, causing them to slow bone production, display RANKL, and secrete M-CSF, which act to increase the number and activity of osteoclasts, resulting in the release of Ca^{2+} by matrix degradation. PTH and vitamin D also act to increase blood Ca^{2+} by stimulating absorption in the intestines and resorption from the kidney tubules.

Calcitonin. This hormone, which is produced in the thyroid gland, acts directly to inhibit the activity of osteoclasts. This will result in the decrease of Ca^{2+} in the blood; however, the hormone does not appear to have a major influence on regulating bone density or structure.

▽ In addition to osteoporosis, skeletal tissues are affected by a wide variety of pathologies that are associated with aging. Many of these diseases, including the three discussed below, result from the perturbations of the basic processes described previously in this chapter.

Achondroplasia is a common cause of dwarfism and results from mutations in a growth factor receptor gene expressed in cartilage cells of the epiphyseal growth plate. The cartilage is thin and disorganized and closure occurs prematurely. Intramembranous growth is not affected and, as a result, bones tend to be short and relatively thick.

Osteopetrosis is a rare condition produced by defects in the activity of osteoclasts. Bone modeling and remodeling are inhibited, and the result is the formation of thick and misshapen bones, which are also weak due to the persistence of woven bone.

Osteosarcomas are malignant tumors that begin in a bone and are characterized by the production of disorganized, woven bone. The tumor invades and replaces normal bone; it may first be detected by a fracture at the site of the tumor. ▼

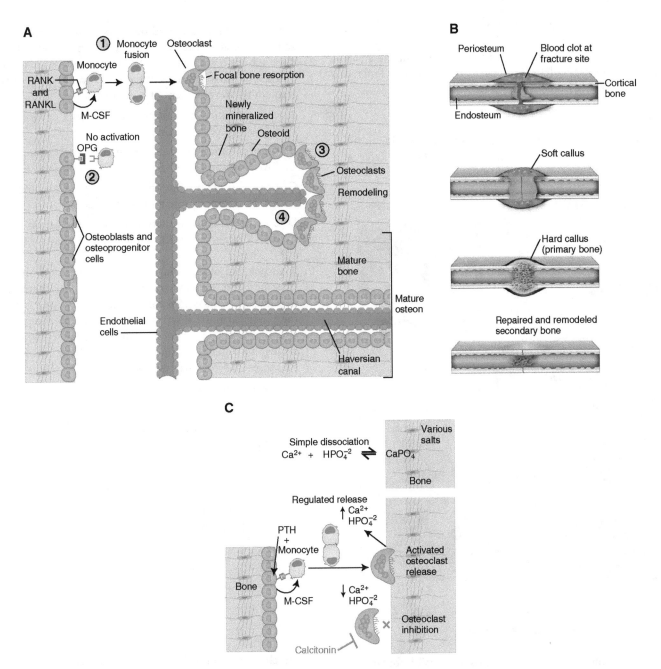

Figure 4-4: Bone maintenance, repair, and physiology. **A.** Bone remodeling. This highly schematic figure represents a Volkmann's canal and adjacent tissue in compact bone to summarize osteoclast formation and activity. (1) Osteoblasts can express the signaling protein RANK-ligand (RANKL) on their surface and secrete macrophage colony-stimulating factor (M-CSF). (2) These two signals stimulate the formation of osteoclasts from monocytes, which express the RANK receptor for RANKL. Alternatively, osteoblasts can secrete osteoprotegerin (OPG) to cover RANKL and reduce the rate of monocyte fusion. (3) Individual osteoclasts will resorb limited amounts of bone, whereas groups of osteoclasts will tunnel into bone, eventually remodeling a volume into a new osteon. (4) The tunnel is lined by osteoprogenitor cells and osteoblasts that fill in the space created with new bone tissue. A blood vessel and associated loose connective tissue also enter as the basis of a Haversian canal to supply the new osteon that forms behind the osteoclasts. **B.** Bone fracture repair. A fracture in compact bone interrupts the blood supply to the central areas, resulting in the formation of a blood clot, the death of osteocytes, and the loss of some additional bone tissue. Signals generated by the clot and cell death stimulate the proliferation of cells in the periosteum and endosteum, which penetrate the clot and produce a soft callus of connective tissue and hyaline cartilage. This is then converted to a hard callus of primary bone via intramembranous and endochondral ossification and eventually remodeled into secondary bone. **C.** Bone regulation of body Ca^{2+}. Bone contains a large reservoir of Ca^{2+} and phosphate. Exchange between bone salts containing Ca^{2+} and phosphate and their soluble ionic forms in extracellular fluid help maintain constant levels of these ions in blood. Parathyroid hormone (PTH) stimulates the release of Ca^{2+} from bone matrix by an indirect mechanism. PTH binds to osteoblasts, causing them to express RANKL and secrete M-CSF, stimulating the production of osteoclasts, which then degrade bone. Calcitonin binds directly to osteoclasts, inhibiting their activity, and thus decreasing the release of Ca^{2+} from bone into the blood.

STUDY QUESTIONS

Directions: Each of the numbered items or incomplete statements is followed by lettered options. Select the **one** lettered option that is **best** in each case.

Questions 1–3

The micrograph below shows an image of bone forming in the diaphysis (shaft) of an embryonic canine limb bone. Refer to the image when answering the following questions.

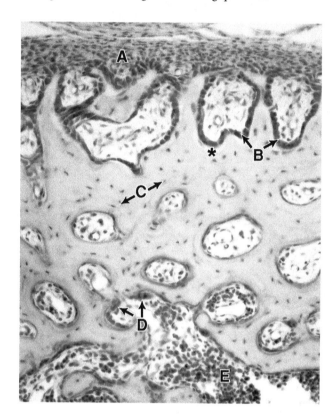

1. What site indicates cells that are most responsible for maintaining the bony matrix in adult bone?

2. What site indicates cells that are most responsible for enlarging the marrow cavity in the diaphysis as long bones grow?

3. What is the predominant type of collagen in the area marked with an asterisk?
   A. Type I collagen
   B. Type II collagen
   C. Type III collagen
   D. Type IV collagen
   E. Type X collagen

4. Osteopetrosis is a rare condition produced by defects in the activity of osteoclasts. Modeling and remodeling normally does not occur in patients with osteopetrosis, which results in thick, misshapen bones. What defect in osteoblasts could result in the development of osteopetrosis?
   A. Gain-of-function mutation in the gene encoding the RANK receptor protein in osteoblasts
   B. Gain-of-function mutation in the gene encoding the RANK-ligand (RANKL) protein in osteoblasts
   C. Loss-of-function mutation in the gene encoding the RANK receptor protein in osteoblasts
   D. Loss-of-function mutation in the gene encoding the RANK-ligand (RANKL) protein in osteoblasts

5. A 60-year-old man is diagnosed with lymphoma and now must undergo bone marrow testing to further characterize and stage the disease. To collect a sample of bone marrow, the physician must insert a large needle into the patient's ileum to collect a marrow aspirate. Using the following key, arrange in order the structures that the large needle must pass through as the marrow is collected.

   1 = skeletal muscle
   2 = endosteum
   3 = inner circumferential lamellae
   4 = marrow
   5 = osteons (Haversian systems)
   6 = outer circumferential lamellae
   7 = periosteum

   A. 1 − 2 − 3 − 7 − 5 − 6 − 4
   B. 1 − 2 − 6 − 3 − 5 − 7 − 4
   C. 1 − 7 − 3 − 5 − 6 − 2 − 4
   D. 1 − 2 − 3 − 5 − 7 − 6 − 4
   E. 1 − 7 − 6 − 5 − 3 − 2 − 4
   F. 1 − 7 − 3 − 5 − 2 − 6 − 4

6. Which of the following structures directly connects the Haversian canal in one osteon to Haversian canals in adjacent osteons?
   A. Canaliculi
   B. Caveoli
   C. Howship's lacunae
   D. Trabeculae
   E. Volkmann's canals

ANSWERS

1—C: Osteocytes are the main resident cells in mature bone. Osteocytes sit in lacunae in the matrix that they maintain.

2—D: Osteoclasts are large, multinucleated cells that are responsible for the removal and turnover of bone matrix. As bones grow and the diaphysis expands in width by intramembranous ossification (as seen in this image), osteoclasts enlarge the marrow cavity.

3—A: The asterisk marks an area of osteoid, the organic components of newly deposited bone matrix. Type I collagen constitutes 90% of the organic material in bone matrix.

4—D: Osteoblasts can express the signaling protein RANKL on their surface, which can bind to RANK receptor on monocytes and cause monocytes to fuse and form osteoclasts. Osteoclasts are required for modeling and remodeling of bone. Loss of RANKL on osteoblasts is expected to result in a decrease in the number of active osteoclasts, and therefore could contribute to osteopetrosis.

5—E: The outer portion of bone consists of a solid mass of compact cortical bone, composed of osteons. The surfaces of compact bone are organized into outer and inner circumferential lamellae, between which are sandwiched osteons. The outer surface is covered with connective tissue periosteum, and the inner surface adjacent to the marrow cavity is lined with endosteum. Thus, a needle would pass through periosteum, outer circumferential lamellae, osteons, inner circumferential lamellae, and endosteum to reach the marrow cavity.

6—E: Osteons are organized around central Haversian canals that run parallel to the long axis of the bone. Volkmann's canals run radially in bone, interconnecting the Haversian canals and linking them to the periosteum and marrow cavity.

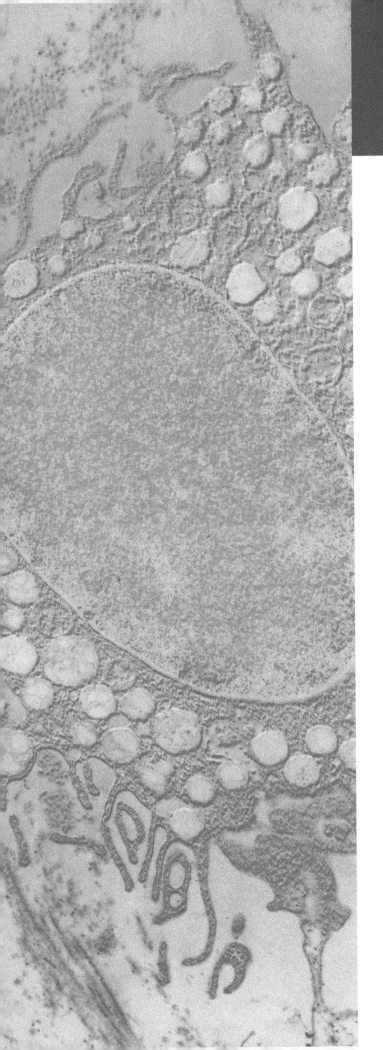

CHAPTER 5

MUSCLE TISSUE

OVERVIEW

The body is composed of four basic tissues: epithelium, connective tissue, muscle tissue, and nervous tissue. Of these four tissues, muscle tissue has a simple but vital role—to contract.

- **Skeletal muscle** powers behavior; it moves limbs, propels the breath, and is involved in producing speech.
- **Cardiac muscle** pumps blood.
- **Smooth muscle** controls blood pressure in vessels and helps to move food through the digestive tract.

Contractions of these three muscle types differ in speed, force, frequency, and extent, but all are initiated by increases in Ca^{2+} concentration, which triggers the movements of the actomyosin filament systems and shortens the cells. The different modes of contraction by the three types of muscle result both from variations in the composition and array of their actomyosin systems and from differences in membrane organization and organelle distribution. The large-scale, regular array of the filaments in skeletal and cardiac muscle produces ordered structures, known as striations, which are visible in the light microscope. Actomyosin arrays in smooth muscle are less abundant, and no special structures are seen on standard sections.

Although contraction is the obvious and main function of muscle, these cells play other roles in the complex economy of the body. The abundant protein in skeletal muscle can be catabolized as required to support maintenance of other tissues. Cardiac muscle cells produce hormones important for the regulation of fluid balance in the body. Smooth muscle in the large blood vessels produces elastic fibers important for their resilience. Myoepithelial cells are a fourth type of contractile cell (see Chapter 1).

This chapter begins with a description of the anatomy of skeletal muscle, followed by a review of the molecular components of contraction common to all muscle and an examination of the specialized features found in skeletal muscle. The discussion then turns to cardiac and smooth muscle. Features of growth and repair are considered for each muscle type.

SKELETAL MUSCLE

This section includes an overview of the components of the molecular machinery that produce contraction in all muscle as well as a discussion of some of the specializations found in skeletal muscle. The anatomy of skeletal muscle will be described first to introduce some specific terminology.

ORGANIZATION OF SKELETAL MUSCLE

Muscles that move the skeleton are composed of long individual cells called **myofibers**. Skeletal muscles are built from myofibers and a set of three supportive connective tissue wrappings: endomysium, perimysium, and epimysium (Figure 5-1A).

Endomysium. A thin extracellular layer consisting of basal lamina and fine reticular fibers that surround each myofiber.

Perimysium. A more dense layer of connective tissue that surrounds groups of myofibers and organizes the myofibers into **fascicles.**

Epimysium. A dense connective tissue that surrounds the entire muscle and is continuous at the ends with tendons. Inside the muscle, the epimysium is continuous with portions of the perimysium surrounding the fascicles. Contraction of myofibers is transmitted through the set of connective tissue wrappings and then to tendons to move a part of the body.

Blood vessels and nerves supplying the muscle course in the connective tissue wrappings. Extensive capillary networks are found in the endomysium surrounding every myofiber. Motor nerve terminals on myofibers are surrounded by specialized regions of the endomysium. Cross-sections of skeletal muscle typically reveal endomysium surrounding each fiber and portions of the perimysium (Figure 5-1D).

STRUCTURE OF SKELETAL MYOFIBERS

Myofibers are large cells that vary in diameter (typically about 100 μm) and length (from millimeters to centimeters). They are formed during development by the fusion of many individual **myoblast** cells, which results in syncytial multinucleated fibers. The plasma membrane of a myofiber is called the **sarcolemma**. The numerous nuclei in each myofiber are found directly beneath the sarcolemma. Almost all of the cytoplasm consists of densely packed **myofibrils** composed of linear chains of **sarcomeres**.

Myofibril. A cylindrical array of actin and myosin filaments, 1–3 μm in diameter, that extends the length of the cell (Figure 5-1B). Mitochondria are occasionally found between myofibrils and usually occupy only a small percentage of the volume of these cells.

Myofibrils have a repeating structure of alternating dark areas (**A bands**) and light areas (**I bands**), which become visible in the microscope by staining or by illumination with polarized light. The letters A and I are derived from polarized light microscopy and represent anisotropic and isotropic, respectively. Each I band is bisected by a thin, dark line called a **Z line**, or **Z disc**.

Sarcomere. The region of a myofibril between two Z lines, which are about 2.5 μm apart in resting muscle (Figure 5-1C). The sarcomere is the contractile unit of skeletal muscle. As sarcomeres contract, Z lines move closer together. Intermediate filaments hold Z lines together in adjacent myofibrils, and the resulting wide-scale registration of sarcomeres causes this muscle to appear striped, or **striated**, as can be clearly seen in longitudinal sections (Figure 5-1E).

GROWTH AND REPAIR OF SKELETAL MUSCLE

Growth. In humans, the maximal number of myofibers per skeletal muscle appears to be established soon after birth. Muscles must gain length, of course, until body growth stops, which is accomplished in part by the fusion of **satellite cells** with myofibers to support growth with additional nuclei. Satellite cells are unfused myoblasts that persist in skeletal muscle after birth, residing in the endomysium around myofibers. Growth of skeletal muscle resulting from exercise may also involve fusion of satellite cells as well as increased synthesis to produce additional myofibrils and larger myofibers.

Repair. Satellite cells serve as stem cells in adult skeletal muscle and may be capable of repairing limited damage by dividing to increase their number and fusing to restore myofibers. Unfortunately, this ability is limited.

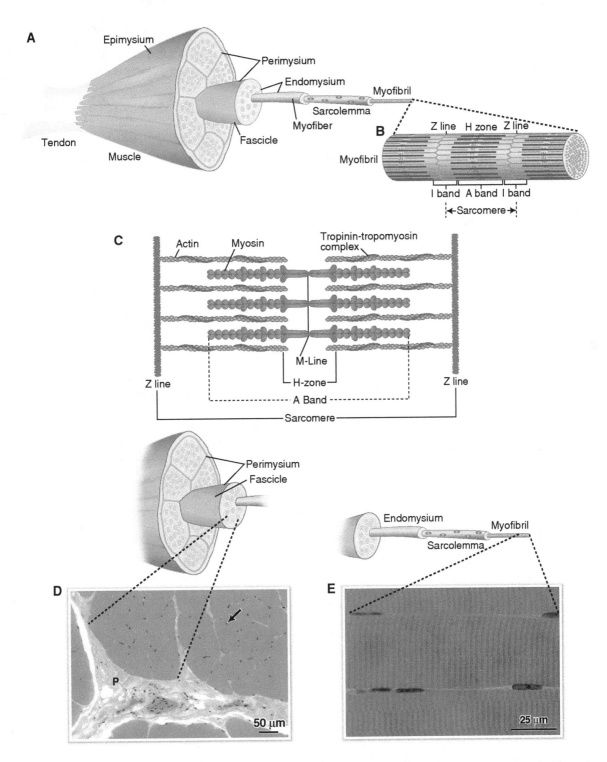

Figure 5-1: Structural features of skeletal muscles and muscle cells. **A.** Connective tissue wrappings in skeletal muscle. The epimysium surrounds the entire muscle and is continuous with tendons at the ends. The perimysium surrounds groups of muscle cells (myofibers) called fascicles. A basal lamina, the endomysium, surrounds each myofiber. **B.** Myofibrils inside myofibers. Myofibrils consist of linear chains of sarcomeres, the region between two Z lines. The I band, associated with each Z line, stains lighter than the A band, which is found on either side of the I band. This arrangement of alternating light and dark regions gives skeletal muscle its striated appearance. **C.** A sarcomere, the functional unit of skeletal muscle. Thin filaments containing actin extend from both faces of the Z lines, forming the I band. They overlap with myosin filaments in the middle of the sarcomere, forming the A band. Muscle contraction involves the relative movements of myosin and actin filaments, which bring the Z lines closer together. **D.** Cross-section of skeletal muscle. A portion of the perimysium (P) traverses near the bottom, and the large individual myofibers appear as polygonal shapes surrounded by the thin endomysium. Myofiber nuclei (*arrow*) are seen just under the endomysium. **E.** Longitudinal section of skeletal muscle. The alternating stripes of I and A bands are obvious, with the thin Z line visible in the middle of the light I bands. Nuclei are found at the edges of the myofibers, which are separated by the endomysium layer.

STRUCTURE AND FUNCTION OF ACTOMYOSIN

Actin, myosin, and related proteins were first discovered in skeletal muscle and are widely expressed in eucaryotic cells. Muscle contractions are extreme examples of basic movements such as those involved in cell division.

ACTIN FILAMENTS

All eucaryotic cells contain members of the actin gene family. Actin is one of the most abundant proteins found in cells.

- **F-actin**, or **thin filaments**, are 5–8 nm in diameter and form spontaneously by the polymerization of monomeric globular actin proteins, **G-actin**, when adenosine triphosphate (ATP) and cations are present.

- Polymerization occurs much faster at one end of F-actin, called the plus end, and this thin filament **polarity** extends to other features of the filament (Figure 5-2A).

- Cells contain many **accessory proteins** that control various features of F-actin assembly, disassembly, and organization.

  - **Tropomyosin** is an elongated molecule that binds to F-actin and stabilizes the filament.

  - **α-Actinin** binds to the plus end of actin filaments in cells, forming cross-linked arrays of widely spaced F-actin (Figure 5-2B).

MYOSIN FILAMENTS

Eucaryotic cells also contain members of the myosin family of motor proteins.

- All myosin filaments involved in muscle contraction are formed from dimers of the **myosin II** class (Figure 5-2C). Myosin monomers of this class contain a **head group** and a long α-helical **tail**. Two small proteins, called **light chains**, are associated with the head group. The tails of two monomers self-associate to form a dimer.

- The antiparallel association of myosin tails produces myosin **filaments**, which grow in a **bipolar** fashion, with head groups found at each end and a headless bare zone in the middle (Figure 5-2D). The resulting structures have a larger diameter than F-actin, about 14 nm. In muscle, these structures are called **thick filaments**.

MYOSIN HEAD MOVEMENTS

The head groups of myosin molecules have both an affinity for binding to F-actin and an adenosine triphosphatase (ATPase) activity.

- Binding and hydrolysis of ATP by a myosin head group results in a cycle of conformational changes coupled to binding and unbinding of the head to actin monomers on F-actin.

- The result of one cycle is that when a myosin head rebinds F-actin, it has moved toward the plus end of the filament to engage a different monomer (Figure 5-2E). The binding of myosin to actin stimulates the ATPase activity of myosin.

FILAMENT SLIDING

Bipolar myosin filaments can interact with two sets of actin filaments, which is the basis of the sliding filament model for muscle.

- Arrays of actin filaments, held together by α-actinin at their plus ends, are found near either terminus of a myosin filament (Figure 5-2F).

- The ATP-driven myosin head movements exert a force that pulls the plus ends of F-actin toward the middle of the myosin filament. Because the thick filaments are bipolar, this myosin-generated force will pull the two arrays of actin toward each other.

- Thus, ATP hydrolysis drives the sliding of actin past myosin filaments and shortens the width of the actin assemblies without changing the length of the filaments. This results in moving the structures tethering the plus ends of the F-actin assemblies closer together.

CONTROL OF FILAMENT SLIDING

Additional components are required to avoid continuous filament sliding in live cells, where ATP is continually present.

- In all three types of muscle, actomyosin interactions are inhibited until Ca^{2+} concentrations are increased above resting levels.

- Accessory proteins on thin filaments accomplish this Ca^{2+}-mediated control for skeletal and cardiac muscle.

- In smooth muscle, control involves both thin and thick filaments.

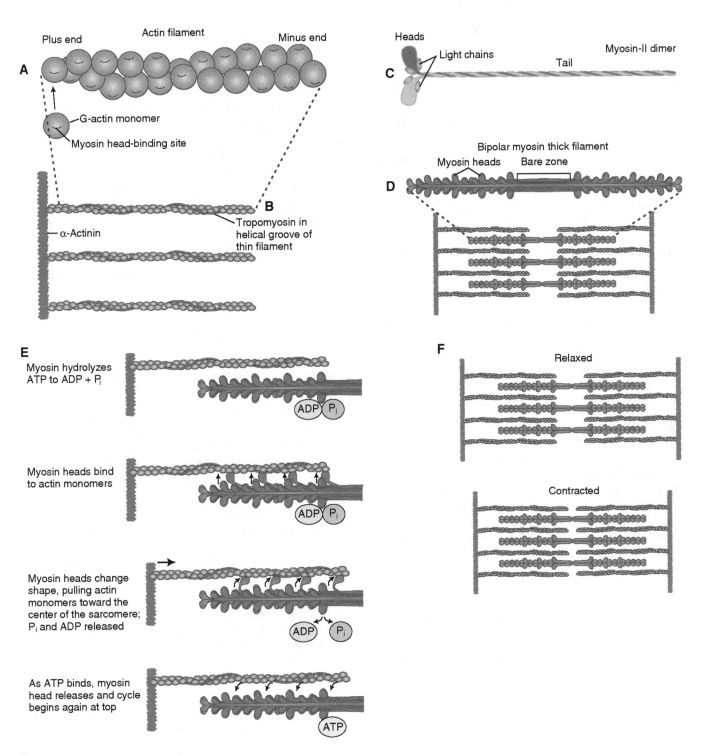

Figure 5-2: Actomyosin components and activities. **A.** Actin filaments assemble from monomers. Under physiologic conditions, actin mono-mers rapidly polymerize and form helical filaments. The addition of monomers to a growing filament is more likely at one end, called the plus end. **B.** Accessory proteins help organize actin filaments. Tropomyosin is an extended rod-like protein that binds along the helical groove of actin filaments and stabilizes them. A second protein, α-actinin, binds to the plus ends of actin filaments and assembles them into widely spaced arrays. **C.** Myosin-II dimer structure. The myosin-II protein has two domains: a compact head group and a long α-helix tail. The tails of two monomers spontaneously associate, forming a dimer. The head groups are associated with small protein subunits, called light chains. **D.** Myosin dimers self-assemble into thick filaments. The dimers associate in an antiparallel fashion, forming a bipolar thick filament. The lower panel shows the arrangement of thick and thin filaments in a single sarcomere. **E.** Myosin heads interact with actin. A cycle of confor-mational changes of the myosin heads is associated with the binding and hydrolysis of adenosine triphosphate (ATP). This cycle involves the binding and unbinding of a myosin head to actin monomers on a nearby thin filament. **F.** Actomyosin interactions produce filament sliding. Each ATP-driven cycle of myosin head movements ratchets the actin filaments toward the middle of the myosin filament. Because myosin filaments are bipolar, both sets of actin arrays slide toward the middle of the myosin filaments and the sarcomere shortens.

SARCOMERE ORGANIZATION IN SKELETAL MUSCLE

The Z line bounding sarcomeres contains α-actinin, which tethers the plus ends of actin filaments (Figure 5-3A). Actin filaments project away from both sides of the Z line and surround myosin filaments in a hexagonal array.

- The **A band** is the area where thin and thick filaments overlap; the **I band** contains only thin filaments.

- Several additional proteins are required to hold the thin and thick filaments in this precise arrangement. For instance, the ends of thick filaments bind to one end of the giant protein, **titin**, which acts like a spring to keep thick filaments centered between Z lines. Other proteins bind at the center of thick filaments or regulate the length of thin filaments. These arrangements ensure that filament sliding is rapid and effective.

MEMBRANE AND CONNECTIVE TISSUE LINKAGES

Protein assemblies that include **dystrophin** link actin filaments of myofibrils to membrane proteins in the sarcolemma (Figure 5-3B). These membrane proteins bind the cell to the endomysium via integrins. These linkages ensure that when the myofibers shorten, the cell shortens, and the force of shortening is transmitted to the connective tissues surrounding the muscle.

▽ In **Duchenne muscular dystrophy**, a progressive X-linked recessive disease, mutations in the gene that codes for dystrophin typically cause a complete loss of this protein, resulting in damage to mature muscle. Satellite cells can produce more fibers but these fibers, lacking dystrophin, are destined to become damaged as well. Many patients with Duchenne muscular dystrophy die by their early twenties of respiratory failure caused by weakening of the muscles that support breathing. ▽

TROPONIN–TROPOMYOSIN COMPLEX

The thin filaments of skeletal muscle also contain **troponin**, a three-subunit protein complex that provides Ca^{2+} regulation of filament sliding.

- When the cytoplasmic Ca^{2+} levels are sufficiently low, the troponin C subunit cannot bind Ca^{2+} and the troponin complex adopts a conformation that allows it to bind both to actin and to tropomyosin, shifting the position of tropomyosin on the thin filament so it blocks the myosin-binding sites on actin (Figure 5-3C).

- When the Ca^{2+} levels are high, the troponin subunit C binds Ca^{2+} and the troponin complex is released from actin, allowing tropomyosin to move so it no longer occludes the myosin-binding site. This allows myosin heads to bind actin, activating the myosin ATPase and initiating filament sliding.

SARCOPLASMIC RETICULUM AND T TUBULES

The control of Ca^{2+} in skeletal muscle involves specializations of the endoplasmic reticulum and plasma membrane (Figure 5-3D).

- Myofibrils are surrounded by the endoplasmic reticulum, known in muscle as the **sarcoplasmic reticulum**. Expansions of the sarcoplasmic reticulum, the **terminal cisternae**, are found at junctions of the A and I bands. The interior of the sarcoplasmic reticulum normally contains a high concentration of Ca^{2+}, pumped in from the cytoplasm by a Ca^{2+} pump, which maintains a low concentration of Ca^{2+} in the cytoplasm.

- The sarcolemma has numerous invaginations that protrude into the cell to form a set of tubules called **transverse**, or **T tubules**. These tubules are found on either side of the Z line at junctions of the A and I bands in close contact with the terminal cisternae, forming an association of membranes called a **triad**.

NEUROMUSCULAR JUNCTION, Ca^{2+} RELEASE, AND CONTRACTION

Every skeletal muscle fiber receives a single synaptic contact, a **neuromuscular junction**, from a single **motor neuron**. A motor neuron can contact several fibers, and the set of fibers it controls is called a **motor unit**.

- When an action potential in a motor neuron arrives at a neuromuscular junction, the neurotransmitter acetylcholine is released from the neuron, which diffuses to the sarcolemma and opens ion channels in the myofiber, causing a local membrane depolarization (Figure 5-3E).

The arrangement of the triads permits activation of all the sarcomeres in a myofiber following stimulation by the motor neuron. The membrane depolarization at the neuromuscular junction rapidly spreads along the sarcolemma and is carried inside the cell by the T tubules, arriving at the triads.

- The depolarization activates Ca^{2+} channels in the terminal cisternae, releasing Ca^{2+} into the cytoplasm. This binds to troponin, changing its configuration and allowing myosin heads to engage actin; filaments are now free to slide and shorten the sarcomeres.

- When the depolarization ends, Ca^{2+} pumps return Ca^{2+} to the inside of the sarcoplasmic reticulum and troponin inhibition is restored; the myosin heads release and cannot reengage actin, and the muscle relaxes.

- All of the myofibers of a motor unit contract maximally when activated. The extent and force of a muscle's contraction depend on the number of motor units activated.

ATP is required both for myosin dissociation from actin and for Ca^{2+} removal. After death, ATP levels eventually decrease, Ca^{2+} levels increase, and myosin heads bind continuously to actin, producing **rigor mortis**.

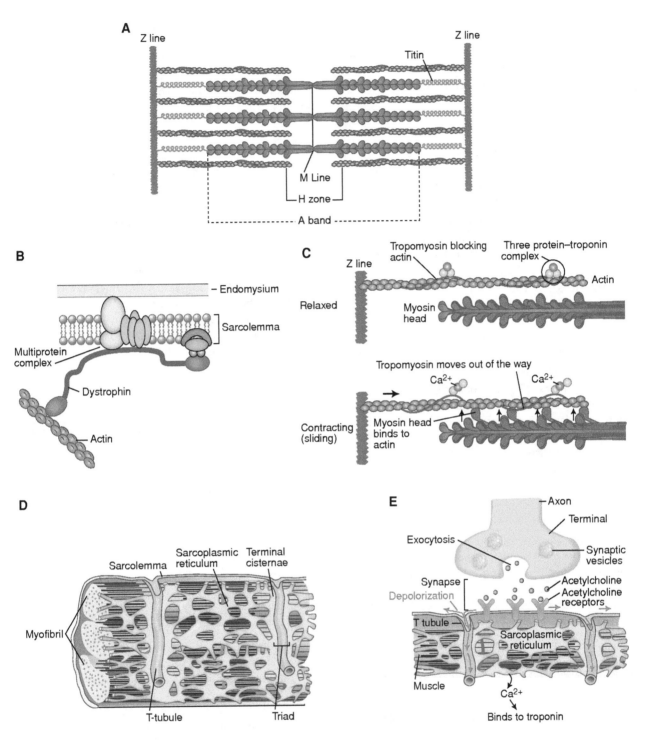

Figure 5-3: Schematics demonstrating skeletal muscle contraction and its control. **A.** A sarcomere. Numerous accessory proteins, such as titin, are required to hold myosin and actin filaments in the proper alignment to allow efficient sliding. **B.** Dystrophin links myofibrils to the sarcolemma. A multisubunit complex, which includes dystrophin, links actin filaments to integral membrane proteins that are bound to the endomysium. This ensures that shortening the sarcomeres shortens the myofiber and that the force of shortening is transmitted to the muscle's connective tissue wrappings. **C.** Troponin regulates myosin binding to actin. The troponin–tropomyosin complex binds to actin in the absence of Ca^{2+}, blocking the myosin heads' access to actin and keeping them inactive (top panel). When Ca^{2+} is present, troponin releases from actin, tropomyosin moves out of the way, the myosin heads bind to actin activating their ATPase, and the filament sliding commences (lower panel). **D.** Sarcoplasmic reticulum and T tubules controlling Ca^{2+}. The sarcoplasmic reticulum surrounds the myofibers. The Ca^{2+} pumps in these membranes remove Ca^{2+} from the cytoplasm. Thin projections of the sarcolemma, called T tubules, come into close contact with expansions of the sarcoplasmic reticulum (called terminal cisternae) and form structures (called triads), which are centered on Z lines. Electrical depolarization of T tubules causes the sarcoplasmic reticulum to release Ca^{2+}. **E.** Neuromuscular junction initiating excitation–contraction coupling. The release of acetylcholine at a motor neuron terminal opens ion channels in the sarcolemma. The depolarization, or excitation, of the membrane is carried inside the myofiber by the T tubules, resulting in Ca^{2+} release from the sarcoplasmic reticulum. This removes the troponin–tropomyosin inhibition, permitting actomyosin filament sliding.

CARDIAC MUSCLE

Cardiac and smooth muscle are constructed from aggregations of smaller cells than those found in skeletal muscle. Figure 5-4A compares the relative sizes and shapes of the three types of muscle cells.

ORGANIZATION OF CARDIAC MUSCLE

Cardiac striated muscle is arranged so that its contractions pump blood from the heart chambers. This is accomplished by linking short, individual cardiac muscle cells into long arrays that encircle the chambers.

Cardiac muscle is not divided into fascicles; thin portions of connective tissue are found at irregular intervals surrounding the extensive set of blood vessels that supply the heart muscle (Figure 5-4B).

Individual cells are surrounded by an endomysium and are closely associated with a network of capillaries.

CARDIAC MUSCLE CELLS

Cardiac muscle cells are about 15 μm in diameter and about 100-μm long; they often are branched and contain one or two nuclei near the center axis. The ends of a cell, or its branches, are linked to other muscle cells by **intercalated discs**, complex junctions that produce large multicellular contractile units rather than the long, single cells found in skeletal muscle (Figure 5-4C). Other differences from the structure of skeletal muscle are as follows:

To provide the ATP needed to energize the continual contraction of the heart, mitochondria are far more abundant in cardiac muscle than in skeletal muscle and occupy about 40% of the cell volume.

When seen with the light microscope, the myofibrils in cardiac muscle cells appear similar to those in skeletal muscle. At the ultrastructural level, myofibrils are closely associated with the numerous mitochondria present, which results in a lower packing density and prevents their regular association into the cylindrical bundles found in skeletal muscle.

Although the components of cardiac muscle sarcomeres are similar to those of skeletal muscle, these cells express isoforms of many proteins that are specific to the heart, including actin and myosin.

T TUBULES AND SARCOPLASMIC RETICULUM

The membrane systems of cardiac cells are organized in a slightly different arrangement from those of skeletal muscle, but the basic strategy of membrane depolarization of T tubules causing Ca^{2+} release from sarcoplasmic reticulum remains the same. T tubules approach myofibrils near Z lines rather than at the A and I junctions. The T tubules are associated with only one protrusion of the sarcoplasmic reticulum, forming a dyad rather than a triad.

INTERCALATED DISCS

The intercalated discs of cardiac muscle represent an important distinguishing feature (Figure 5-4B and C). These complex cellular junctions provide both mechanical and electrical connections between adjacent cells. Transverse portions of intercalated discs contain **desmosomes** and **adherens junctions** and bond the two cells together. The desmosomes are associated with intermediate filaments, and the adherens junctions are associated with the ends of actin filaments of sarcomeres and function as hemi Z lines. Both structures link cells together via cadherin proteins. The longitudinal portions of intercalated discs contain **gap junctions**, which transmit the membrane depolarization of one cell to its "neighbor" cell and allow these large multicellular assemblies to contract as a unit (Figure 5-4D).

COORDINATION OF HEART MUSCLE CONTRACTION

The rhythmic contractions of the heart are initiated and transmitted by groups of specialized cardiac muscle cells (see Chapter 7). Localized membrane depolarization of myocytes in the heart chambers are rapidly transmitted to neighboring cells by the gap junctions present in intercalated discs, and the resulting contraction sweeps along the chamber walls, expelling blood.

GROWTH AND REPAIR OF CARDIAC MUSCLE

Growth. As with skeletal muscle myofibers, the maximal adult number of cardiac muscle cells is achieved soon after birth. Growth of the heart after that occurs by cell hypertrophy and not by cell division.

Repair. Cardiac muscle stem cells are apparently poorly responsive to signals produced by damage. Thus, when heart muscle cells die (e.g., following blockage of a coronary artery), they are replaced by connective tissue scars, resulting in reduced function. Because **coronary artery blockage** is common, extensive research effort is being directed toward identifying cells that would serve as stem cells and be able to replace damaged cardiac muscle.

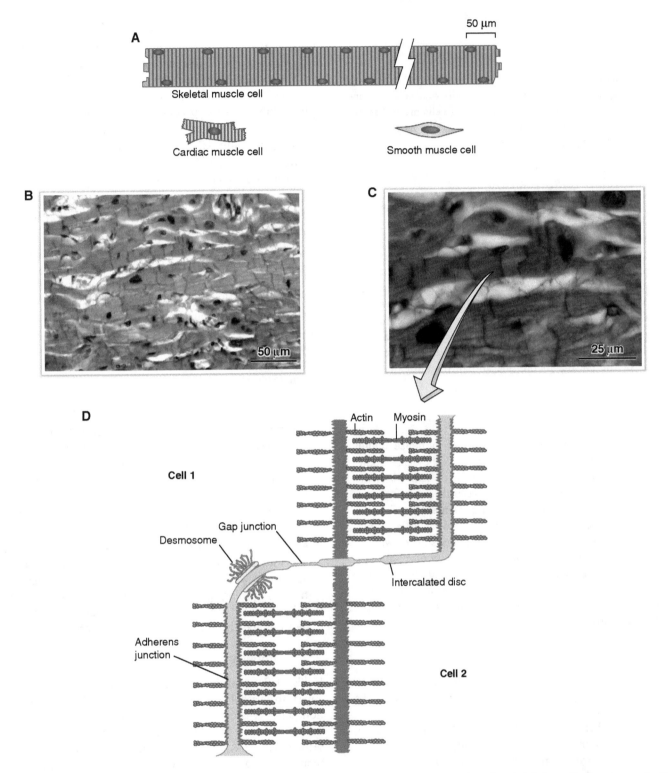

Figure 5-4: Cardiac muscle structure and function. **A.** Comparison of the three types of muscle cells: skeletal, cardiac, and smooth. The extremely large skeletal myofibers are long, multinucleated cylinders, with the many nuclei seen at the periphery of the fibers. Cardiac muscle cells are much smaller and may be branched. Each cell contains one or two nuclei embedded deep in the cytoplasm. Smooth muscle cells are tapered, and their single nucleus is also centrally located. **B.** Longitudinal section of cardiac muscle. The individual cells, linked by intercalated discs, form long multicellular chains. **C.** Intercalated discs in cardiac muscle shown at higher magnification than in part B. Intercalated discs join the ends of cells and their branches to each other and appear as straight or jagged thin lines. **D.** Intercalated discs linking cells physically and electrically. Transverse portions of intercalated discs contain both desmosomes and adherens junctions, which serve as hemi Z lines linking the contractile apparatus of the two cells. Gap junctions found on the lateral faces of intercalated discs provide electrical continuity between cells. The intercalated discs transform individual cardiac muscle cells into a structural and functional syncytium.

SMOOTH MUSCLE

ORGANIZATION OF SMOOTH MUSCLE

Smooth muscles are often arranged as sheets of cells encircling a lumen (e.g., in the gut) (Figure 5-5A). Individual cells are surrounded by a basal lamina, and groups of cells are linked by adherens junctions and encased in a matrix of extracellular material containing abundant reticular fibers.

SMOOTH MUSCLE CELLS

Smooth muscle cells are spindle shaped, about 5 µm in diameter near the midpoint, and can vary in length from 20 µm to 500 µm, depending on the organ in which they are found. Smooth muscle cells contain a single centrally located nucleus.

Smooth muscle cells are not striated, and their cytoplasm has no distinctive features. They contain abundant desmin filaments. Vascular smooth muscle cells also contain vimentin filaments. These intermediate filaments are associated with structures called **dense bodies**, which are found on the inner surface of the plasma membrane and throughout the cytoplasm.

DENSE BODIES AND ACTOMYOSIN

Actomyosin in smooth muscle is organized by **dense bodies** rather than by Z lines (Figure 5-5B and C). The dense bodies contain α-actinin and tether the plus ends of actin filaments. Myosin filaments are interspersed between the actin filaments and, when activated, use the sliding filament mechanism described for actomyosin in skeletal muscle to ratchet actin filaments. The force generated by the sliding moves the network of membrane bound and cytoplasmic dense bodies together, shortening the cell. Contractions of smooth muscle are slower and less powerful than the contractions of striated muscle, in part because the ratio of myosin to actin is four- to fivefold lower in smooth muscle.

CAVEOLAE

In addition to lacking sarcomeres, smooth muscle lacks T tubules. Plasma membrane specializations, called caveolae, are likely to be involved in Ca^{2+} regulation in these cells (Figure 5-5C). Caveolae are **cholesterol-rich membrane domains** containing select sets of membrane proteins, including Ca^{2+} pumps, Ca^{2+} channels, and hormone receptors. It is thought that caveolae may be responsible for the direct entry of Ca^{2+} from outside the cells via ion channels and its removal by pumping. It is also likely that caveolae contain signaling complexes that trigger Ca^{2+} release or reuptake from smooth muscle endoplasmic reticulum.

ROLE OF Ca^{2+} AND CALMODULIN IN SMOOTH MUSCLE

The Ca^{2+}-mediated control of myosin ATPase in smooth muscle depends on the Ca^{2+}-binding protein **calmodulin** rather than on a troponin complex. When cytoplasmic Ca^{2+} levels are increased by external signals or internal activity, the Ca^{2+}-calmodulin complexes produced activate smooth muscle actomyosin by two mechanisms (Figure 5-5C).

- **Ca^{2+}-calmodulin stimulates a kinase** that phosphorylates myosin light chains. This phosphorylation is required for the myosin head to bind to actin.
- **Ca^{2+}-calmodulin binds to thin filaments**, displacing the protein caldesmon, which prevents smooth muscle thick filaments from interacting with actin.

Once activated, smooth muscle contraction tends to be slow because the conformational changes of the myosin heads that produce sliding of smooth muscle filaments require more time than those of striated muscle. Although they are slow, these contractions can produce dramatic shortenings, distorting the nucleus into a "corkscrew" shape (Figure 5-5A and B). Contractions may also be long lasting. Vascular smooth muscle must remain at least partially contracted throughout life to maintain adequate blood pressure.

SMOOTH MUSCLE CONTRACTILE PATTERNS

- **Visceral muscle** consists of groups of smooth muscle cells that are electrically coupled via gap junctions and tend to act as a single unit. The cells are often arranged in sheets surrounding the lumen of an organ. Such muscles may be spontaneously active and also respond to autonomic innervation and hormonal signals. The gap junctions between cells mediate widespread contraction in visceral muscles.
- **Multiunit smooth muscle** resembles the motor units of skeletal muscle. An example is the bundle of smooth muscle associated with skin hairs that raise the hairs when contracted, forming "goose bumps." These cells are not electrically coupled and can produce graded contractions, depending on how many cells are stimulated.

GROWTH AND REPAIR OF SMOOTH MUSCLE

Unlike their striated counterparts, many smooth muscle cells appear to be capable of dividing and replacing lost cells. In addition, other cells have been shown to be capable of dividing to produce progeny that differentiate into smooth muscle. This is seen, for example, when blood vessels expand into regenerating tissue or into a tumor. Pericyte cells of the vessels and smooth muscle cells themselves divide and become smooth muscle in newly formed vessels. However, in other cases, such as the urinary bladder, replacement of smooth muscle after damage or surgery is often insufficient or not properly organized, and this can produce long-term functional problems.

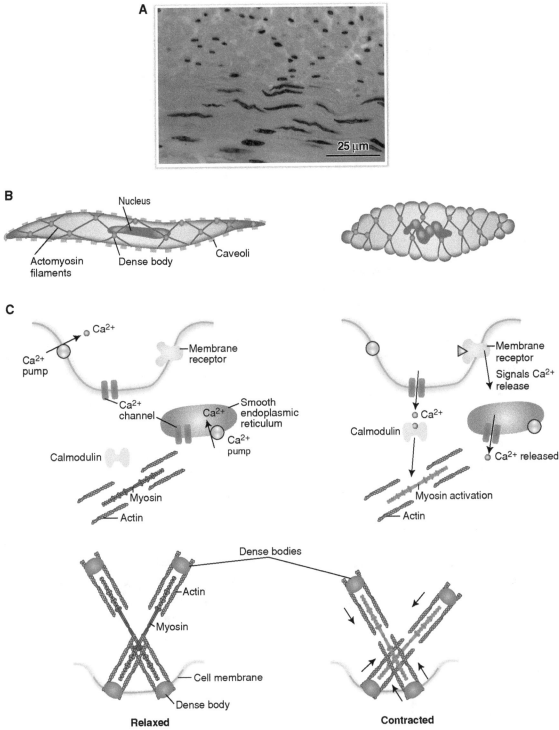

Figure 5-5: Structure and function of smooth muscle. **A.** Smooth muscle from the wall of the gut. A group of cells cut longitudinally is seen at the bottom of the image. Some nuclei have a characteristic corkscrew shape, indicating that these cells were contracted when fixed. Cells cut in cross-section are seen at the top of the image. The nuclei are centrally located when present in a section, and the tapered nature of these cells can be appreciated from the variations in circumference. **B.** Dense bodies organizing contraction of smooth muscle. Dense bodies are found on the inside of the plasma membrane and in the cytoplasm. These structures contain α-actinin and organize actin filaments, which are associated with smooth muscle myosin. When the actomyosin slides, the force that is generated pulls dense bodies together, shortening and squeezing the cells, which often distorts the nucleus. **C.** Regulation of smooth muscle contraction. Caveolae are invagination of the plasma membrane that contain Ca^{2+} channels and Ca^{2+} pumps, which regulate Ca^{2+} entry and removal (*top*). Caveolae may also signal release and reuptake of Ca^{2+} from the endoplasmic reticulum. An increase in Ca^{2+} concentration in smooth muscle results in the protein calmodulin binding Ca^{2+}. Ca^{2+}-calmodulin then activates a kinase that phosphorylates light chains of smooth muscle myosin, which activates the myosin adenosine triphosphatase (ATPase) activity. Calmodulin also removes an inhibitory protein from actin. These two actions result in the sliding of actin past myosin filaments, pulling on dense bodies and contracting the cell (*bottom*).

STUDY QUESTIONS

Directions: Each of the numbered items or incomplete statements is followed by lettered options. Select the **one** lettered option that is **best** in each case.

1. What is the function of the sarcoplasmic reticulum in skeletal muscle?

   A. Ca^{2+} release from the transverse tubules during muscle relaxation

   B. Cellular storage of Ca^{2+}

   C. Cellular storage of glycogen

   D. Degradation of cellular glycogen

   E. Transport of Ca^{2+} into the terminal cisternae during muscle contraction

2. A 66-year-old man who lives alone has a severe myocardial infarction and dies during the night. The medical examiner's office is called the following morning and describes the man's body as being in rigor mortis. The state of rigor mortis is due to which of the following?

   A. Absence of adenosine triphosphate (ATP) preventing detachment of the myosin heads from actin

   B. Enhanced retrieval of Ca^{2+} by the sarcoplasmic reticulum

   C. Failure to disengage tropomyosin and troponin from the myosin active sites

   D. Increased lactic acid production

   E. Inhibition of Ca^{2+} leakage from the extracellular fluid and sarcoplasmic reticulum

3. Duchenne muscular dystrophy is a progressive X-linked recessive disease caused by mutation in the gene that codes for the protein dystrophin. This disease results in damage to mature muscle and is often fatal. What is the normal function of dystrophin in healthy muscle?

   A. Holds thick and thin filaments in the proper alignment to allow efficient sliding

   B. Links actin filaments in myofibrils to membrane proteins in the sarcolemma

   C. Links sarcomeres together in myofibrils

   D. Promotes regeneration of damaged myofibers

   E. Sequesters Ca^{2+} when contraction is complete

4. Neurotransmitter released from motor nerve terminals sets off a sequence of events that ultimately results in contraction of skeletal muscle. Arrange in the correct order the events that occur when neurotransmitter is released.

   1 = depolarization of the sarcolemma

   2 = binding of Ca^{2+} to troponin

   3 = depolarization of T tubules

   4 = shift in the position of tropomyosin to expose myosin-binding sites on actin

   5 = release of Ca^{2+} from the sarcoplasmic reticulum

   6 = repeated binding and release of myosin to actin

   A. 1 – 2 – 3 – 4 – 5 – 6

   B. 1 – 3 – 5 – 4 – 2 – 6

   C. 1 – 5 – 3 – 2 – 4 – 6

   D. 1 – 3 – 5 – 2 – 4 – 6

   E. 1 – 2 – 4 – 3 – 5 – 6

5. Gap junctions allow the signal to contract and spread in a wave from one cardiac muscle cell to another, synchronizing contraction. Gap junctions are located in which of the following subcellular structures?

   A. Caveolae

   B. Dense bodies

   C. Dyads

   D. Intercalated discs

   E. Sarcoplasmic reticulum

   F. Transverse tubules

   G. Triads

6. Which cellular structure is thought to regulate Ca^{2+} in smooth muscle cells?

   A. Caveolae

   B. Dense bodies

   C. Intercalated discs

   D. Sarcoplasmic reticulum

   E. T tubules

ANSWERS

1—B: Ca^{2+} is responsible for the coupling of excitation and contraction in skeletal muscle. The sarcoplasmic reticulum is a modified endoplasmic reticulum. Ca^{2+} is concentrated in the lumen of the sarcoplasmic reticulum (B). Glycogen is stored as droplets in the cytoplasm (C). The transverse tubule system, or T system, is an extension of the cell membrane of the myofiber (sarcolemma). The T system allows for simultaneous contraction of all myofibrils because it encircles the A-I bands in each sarcomere of every myofibril (A). It is important to note that cardiac muscle also has a T system, although it is not as elaborate and well organized as that found in skeletal muscle (e.g., dyads are present rather than the triads of skeletal muscle, and there are fewer T tubules in the atrial versus ventricular muscle).

2—A: There is a small amount of production of adenosine triphosphate (ATP) after death through anaerobic and phosphagen pathways. However, there is insufficient ATP to induce the detachment of the myosin heads from actin. Ca^{2+} continues to leak from the extracellular fluid and the sarcoplasmic reticulum (E); however, the sarcoplasmic reticulum is no longer able to retrieve the Ca^{2+} (B). Tropomyosin and troponin are disengaged from the myosin active sites (C). Lactic acid is produced during rigor mortis through anaerobic pathways. The high levels of lactic acid cause deterioration of the skeletal muscle and end the state of rigor mortis (D).

3—B: Dystrophin is part of a multisubunit complex that links actin filaments in myofibrils to integral membrane proteins in the sarcolemma, which are also bound to the endomysium. This ensures that when the sarcomere shortens, the myofiber shortens, and that the force of shortening is transmitted to the muscle's connective tissue wrappings.

4—D: When cytoplasmic Ca^{2+} levels are low, tropomyosin blocks the myosin-binding site on actin. Following an action potential, membrane depolarization at the neuromuscular junction rapidly spreads down the T tubules, triggering release of Ca^{2+} from the sarcoplasmic reticulum. Ca^{2+} binds to troponin, releasing the troponin complex from actin, which allows tropomyosin to shift its position on actin and expose the myosin-binding site. This allows myosin heads to bind to actin, activating the myosin adenosine triphosphatase (ATPase). The repeated binding and release of myosin to actin is responsible for filament sliding and muscle contraction.

5—D: Intercalated discs are complex cellular junctions that provide both mechanical and electrical connections between adjacent cardiac muscle cells. The transverse portions of intercalated discs bond the two cells together and function as hemi Z lines to link the contractile apparatus of the two cells. The longitudinal portions of intercalated discs contain gap junctions, which transmit depolarization to a neighboring cell, synchronizing contraction.

6—A: Smooth muscle cells lack T tubules. Instead, membrane specializations called caveolae are thought to be involved in Ca^{2+} in these cells. Caveolae are invaginations of the plasma membrane that contain Ca^{2+} pumps, Ca^{2+} channels, and hormone receptors. It is thought that caveolae may be responsible for the direct entry of Ca^{2+} from outside the cells via ion channels and its removal by pumping. They may also signal release and reuptake of Ca^{2+} from the smooth endoplasmic reticulum.

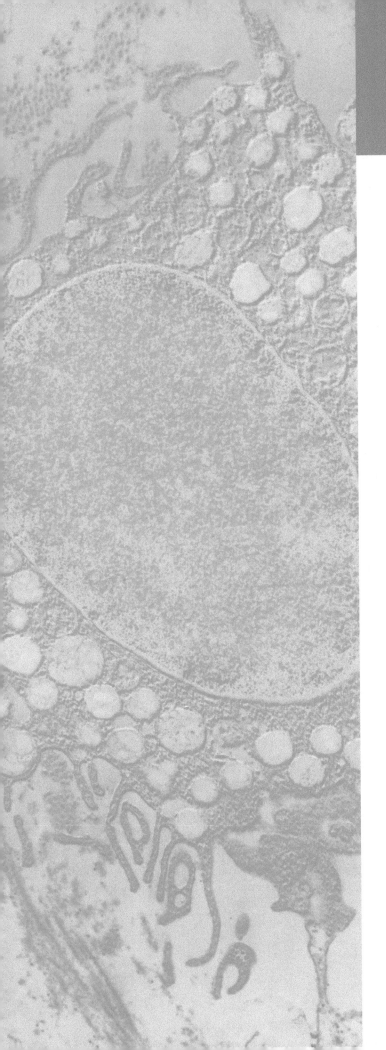

CHAPTER 6

NERVOUS TISSUE

OVERVIEW

The nervous system is comprised of cellular networks that collect and integrate information, direct behavior, and create concepts and memories. You are reading these words using input from the sensory receptors in your eyes to acquire visual signals for interpretation by your brain in concert with the output of motor systems to control eye movements required to track the words. At the same time, other components of your nervous system are managing "housekeeping" functions, including maintaining posture, breathing, and moving food through your digestive system. This is a huge task; when a body is at rest, the brain by itself is responsible for one-fifth of the body's oxygen consumption.

The nervous system consists of two kinds of cells, neurons and glial cells. **Neurons**, or nerve cells, are the functional units of information and signal processing, and **glial** cells provide essential support to neurons. There are approximately 10^{11} neurons and 10^{12} glial cells in the body. Neurons are electrically excitable cells linked together into computational circuits, with each cell having hundreds or thousands of specific connections. Simple circuits, for example, can involve only sets of the few cells that are required to make the leg extend when the patellar tendon is stretched. The structures involved in more complex activities such as reading likely involve at least hundreds of circuits and millions of cells.

The basic organization of the nervous system is described from both anatomic and functional perspectives. The nervous system has two **anatomic** realms.

- The **peripheral nervous system (PNS)** receives sensory information from the outside world and from conditions within the body. It also transmits commands to effectors such as muscles and glands.

- The **central nervous system (CNS)**, which consists of the brain and the spinal cord, integrates sensory inputs and produces coordinated outputs. The sensory systems of the eye and ear are also part of the CNS.

The nervous system also has two **functional** divisions; one includes a subjective component.

- The **somatic** or **voluntary** nervous system is comprised of central elements that produce output that can be controlled by thinking.

- The **autonomic nervous system (ANS)** consists of both central and peripheral elements that control involuntary activities such as gut motility. The **ANS** has **parasympathetic** and **sympathetic** divisions that often have opposed actions.

The nervous system is often compared to a digital computer because both are involved with information processing. Digital computers are built to precise specifications, with the location of each component and its connections predetermined. However, there is far too little genetic information to direct complete specification of the nervous system, and a selective strategy is used for its construction. After the developmental period, which extends for years after birth, subtle cellular changes must be involved in memory formation and learning; any larger plasticity required for regeneration is severely limited, however, and damage to nervous tissue is especially debilitating.

NEURONAL STRUCTURES

Neurons receive, process, and relay signals. Almost all neurons consist of three parts: **dendrites**, a **cell body**, and an **axon**. The many variations displayed by these components make neurons the most diverse cells in the body in terms of size and shape. Figure 6-1A provides a generalized view of the arrangement of these parts found on the three common neuronal types. Neurons are polarized cells; signals arrive at dendrites and move to the cell body and then to the axon, where they are transmitted to the next cell. **Synapses** are specialized membrane regions that receive and transfer signals from cell to cell, typically neuron to neuron or neuron to effector cell.

In this chapter, we will first consider the structural elements of neurons and then briefly describe how these elements are involved in signaling. A wide variety of staining methods have been developed to visualize nervous tissue, including the Golgi method, which stains some cells completely and leaves neighbor cells faint; silver stains that react with axons and dendrites; and Nissl stains that identify the abundant endoplasmic reticulum in cell bodies.

DENDRITES

Dendrites are highly branched membrane extensions of the cell body. The branches are typically short, although some cells (e.g., the Purkinje neurons in the cerebellum of the CNS) have impressively large and complex dendritic processes (Figure 6-1B). Dendrites serve as the major site for receiving and integrating incoming signals, and thus the shape and extent of dendrites are the major determinants of a neuron's function.

Many dendrites of neurons found in the cerebellum and cerebral cortex contain small, often mushroom-shaped, membrane protrusions called **dendritic spines**.

- Spines are specialized to receive synaptic input from other neurons.

- Spines are dynamic structures that can rapidly change shape, grow, and appear or disappear. It is assumed that the plasticity of these spines is an important component of learning and memory.

CELL BODY

The nucleus and much of the metabolic and synthetic machinery of neurons is housed in the cell body, or **perikaryon**. The diameter of a cell body ranges from 5 μm for granule cells in the CNS, which are among the smallest cells in the body, to 100 μm in large cells. However, most of a neuron's mass is in its dendrites and axon. A typical neuronal cell body has the following characteristics:

- The nucleus stains pale and usually shows a prominent nucleolus.

- Because most of the components of dendrites and axons, including their extensive membrane coverings, are produced in the cell body, it is packed with rough endoplasmic reticulum (RER) and an active Golgi apparatus. The RER in neurons is called **Nissl substance**, or Nissl bodies, because it is the target of the Nissl stain, which is used to detect neuronal cell bodies (Figure 6-1C).

- Neurons contain a special class of intermediate filaments, called **neurofilaments**, which originate in the cell body and extend into axons and dendrites. These filaments are assumed to provide structural support to these processes; the diameter of the cell's axon is related to its production of neurofilaments.

▽ Improper assembly or overproduction of intermediate filaments in motor neurons results in the destruction of filaments in the fatal disease **amyotrophic lateral sclerosis**. ▽

The most common shapes for cell bodies are round or oval, like the small **granule cells** of the cerebral cortex, or triangular like the **pyramidal cells**, which tend to be larger (Figure 6-1D). The cell bodies of some neurons receive a significant number of synaptic contacts, although this is uncommon.

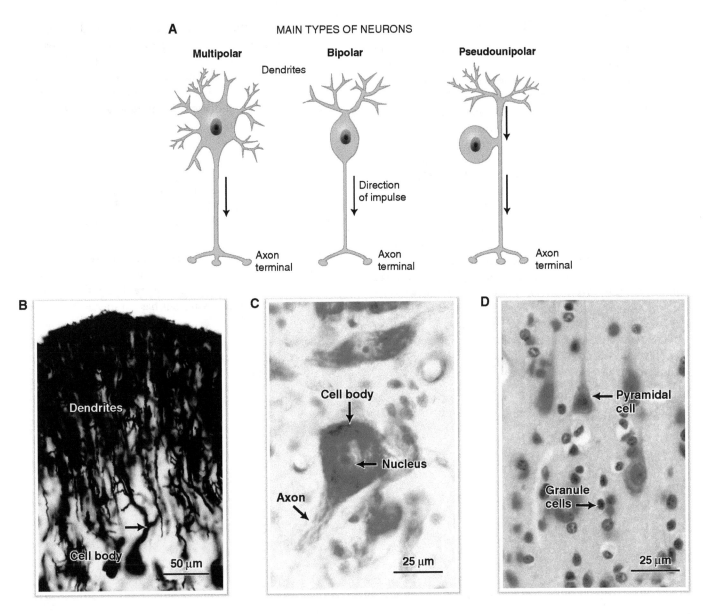

A   MAIN TYPES OF NEURONS

Figure 6-1: Structural features of neurons. **A.** Three main types of neurons, as defined by arrangement of the cell body and its processes. In multipolar neurons, the most common type, many dendrites arise independently from the cell body. Bipolar neurons have a single dendrite and axon and are found in the sensory systems (e.g., the retina). Pseudounipolar neurons begin life as bipolar cells, but the axon and dendrite fuse during development. Pseudounipolar neurons are found in sensory and cranial ganglia. **B.** Dendrites on Purkinje neurons in the cerebellum. These cells typically have a single dendrite (*arrow*), which branches repeatedly and forms dense, planar arbors that project toward the surface of the cerebellum. (Silver stain) **C.** Cell body of a motor neuron in the gray matter of the spinal cord. The Nissl stain reveals accumulations of rough endoplasmic reticulum in the cell body and a proximal axon in these large cells. (Nissl stain) **D.** Pyramidal and granule cells in the cerebrum. The cerebral cortex contains layers occupied by similarly shaped cells. In this section, large pyramidal neurons and smaller granule cells are seen in adjacent areas. (*continued on page 73*)

AXONS

Axons, or **nerve fibers**, connect the cell body to its terminal synapses and the signaling target of the neuron. They can be short, as in a few tens of microns, or more than 1-meter long in the case of pyramidal neurons, which send signals from the brain to the lower spinal cord. The diameter of an axon is constant along its length. Unlike dendrites, most axons are unbranched, although some may produce a few collateral branches. Most neurons contain a single axon, but there are neurons of sensory and sympathetic ganglia that have more than one axon. Figure 6-1E shows an example of the initial portion of a single axon in a developing brain.

■ The **axoplasm**, or the cytoplasm of the axon, is filled with microtubules, neurofilaments, and a subset of components produced in the cell body. Axons require transport machinery to distribute these components along their length. **Axoplasmic transport** depends on polarized microtubules extending in staggered arrays from the cell body to the end of the axon process. Microtubules serve as guides for the adenosine triphosphate (ATP)-energized molecular motors **kinesin** and **dynein**, which move mitochondria, vesicles, and other components up and down the axon.

■ The region of the axon adjacent to the cell body is known as the **axon hillock**. A barrier prevents general diffusion of material from the cell body into the axon; Nissl substance (i.e., the RER), for example, extends only partially into the hillock (Figure 6-1C). The axon hillock is also the site of action potential initiation. Many long axons are wrapped with myelin produced by glial cells.

SYNAPSES

The synapse is a specialized signaling contact between an axon terminal and its target. The targets are dendrites (or more rarely the cell body or axon) of another neuron or other cells such as muscle fibers (Figure 6-1F).

■ The signaling at most vertebrate synapses involves the release of a chemical messenger toward receptors on the target cell. This arrangement ensures that information flow is polarized; the receiving cell has no means of sending a signal back.

■ Some axon terminals produce electrical connections with targets via gap junctions, but this direct electrical signaling is rare in vertebrates.

E

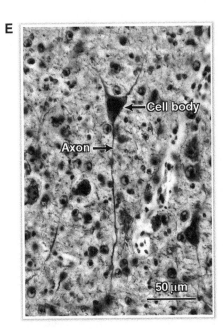

F

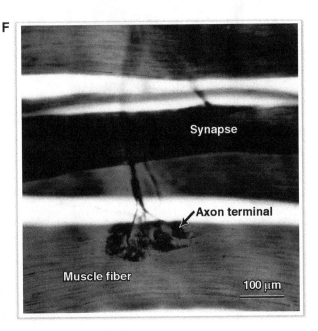

Figure 6-1: (*continued*) **E.** An axon on a pyramidal neuron in a developing chick brainstem. Section courtesy of Thomas Parks, PhD, University of Utah School of Medicine, Salt Lake City, Utah. (Silver stain) **F.** Synapses of motor neurons on skeletal muscle. Two synapses are seen in this preparation from lizard tissue. Individual axon terminals, called motor end plates on these cells, can be seen in the lower synapse. (Gold formate stain)

NEURONAL ACTIVITIES

The transmission of signals by neurons is accomplished by changes in their **resting membrane electric potential**. Signals received at **synaptic inputs** can either increase or decrease the magnitude of the membrane potential on the dendrites and cell body, largely via changes in **ion channels**. If the potential is decreased sufficiently, an explosive, self-perpetuating depolarization, called an **action potential**, is initiated at the axon hillock. The action potential then travels to its termini and activates synaptic output to signal the next cell. In most cases, synaptic signaling involves the release of chemical neurotransmitters. Summation of potential changes spreading along dendrites toward the axon determines the frequency of action potential firing, and hence the signaling activity of the neuron.

MEMBRANE POTENTIAL AND SIGNAL CONDUCTION

Neurons are well supplied with sodium pumps (see Chapter 1). The action of these pumps keeps the cytoplasm of neurons high in K^+, low in Na^+, and the cells typically have an electrical potential across the plasma membrane of −60 to −70 mV relative to the outside.

- There are numerous ion channels specific for Na^+, K^+, or Ca^{2+} present in the neuronal membranes. In electrically excitable cells (e.g., neurons), some are **voltage-gated** ion channels with the membrane potential regulating their openings.

- A local depolarization of the membrane toward 0 mV will cause voltage-gated Na^+ channels in the area to open. Na^+ enters, causing a further depolarization that spreads to adjacent membrane areas. This spread of depolarization occurs along membranes of dendrites and the cell body (Figure 6-2A).

If the concentration of voltage-gated Na^+ channels is sufficiently high, the membrane potential can approach +50 mV within a few milliseconds. This explosive change in membrane potential is called an **action potential**, which can propagate quickly along membranes and serves as the unit of signal conduction in the nervous system. High concentrations of Na^+ channels are found in a region of the axon hillock, called the **initial segment**, and along the remainder of an axon, allowing axons to transmit signals rapidly via action potentials.

- On axons covered with myelin, the Na^+ channels are clustered at intervals, called **nodes of Ranvier**, where the myelin wrapping is thin (Figure 6-2A and B). The membrane depolarization in axons is large enough to allow action potentials to jump from node to node; this **saltatory conduction** is extremely rapid.

- After the peak of the action potential passes, several events quickly return the membrane potential to −60 to −70 mV. Na^+ channels close automatically, stopping Na^+ entry; voltage-gated K^+ channels open, allowing positive charges to escape, and Na^+ pumps restore the normal ionic balance.

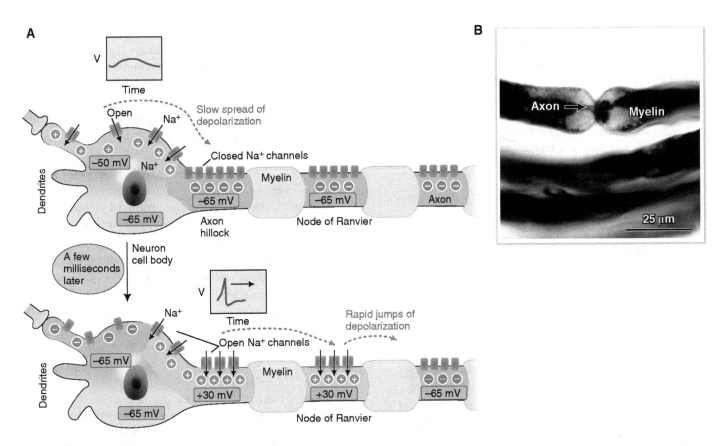

Figure 6-2: Signal conduction and neuronal circuits. **A.** Transmission of membrane depolarization. The upper panel shows membrane depolarization in a dendrite causing voltage-gated Na⁺ channels to open, which slowly spreads the 10–15 mV depolarization toward the cell body. The lower panel shows the depolarization reaching the initial segment of the axon hillock, where voltage-gated Na⁺ channels are tightly packed. A large depolarization results, producing an action potential that jumps across an insulating myelin segment to the next patch of free axonal membrane at a node of Ranvier. **B.** Node of Ranvier. A node is formed at the junction of two Schwann cell myelin wrappings on a single peripheral nerve fiber. (Osmium stain) (*continued on page 77*)

SYNAPTIC ACTIVITY AND INTEGRATION

The cytoplasm of **presynaptic terminals** of axons is filled with mitochondria, **synaptic vesicles** that contain the **neurotransmitter** specific for that cell, and machinery for recycling synaptic vesicles (Figure 6-2C).

- Synaptic vesicle membranes contain proteins involved in vesicle attachment, membrane fusion, exocytosis, and endocytosis for releasing and recycling vesicles. The presynaptic membranes contain Ca^{2+}-specific ion channels and docking sites for synaptic vesicles.

- When an action potential reaches the presynaptic areas at the end of an axon, the membrane depolarization causes voltage-gated Ca^{2+} channels to open. The increased Ca^{2+} concentration triggers the exocytosis of docked synaptic vesicles, releasing their stored neurotransmitter into the **synaptic cleft**, a thin space separating the terminal from the adjacent postsynaptic cell.

- The neurotransmitters diffuse across the cleft and bind to receptors found in clusters on the **postsynaptic membrane**.

The postsynaptic receptors are **ligand-gated ion channels** that open after binding neurotransmitter molecules.

- If the neurotransmitter is **excitatory**, as glutamate is in the CNS, the ion channels of the receptors will allow cations, usually Na^+, to enter the postsynaptic membrane, causing a local depolarization and opening nearby voltage-gated Na^+ channels.

- If the neurotransmitter is **inhibitory**, as glycine or γ-amino-butyric acid (GABA) are in the CNS, the ion channels of the receptors allow either Cl^- into the cell or K^+ out of the cell, either of which will increase the negativity of the membrane potential and decrease the probability that voltage-gated Na^+ channels in the area will open.

A neuron in the CNS may receive many thousands of synapses, both excitatory and inhibitory. Activity at these sites produces local changes in membrane potential, called depolarizations or hyperpolarizations, which interact over the complex geometry of the dendrites and cell body. The result is summed at the axon hillock, and this integration determines the rate at which action potentials are sent down the axon to the neuron's targets.

COMPUTATIONS BY NEURONAL CIRCUITS

The **four functional neuronal classes** are assembled into computational circuits and utilize signal conduction and synaptic transmission to initiate and regulate behavior.

- **Motor (efferent)** neurons control effector organs such as muscles or glands, and are located primarily in the CNS.

- **Sensory (afferent)** neurons receive stimuli from sensors monitoring internal or external stimuli. The cell bodies of sensory neurons may be either in the CNS or in the peripheral ganglia.

- **Projection neurons**, often large pyramidal cells in the CNS, connect different regions, which may be far apart.

- **Interneurons** establish local circuits by connecting other nearby neurons in the CNS or in the peripheral ganglia.

The **patellar tendon reflex** (also known as the knee-jerk reflex) provides a familiar example of a simple circuit (Figure 6-2D).

- A tap on the tendon stretches the quadriceps muscle, which is located on the anterior aspect of the thigh. Pseudounipolar sensory neurons of dorsal root ganglia attached to muscle spindles sense this stretch, and action potentials are sent toward the spinal cord.

- Axons of these cells bifurcate and send these action potentials to two types of cells in the spinal cord. Axons of one branch maintain excitatory synapses on motor neurons that connect to the quadriceps muscle, which then signal this muscle to contract. Axons of the other branch maintain excitatory synapses on interneurons, which in turn activate inhibitory synapses on motor neurons that synapse with the hamstring muscles on the posterior thigh, causing this opposing muscle group to relax.

- The result is a rapid jerk forward of the lower leg.

This simple circuit is part of a system that controls aspects of posture and walking. Such circuits are established early in life and normally do not change. Other more complicated circuits in the CNS involved in learning and memory must involve subtle changes to both the structure and biochemistry of axons and dendrites.

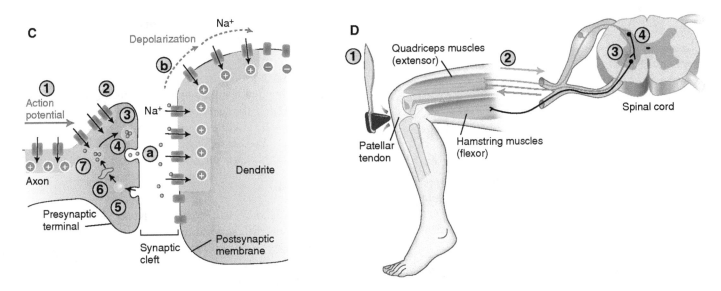

Figure 6-2: (*continued*) **C.** Excitatory synaptic transmission. Presynaptic events: (1) An action potential reaches the presynaptic area of a terminal, where it depolarizes the membrane and opens voltage-gated Ca^{2+} channels. (2) The increased concentration of Ca^{2+} causes docked synaptic vesicles (3) to fuse with the membrane and release their contents of neurotransmitter (4) into the synaptic cleft. (5) The fused vesicle membrane is internalized, (6) processed through endosomes, (7) reloaded with neurotransmitter via cotransporters in the vesicles, and the loaded vesicles reattach to the membrane. Postsynaptic events: (a) Neurotransmitter molecules diffuse across the synaptic cleft and bind to ligand-gated ion channels on the postsynaptic membrane. These channels open, allowing cations (e.g., Na^+) to enter. (b) This depolarizes the membrane, opening adjacent voltage-gated Na^+ channels and spreading the depolarization away from the synapse. **D.** Patellar tendon reflex: (1) Stretching the patellar tendon lengthens the quadriceps muscle, which is detected by muscle-spindle fibers. (2) Sensory nerves (*orange*) are activated and send action potentials to the spinal cord. (3) Branches of each axon synapse with motor neurons (*red*), which then stimulate the quadriceps. (4) Other branches synapse with an interneuron (*blue*) that inhibits motor neurons (*black*) that control the hamstring muscles, allowing them to relax so that the stimulated quadriceps can contract. This reflex will normally serve to maintain a constant length of the quadriceps, a component of systems controlling standing posture or movement.

GLIAL CELLS

Glia are the principal support cells of nervous tissue, and they are about 10 times more numerous than neurons. Glia fill spaces between neurons, wrap axons with myelin, surround blood vessels, and line internal and external surfaces in the CNS. Damage to the CNS can result in the division of glial cells in adults; however, it is with rare exception that adult neurons divide.

GLIA OF THE CENTRAL NERVOUS SYSTEM

There are two types of glia found in the CNS: astrocytes and oligodendrocytes. They are both considered **macroglia** and are derived from **neuroectodermal** tissue. **Microglia** are derived from bone marrow progenitor cells and are related to macrophages.

ASTROCYTES Astrocytes are the most common glia cells. Astrocytes have small, round cell bodies that support numerous complex processes that extend between neurons and their processes. Some astrocytes maintain expanded processes, called **end feet**, which surround portions of capillaries or line the pia, a loose connective tissue that covers the CNS (Figure 6-3A). This lining of the brain and spinal cord is called the **glial limiting membrane**, or **glial limitans**. Astrocytes contain intermediate filaments produced from **glial fibrillary acidic protein** (GFAP) subunits, and can be identified using antibodies that react against this protein. These cells adopt various shapes and are sometimes classified as protoplasmic or fibrous.

Astrocytes surround synaptic areas, and are involved in supporting neuronal activity, as follows:

- Astrocytes actively **absorb extracellular K$^+$** released from neurons by electrical activity and neurotransmitters that escape from the synaptic cleft. The major excitatory transmitter in the CNS is glutamate. Astrocytes import glutamate as it is released from synapses or as it diffuses from the blood and convert it to glutamine, which is then released and imported by neurons. This is a critical activity because excess extracellular glutamate can injure or kill neurons. Astrocytes are linked together via gap junctions. It is possible that astrocyte networks may serve a role in modulating neural activity by regulating important components of the extracellular environment.

- Astrocytes **produce and export several factors important for neuron development and survival**. These factors include nerve growth factor, brain-derived neurotrophic factor, and glial-derived neurotrophic factors. Such factors may prove to be therapeutic for neurodegenerative conditions, including Parkinson's disease and Alzheimer's disease.

▽ Astrocytes respond to brain injuries by proliferating, helping to phagocytize material, and by forming scars to fill in empty spaces. These scars, however, may interfere with subsequent neuronal regeneration. **Astrocytomas**, the tumors that arise from astrocytes, account for about 75% of brain tumors in adults. These tumors are not common; however, rapidly growing astrocytomas usually are untreatable and often are rapidly fatal. ▽

OLIGODENDROCYTES Oligodendrocytes are small cells responsible for **forming and maintaining the myelin sheath** that wraps in a helix around segments of axons in the CNS. This process requires considerable membrane synthesis, and active oligodendrocyte cell bodies contain abundant RER and an expanded Golgi apparatus. Each oligodendrocyte can produce 40–50 segments of myelin and interact with several different axons (Figure 6-3B).

▽ Oligodendrocytes may help **maintain CNS neurons** and promote their survival and the limited regeneration after damage by expressing nerve growth factor and neurotrophin-3. The cell surface of oligodendrocytes contains chondroitin sulfate proteoglycans and other molecules that inhibit axon growth. These molecules may play a role in guiding axons during development, and they may also inhibit long-range regeneration of damaged axons in adults. Oligodendrocytes are the **source of about 10% of brain tumors in adults**. ▽

MICROGLIA Microglia normally account for less than 10% of total glia in CNS. Microglia are difficult to identify on sections without the use of specific antibody reagents.

▽ Damage to the CNS causes microglia to increase in number, both by local cell division and by invasion of bone marrow-derived cells through blood vessels that become leaky after injury. Microglia can adopt various shapes, depending on whether they are quiescent or responding to a lesion such as a hemorrhage or tumor. Like macrophages, microglia phagocytize dying cells and debris. Under some conditions, microglia can release toxic molecules and glutamate, which may contribute to the pathology of **stroke, Alzheimer's disease**, and **acquired immune deficiency syndrome (AIDS)**. ▽

GLIA OF THE PERIPHERAL NERVOUS SYSTEM: THE SCHWANN CELL

Schwann cells are the only glial cells of the PNS, and they perform functions of astrocytes, oligodendrocytes, and microglia. Like astrocytes, they can send out processes to surround cell bodies and their processes. In peripheral ganglia, these cells may be called **capsule** or **satellite cells** when they surround cell bodies (Figure 6-3C). Like oligodendrocytes, Schwann cells ensheathe axons and can wrap them with myelin. All PNS axons are associated with Schwann cells (Figure 6-3D). If the axons are less than 1 μm in diameter, they are ensheathed by Schwann cells; larger axons receive myelin wrappings. A single Schwann cell can ensheathe several small axons; however, they form only one myelin segment on one large axon, unlike oligodendrocytes, which can form many myelin segments on several axons. Like microglia, Schwann cells can also act as phagocytes to remove damaged cells and debris.

A

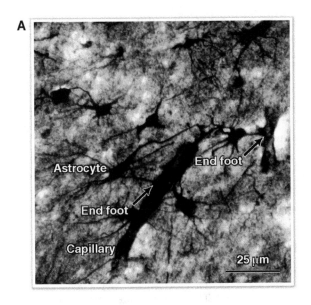

B

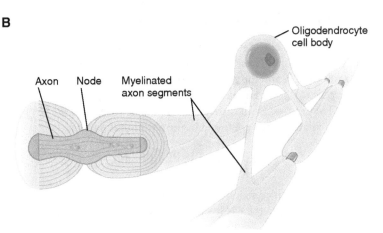

C

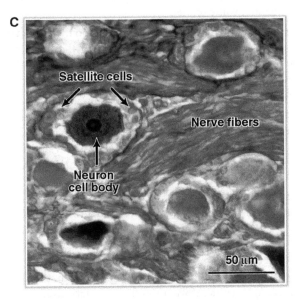

D

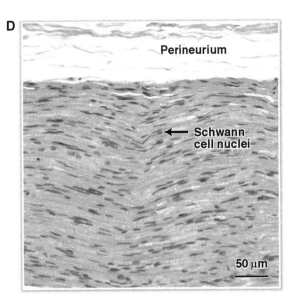

Figure 6-3: Glial cells. **A.** Astrocytes in the central nervous system (CNS). These multibranched cells are in contact with each other and also project end feet onto the surface of capillaries. (Cajal reduced silver stain) **B.** Oligodendrocytes produce myelin in the CNS. Oligodendrocyte processes form segmental myelin wrappings along axons. These cells can form dozens of such segments on several different axons. **C.** Satellite cells in a dorsal root ganglion. Schwann cells surround the large pseudounipolar cell bodies of the sensory neurons; in this case, they usually are called satellite cells. Shrinkage occurred in this preparation, separating the small satellite cells from the large neuronal cell bodies. (Silver stain) **D.** Schwann cells surround axons in the peripheral nerves. A fascicle of axons runs horizontally across this section, below the perineurium. Most of the nuclei visible in the nerve are those of Schwann cells, which will form individual myelin segments on single large axons, or enclose several small axons in cellular infoldings. Nerves also contain blood vessels surrounded by loose connective tissue, which will contribute occasional nuclear profiles.

ORGANIZATION OF THE NERVOUS SYSTEM

During development, neurons migrate to specific areas and aggregate in groups. These groups become the **nuclei** and the **cortex** of the CNS and the **ganglia** of the PNS. Long axons connecting cells to distant targets fasciculate and travel together. In the CNS, a group of fasciculated axons is called a **tract**; in the PNS, fasciculated axons are found in the **peripheral nerves**.

NERVOUS TISSUE OF THE PERIPHERAL NERVOUS SYSTEM

GANGLIA A ganglion contains a compact group of neuronal cell bodies found outside the CNS. Small satellite or Schwann cells usually surround these cell bodies.

- **Sensory ganglia** transmit afferent signals received from peripheral receptors into the CNS. They are either associated with cranial nerves, the **cranial ganglia**, or with inputs to the spinal cord, the **dorsal root ganglia**. These ganglia are surrounded by a dense irregular connective tissue **capsule**, and contain the cell bodies of pseudounipolar neurons; they do not have synapses (Figure 6-4A).

- **Autonomic ganglia** are relay stations for efferent neurons of the CNS. Axons from neurons in the brain or spinal cord, called **preganglionic** neurons, synapse on the principal neurons in an autonomic ganglion, which send **postganglionic** axons out of the ganglion to targets that are usually muscles or glands. Autonomic ganglia may contain interneurons. As mentioned earlier, the ANS has two branches, sympathetic and parasympathetic. **Sympathetic ganglia** tend to be at some distance from the organs that they innervate, whereas **parasympathetic ganglia** are often embedded in their target organ and may not have their own capsule. The actions of these two branches are often opposed in a complementary sense—the parasympathetic system increases blood flow to the intestines after a meal by relaxing smooth muscle in vessels, and the sympathetic system decreases this flow if blood is needed elsewhere.

- **Enteric ganglia** (more commonly called **plexi**) are small groups of neurons that are present in the digestive tract from the lower esophagus to the rectum (Figure 6-4B). This **enteric nervous system** contains more neurons than the spinal cord. Although the ganglia receive input from the vagus nerve and are often classified as part of the parasympathetic branch of the autonomic nervous system, the enteric nervous system appears to operate largely on its own to regulate gut motility and other activities (see Chapter 12). It is sometimes considered a "second brain."

NERVES Peripheral nerves contain axons of sensory and motor neurons. Individual axons range in diameter from less than 1 μm to 30 μm and are associated with Schwann cells. Peripheral nerves contain three types of coverings: epineurium, perineurium, and endoneurium (Figure 6-4C).

- The **epineurium** is the outer covering of large nerves and nerve trunks, and consists of dense irregular connective tissue with many collagen fibers arranged longitudinally. At the site where a nerve enters or exits a ganglion, its epineurium is continuous with the ganglion's capsule.

- The **perineurium** surrounds **fascicles** of associated axons that course inside the nerve. This layer is composed of **perineural** cells, specialized fibroblasts linked by tight junctions.

- The **endoneurium** is a very thin connective tissue containing reticular fibers secreted by the Schwann cells that surround axons and their associated Schwann cells.

NERVOUS TISSUE OF THE CENTRAL NERVOUS SYSTEM

GRAY AND WHITE MATTER The CNS is divided into two main components based on the number of neuronal cell bodies present: gray matter and white matter.

- **Gray matter** contains neuronal cell bodies and their dendrites and synapses, and is found in the center of the spinal cord and in the layers of the cerebral cortex (Figure 6-4D). Many of the axons present in gray matter are unmyelinated.

- **White matter** contains axons, both myelinated and unmyelinated, and only a few neuronal cell bodies. Because of the abundance of axons and myelin, this area has a shiny, white appearance in fresh specimens. Both gray and white matter contain astrocytes, oligodendrocytes, and microglia.

NUCLEI, TRACTS, AND CORTEX

- A **nucleus** in the CNS is a collection of neuronal cell bodies (Figure 6-4E). The neurons are usually born over a short time period, travel together to populate the area, and are involved in a related set of functions. For example, the sensory **trigeminal nucleus**, which is located in the lower brainstem, receives information from the face about touch, temperature, and pain. Nuclei usually contain both projection neurons that connect to cells in other CNS nuclei or PNS ganglia as well as local interneurons.

- **Tracts** are collections of axons of projection neurons that travel together in the CNS (Figure 6-4E).

- The **cortex** is a region in the brain that is organized into layers of neurons. The cerebral cortex is involved with functions that include thought, memory, and language, and contains six layers of cells and their processes. The cerebellum contains only three layers. The layers have characteristic patterns of afferent and efferent connections but are not readily apparent on many sections.

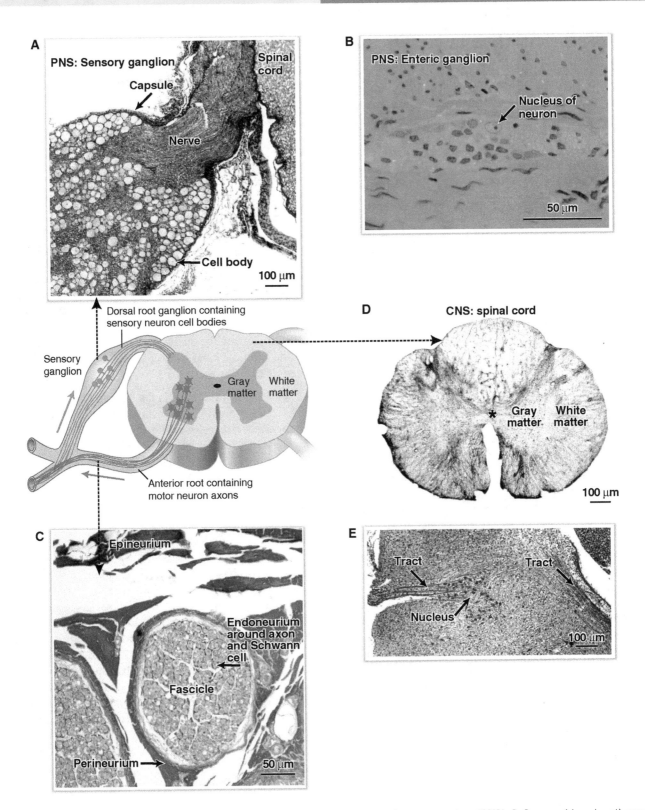

Figure 6-4: Structural features of the peripheral nervous system (PNS) and the central nervous system (CNS). **A.** Sensory (dorsal root) ganglion in the PNS. The ganglion is surrounded by dense irregular connective tissue, stained blue, and contains the cell bodies of pseudounipolar neurons. The cells can be seen at higher magnification in Figure 6-3C. A bundle of nerve fibers running toward the spinal cord exits the ganglion at the top. (Azan stain) **B.** Auerbach's plexus in the wall of the colon. This unencapsulated, enteric ganglion sits between smooth muscle layers and contains neurons and smaller Schwann cells. These ganglia comprise the enteric nervous system, which regulates motility and other activities of the digestive tract. **C.** Cross-section of a peripheral nerve showing three connective tissue sheaths. **D.** Cross-section of the spinal cord. The Nissl stain reveals the distinction between gray matter containing the bulk of the cell bodies and white matter consisting largely of axons and glial cells in the CNS. A simple layer of ependymal cells lines the central canal (*). (Nissl stain) **E.** Nucleus and tracts in the CNS. This section is taken from the developing brainstem of a chick and shows the early formation of a nucleus, a collection of large neuron cell bodies. It is associated with a bundle of axons, a tract, and additional tracts are seen coursing to other forming brain structures. Section courtesy of Thomas Parks, PhD, University of Utah School of Medicine, Salt Lake City, Utah. (Silver stain)

SUPPORT TISSUES OF THE CENTRAL NERVOUS SYSTEM

The CNS is protected by the skull and vertebral column, which are lined by the **meninges**, a set of connective tissue layers that provide additional protection. The brain contains **ventricles**, four spaces lined with **ependymal** cells that communicate with the **central canal**, which continues down the spinal cord early in development. **Cerebrospinal fluid (CSF)** is produced in the ventricles and bathes and cushions CNS cells. The blood–brain barrier of the CNS capillaries limits access of blood components to this tissue.

MENINGES

- **Dura mater.** This outermost layer of dense connective tissue is attached to the inside of the skull, and is separated from vertebral bones by an **epidural space** that contains loose connective tissue. A layer of squamous mesothelial cells lines the smooth inner surface of the dura in the brain. Both surfaces have a mesothelial lining in the spinal cord. The dura contains blood vessels and sympathetic nerve fibers (Figure 6-5A).

- **Arachnoid and pia mater.** These two layers may be considered a unit, called the **leptomeninges**, which encloses cushions of CSF to protect the CNS.

The outer and inner surfaces of the **arachnoid** are covered with a mesothelium, with the outer surface adjacent to the dura. Portions of the inner surface of the arachnoid extend inward in complex, thin branching protrusions called **trabeculae**, which eventually are in contact with the pia mater. The **subarachnoid space** between the trabeculae fills with CSF via openings in the fourth ventricle. This CSF flows through the subarachnoid space, down to the spinal cord, and up to the top of the brain. CSF is returned to the bloodstream via protrusions of the arachnoid into the dura, called **arachnoid villi**, which are associated with porous venous sinuses.

The **pia mater** is composed of loose connective tissue facing the arachnoid and lined on the inside with mesothelial cells (Figure 6-5B). It contains numerous blood vessels and sympathetic nerves. Blood vessels enter and exit the CNS in the **perivascular spaces**, covered by extensions of the pia mater. The pia mater covers the surface of the CNS and is lined on its internal surface by the **glial limiting membrane**, elaborated by astrocytes.

EPENDYMA

- **Ependymal cells.** These cuboidal or columnar epithelial cells are derived from the neuroepithelium of the neural tube and form a simple layer, the **ependyma**, which lines the ventricles and the central canal of the CNS and faces the CSF (Figure 6-5C). The ependymal cell layer allows exchange of molecules less than 100,000 molecular weight between the extracellular fluid around cells in the CNS and the CSF. Cilia on the ependymal cells help circulate the CSF.

- **Choroid plexus and CSF.** The choroid plexus both produces and regulates the composition of CSF. It is formed by convoluted extensions of the pia mater that are covered with ependymal cells and push into the ventricles at several sites (Figure 6-5D). The underlying pia is rich in leaky capillaries. Ependymal cells move fluid from the capillaries into the ventricles to form CSF and also to remove some components (e.g., metabolic wastes) by transporting them back toward the bloodstream. Some CSF is also derived from fluid released from perivascular spaces and capillaries in the brain itself. The CSF resembles blood plasma but is greatly depleted in proteins.

▽ A chronic **increase in the volume of CSF** can occur by a reduction in CSF absorption by the arachnoid villi or by a blockage of its release into the subarachnoid space in the fourth ventricle. This results in **hydrocephalus**, an increase in pressure inside the CNS. In infants, the skull expands to accommodate normal growth, and the increased pressure results in head enlargement and eventually to learning disorders. In adults, the increased pressure produces a variety of symptoms as cells die, including **dementia**. ▽

BLOOD VESSELS

The CNS requires an extensive blood supply to support its constantly high metabolic rate. Numerous capillaries are found in the parenchyma, partially covered by the end-feet of astrocytes. Capillaries in the CNS are responsible for maintaining a **blood–brain barrier**, which prevents the efflux of many components in the blood from these capillaries. The endothelial cells of the capillaries are responsible for this barrier. Tight junctions seal spaces between the cells, with only a small amount of transcytosis occurring across the endothelial walls. This protects the tissue from harmful bloodborne toxins and also prevents entry of medically useful agents, including antibiotics or chemotherapeutic drugs.

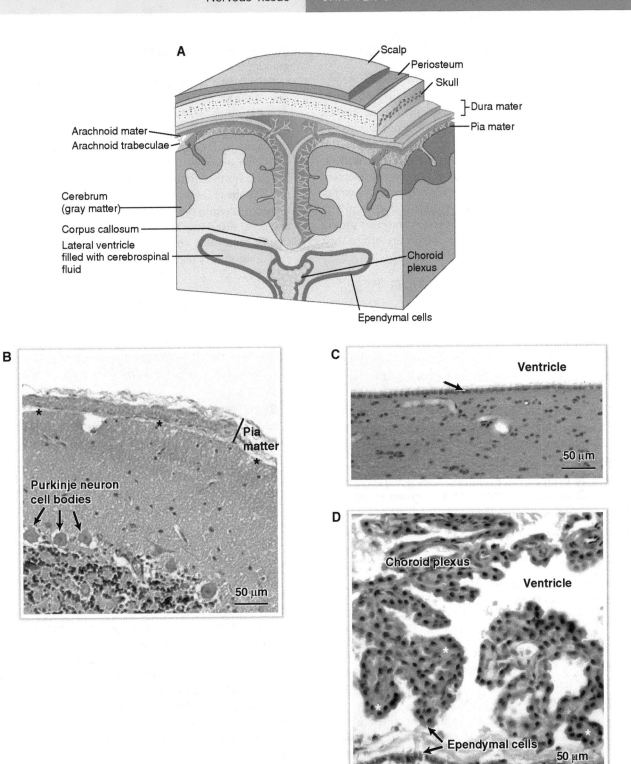

Figure 6-5: Support tissues of the central nervous system (CNS). **A.** The meninges lining the outside of the brain and spinal cord. This three-layered set of outer wrappings contains blood vessels that supply CNS tissue and also form channels, the subarachnoid spaces, filled with cerebrospinal fluid. **B.** The pia mater on the surface of the cerebellum. This band of loose connective tissue includes a simple layer of mesothelial cells facing the cerebellum. A vein in the pia mater filled with red blood cells sits above the mesothelium to the left. Not visible is the glial limiting membrane (*), a set of processes that line the surface directly below the mesothelium of the pia mater. A row of Purkinje neuron cell bodies can be seen, but their dendrites are not obvious with the stain used in this preparation. **C.** Ependymal cells lining a brain ventricle. The ependymal layer (*arrow*) appears as a simple cuboidal epithelium in this section. Cilia projecting from the surface of these cells appear as wisps projecting into the ventricular lumen. The lumens of small blood vessels can be seen below the ependymal layer. **D.** Choroid plexus. This structure is formed from extensions of the pia mater into ventricles that are covered by ependymal cells. The pia contains an extensive blood supply and many capillaries (*). A layer of ependymal cells lines the ventricle wall at the bottom.

DEVELOPMENT OF NERVOUS TISSUE AND REACTIONS TO INJURY

Neurons and glia develop very early from stem cells produced in the embryonic ectoderm. After cell migration, neuronal circuits are established via selective processes that use trophic factors to maintain connections. Although glial cells can divide and restore lost cells, most injured neurons die and their replacement in the CNS is very rare. Small populations of neuronal stem cells have been identified in adults, but the environment of damaged nervous tissue may make it difficult for circuits to reconnect, even if cells are available. Severed axons that project outside the CNS can regrow new processes and return to their original peripheral targets.

ORIGINS AND DEVELOPMENT OF NERVOUS TISSUE

Nervous tissue arises in the embryo from the following regions of ectoderm (Figure 6-6A):

- Most of the CNS is derived from a structure called the **neural plate,** which rolls up into the **neural tube** and then differentiates to form the CNS, including neurons, glia (except for microglia), and the ependymal cells.

- Separate portions of the ectoderm, called the **cranial placodes,** lie outside the neural plate and are the source of some of the sensory nuclei in the brain and components of the eye and ear.

- Most of the neurons and all of the glia (Schwann cells) of the PNS originate from the **neural crest,** a group of cells that arise on either side of the neural tube.

Neurons migrate from the neuroepithelium of the neural tube, or from the neural crest and placodes, and form aggregates in what will become their final location in the CNS or PNS.

- The neurons in these aggregates are genetically programmed to send processes to other regions of the developing embryo to begin the process of forming neural circuits.

- Neurons that produce successful contact with appropriate targets often receive trophic factors, such as members of the neurotrophin family, which maintain these cells; neurons that fail to make such contacts and do not receive trophic support are programmed to die (Figure 6-6B).

- Subsequent **signaling activity** may then fine-tune the structure of these forming circuits. This selective strategy involving trophic factors and activity allows the construction of as many as 10^{14} synapses in the brain using information from a subset of the 25,000 genes in the human genome.

Glial cells migrate with neurons and provide support, including the production of trophic factors. **Radial glia** are present in the developing brain and spinal cord, and serve as scaffolds for guiding the migration of developing neurons. Some of these radial glia later transform into specialized glia (e.g., the Müller cells of the retina), and others apparently convert into astrocytes.

A family of stem cells arises in the early ectoderm and is involved in producing neurons and glia. Most neurons differentiate early in development and permanently exit the cell cycle. Glial cells maintain the ability to divide in response to injury, but neurons that die are seldom replaced. The discovery of small populations of neuronal stem cells in the adult brain offers the hope for treating the debilitating losses associated with stroke and other diseases of the CNS. Neuronal stem cells are found in the **hippocampus,** an area involved in learning and memory. Cell division may play a role in the plasticity required for these activities.

INJURIES INVOLVING THE NERVOUS SYSTEM

It is possible for neurons to recover from injuries that sever their axons by growing new processes, and is far more common in the PNS than in the CNS. The distal portion of the neuron, including the synapses, reacts differently from the proximal portion (Figure 6-6C).

- **Distal reactions.** This portion of the axon and its synapses usually degenerate quickly in a process called **Wallerian degeneration.** The myelin disappears and phagocytic cells in the area remove the resulting debris.

- **Proximal reactions.** The cell bodies of injured neurons undergo **chromatolysis,** which includes loss of Nissl substance (i.e., RER), movement of the nucleus toward the plasma membrane, and swelling of the cell body. If the axon regenerates and reconnects and trophic support is returned, these changes will reverse.

Unfortunately, most damaged connections in the CNS are not reestablished. The cell body may survive if its axon has collateral branches; however, in many cases, the cell body dies after an axon is severed. Nearby astrocytes proliferate to replace lost axons and cell bodies, leaving a glial scar. The scar and inhibitory molecules on the surfaces of oligodendrocytes appear to prevent replacement of most connections, even if there are neurons available for activation and growth.

Neurons in the PNS will often regenerate a new axon. Successful reconnections occur if short distances are involved or if proliferating Schwann cells provide pathways to guide the growing axon back to its target.

- **Transsynaptic reactions.** In most cases, only injured neurons undergo chromatolysis and cell death. However, in some instances, the presynaptic and postsynaptic partner of an injured cell may suffer the same fate.

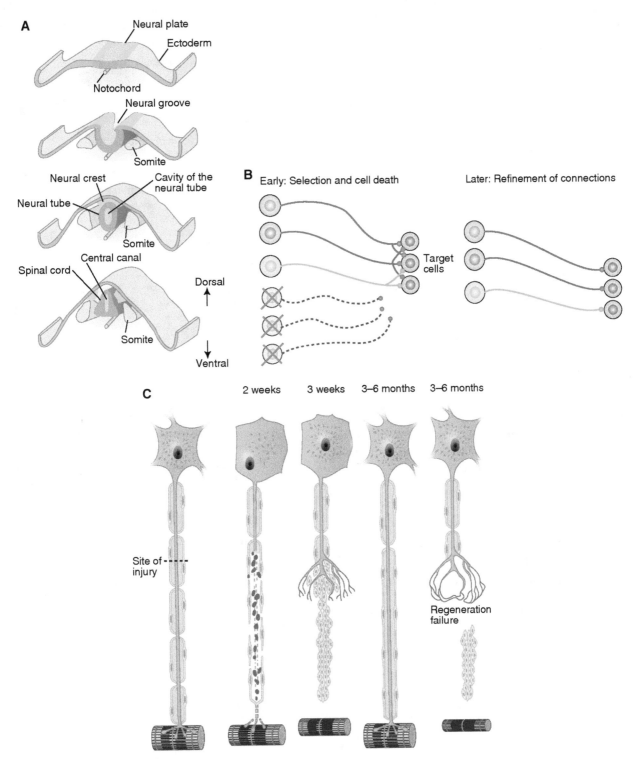

Figure 6-6: Nervous system development and reaction to injury. **A.** Development of the nervous system from the neural tube and neural crest. **B.** Selection and refinement of connections. A group of nerve cells born at the same time will be programmed to send axonal projections toward target cells. The axons that arrive early and form successful connections receive trophic support from the targets. These neurons survive a period of cell death that occurs naturally for all groups of neurons sometime after their birth. Neurons that fail to make enough connections die. This selection establishes the initial wiring of neural circuits. The early connections tend to be widely dispersed. Subsequent electrical and synaptic activity in the circuit eliminates some synapses and strengthens others, reducing the spread of the initial connections and producing functional circuits. **C.** Reactions of neurons to injury. When a central or peripheral axon is severed, its distal portion degenerates and the myelin wrappings are lost. The cell body undergoes chromatolysis; the Nissl substance (rough endoplasmic reticulum) is greatly reduced, the nucleus moves eccentrically, and the cell body swells. In the peripheral nervous system, axons will eventually begin growing from the proximal stump. If newly proliferated Schwann cells provide proper guidance channels, the regenerating axon can find its target and reinnervate it. In the central nervous system, many neurons die after injury, and barriers produced by glial scars and inhibitory molecules on oligodendrocytes often interfere with axon regrowth, even when the neuron survives.

STUDY QUESTIONS

Directions: Each of the numbered items or incomplete statements is followed by lettered options. Select the **one** lettered option that is **best** in each case.

1. The blood–brain barrier protects the brain from exposure to large toxic molecules, but presents a serious hurdle for delivering therapeutic drugs to the central nervous system. What is the cellular basis of the blood–brain barrier?

   A. End feet of astrocytes covering brain capillaries

   B. Fenestrations between brain capillary endothelial cells

   C. Gap junctions between brain capillary endothelial cells

   D. Tight (occluding) junctions between brain capillary endothelial cells

   E. Tight (occluding) junctions between ependymal cells

   F. Tight (occluding) junctions between microglia

2. A 33-year-old woman is referred to the neurology clinic because she has experienced weakness of the eye muscles for the past 2 months. Subsequently, she has experienced diplopia (double vision) and difficulty swallowing, and her speech is slurred. Physical examination shows bilateral ptosis (drooping eyelid), unstable gait, and shortness of breath. Laboratory studies show autoantibodies to the acetylcholine receptor, the receptor for the transmitter that mediates synaptic transmission between motor neurons and skeletal muscle. In which of the following structures do these antibodies bind?

   A. Axon hillock of motor neurons

   B. Dendrites of motor neuron

   C. Presynaptic motor neuron membrane

   D. Postsynaptic muscle membrane

   E. Synaptic vesicles in motor neurons

3. What event results in the generation of action potentials in a typical neuron?

   A. Closing of Na^+ channels

   B. Hyperpolarization of the neuron membrane

   C. Opening of K^+ channels

   D. Opening of Na^+ channels

   E. Pumping of Na^+ from the neuron

4. Neuronal dysplasias represent a family of clinical conditions in which neurons are found in inappropriate locations in the brain. Some dysplasias may result from the absence or defect in which type of cell?

   A. Astrocytes

   B. Microglial cells

   C. Oligodendrocytes

   D. Radial glia

   E. Schwann cells

5. Glutamate is an abundant and important neurotransmitter in the central nervous system (CNS). However, damage to CNS neurons, perhaps as a result of trauma or stroke, can cause excessive amounts of glutamate to be released. This surge of glutamate can radiate from the site of trauma and kill nearby neurons, a phenomenon known as excitotoxicity. Which cells in the CNS normally import glutamate following synaptic transmission and prevent the buildup of extracellular glutamate in an intact healthy brain?

   A. Astrocytes

   B. Ependymal cells

   C. Microglia

   D. Oligodendrocytes

   E. Schwann cells

ANSWERS

1—D: The barrier function of the blood–brain barrier is formed by tight (occluding) junctions (zonulae occludens) between the endothelial cells that comprise the lining of brain capillaries. Adding to the impermeability are the nonfenestrated nature of the capillary endothelium and the paucity or absence of pinocytotic vesicles that represent the physiologic pores seen in other endothelia. Astrocytes form end-foot processes around the brain capillaries that induce and maintain the blood–brain barrier, but do not form the barrier itself.

2—D: The patient has myasthenia gravis, an autoimmune disease in which the body produces antibodies to receptors for the neurotransmitter acetylcholine (ACh), which are expressed on the postsynaptic membrane of muscles. In normal individuals, motor nerve terminals release ACh, which binds to ACh receptors on muscle cells, resulting in depolarization and muscle contraction. Because of the autoantibodies, the number of receptors is reduced in myasthenia gravis, which results in reduced synaptic transmission and consequently in reduced muscle activation.

3—D: A local depolarization of the neuronal membrane toward 0 mV (e.g., from synaptic activation) will cause voltage-gated Na^+ channels in the area to open. Na^+ enters the neuron, causing a further depolarization to spread to adjacent membrane areas. If the concentration of voltage-gated Na^+ channels is sufficiently high, the membrane potential can approach $+50$ mV in a few milliseconds. This explosive change in membrane potential is called action potential.

4—D: Radial glia are present in the developing brain and spinal cord and serve as scaffolds for guiding the migration of developing neurons. The absence or defect in radial glia could cause aberrant migration of central nervous system neurons during embryonic development, resulting in neuronal dysplasia.

5—A: Astrocytes surround synaptic areas in the central nervous system and actively absorb extracellular K^+ released from neurons by electrical activity and neurotransmitters, such as glutamate, that escape from the synaptic cleft.

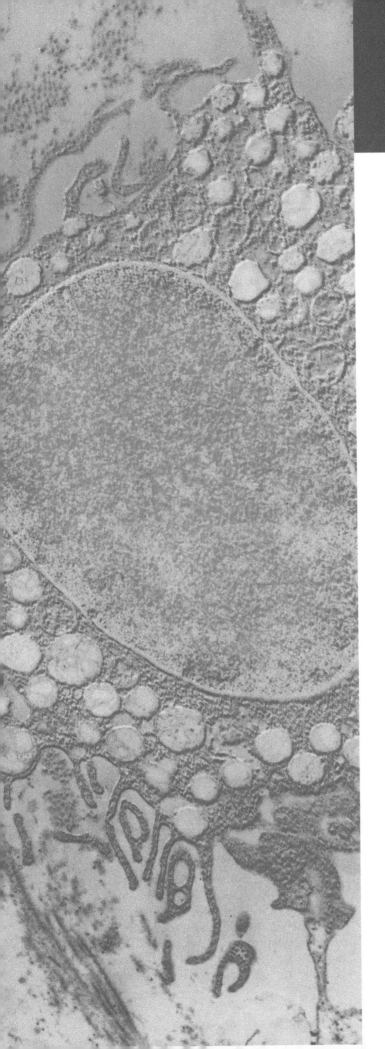

CARDIOVASCULAR SYSTEM

OVERVIEW

The cardiovascular system is the body's principal chemical and cellular transportation network. It supplies nutrients and O_2 to tissues to support their metabolism by continuously circulating blood through the intestines and lungs. It plays a fundamental role in homeostasis by maintaining the composition of water and electrolytes in extracellular fluid and removing metabolic wastes by circulating blood through the kidneys. It is the delivery route for immune system cells that defend the body against pathogens that manage to breech barrier epithelia. The cardiovascular system also distributes the vast array of signaling molecules, such as hormones and cytokines, which orchestrate much of our physiology and influence our behavior.

The cardiovascular system is comprised of the heart, which is a pump, and the vessels, which carry blood on circuits around the body. Both the heart and the large conducting vessels are comprised of a concentric, three-layered structure. The corresponding layers of the heart and vessels share similar composition and functions. Endothelial cells line the innermost layer of the heart and vessels and are supported by a layer of loose connective tissue. The middle layer of both structures is comprised principally of muscle; cardiac muscle powers the pumping of the heart, and smooth muscle controls the diameter of vessels.

The outer layer of the heart and vessels is constructed from connective tissue, and each serves to facilitate proper interactions with surrounding structures. In contrast, the exchangers (capillaries and postcapillary venules) consist mainly of a single layer of endothelial cells surrounded by a basal lamina. Endothelial cells also line the lymphatic vessels, which return a portion of the fluid that escapes from capillaries and venules back to the circulatory system.

HEART

This section begins with a description of the heart followed by components in the order taken by circulating blood: arteries, arterioles, capillaries, venules, and veins.

FUNCTION OF THE HEART

The major function of the heart is to provide the propulsive force to circulate blood. However, the heart also functions as an endocrine organ; for example, cardiac muscle cells located in the atria produce the hormone **atrial natriuretic peptide (ANP)**. The release of this hormone is stimulated by high blood pressure in the atria. ANP binds to receptors found at many sites in the body and produces several effects, the sum of which results in the reduction of blood pressure and the increase of sodium excretion into the urine.

STRUCTURE OF THE HEART

The heart contains four chambers consisting of **two atria** and **two ventricles**.

- **Atria.** Receive blood from the body (right atrium) or the lungs (left atrium). The atria then contract to force the blood into the more muscular **ventricles** to which they connect (Figure 7-1A).
- **Ventricles.** Receive blood from the two atria. Output of the right ventricle is directed to the lungs via the pulmonary artery, whereas the left ventricle supplies blood to other organs of the body via the aorta.

The chambers, like the large blood vessels, are composed of three layers, from deep to superficial: endocardium, myocardium, and epicardium (Figures 7-1B–E). The **cardiac skeleton**, composed of dense connective tissue, surrounds the valves and provides attachments for the muscle of the four chambers.

ENDOCARDIUM Endothelial cells line the heart chambers and all subsequent blood vessels to provide an **antithrombotic** surface that maintains blood as a fluid by preventing clotting. The loose connective tissue immediately adjacent to the endothelial cells may contain some smooth muscle along with fibroblasts and elastic and collagen fibers; in some sites, this tissue may be expanded as a **subendocardial layer** that contains small blood vessels and nerves.

- Noting differences in the thickness of the three cardiac layers aids identifications. The endocardium is relatively thick in the atria and usually thicker than the epicardium, whereas in the ventricles, the endocardium is the thinnest layer.

MYOCARDIUM Multicellular assemblies of **cardiac muscle cells**, linked by **intercalated discs**, are the main feature of the myocardium in both the atria and the ventricles. These assemblies are arranged helically around the chambers to expel blood efficiently upon contraction. This overall orientation is not apparent in typical histologic sections, where the alignment of nearby cells can vary widely.

- The muscle cells are covered by a thin endomysium and are in close proximity to numerous capillaries.
- These elements are fragile, and the muscle cells may become widely separated during the process of preparing sections.
- The myocardium is an extremely thick layer in the ventricles and is much thinner in the atria.

EPICARDIUM The epicardium contains large blood vessels surrounded by collagenous connective tissue and adipocytes.

- A simple layer of mesothelial cells, which secrete a serous fluid that lubricates the surface, covers the epicardium.
- The ventricular epicardium is relatively thicker than the epicardium in the atria and contains a significant amount of adipose tissue, which may help cushion the heart as it moves in the chest.

CARDIAC SKELETON Dense irregular connective tissue, often called the skeleton of the heart, provides attachments for cardiac muscle and supports the **heart valves**, which prevent backflow from the chambers when they contract (Figure 7-1A). This tissue is found in the superior region of the interventricular septum and as rings surrounding the four valves.

SPECIALIZED ELECTRICAL TISSUE

The heart contains sets of specialized cardiac muscle cells that generate and distribute the electrical signals that initiate contractions of the chambers.

- The **sinoatrial (SA)** and **atrioventricular (AV) nodes** are found in the right atrium (Figure 7-1A). These nodes are natural pacemakers consisting of small cardiac myocytes that spontaneously depolarize at regular intervals to generate the heart rhythm; the SA node normally dominates. The rate of firing of these nodes can be influenced by nearby sympathetic and parasympathetic nerves.
- Depolarization is transmitted from the AV node to the base of the ventricles by a bundle of myocytes that are specialized for electrical transmission and linked by gap junctions. The initial AV bundle penetrates the cardiac skeleton and forms several branches. **Purkinje fibers** are transmission elements found in the **subendocardial region** of ventricles (Figure 7-1A and F). These fibers are recognizable by their large size; they contain scant, poorly organized myofibrils. Purkinje fibers lack T tubules; thus their depolarization does not lead to contraction. They enter the myocardium of the ventricles, make contact with normal cardiac muscle via gap junctions, and initiate contraction in those cells. The contractions initiated by Purkinje fibers efficiently expel blood from the ventricles.

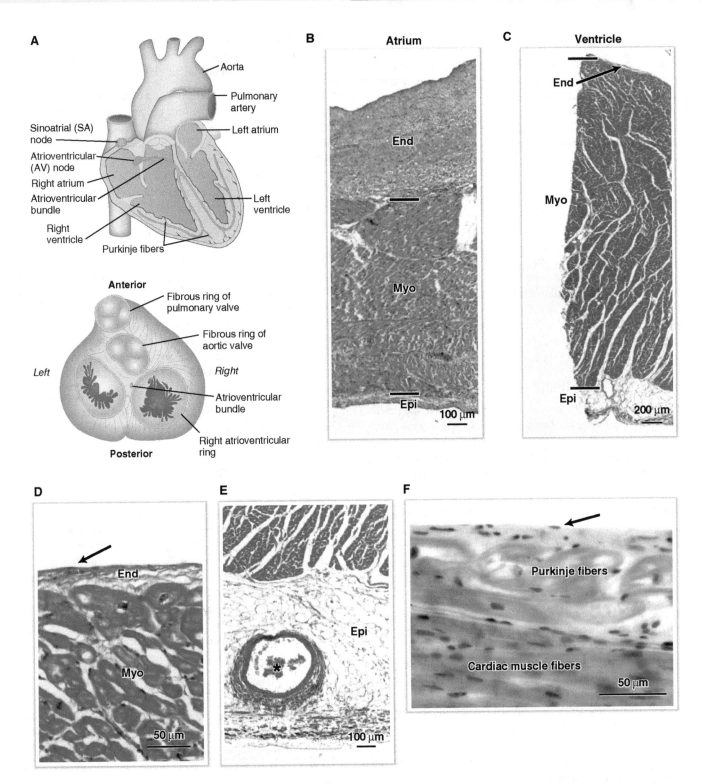

Figure 7-1: The heart. **A.** (Top) Heart chambers shown in an anterior view. The electrically active tissue is indicated in orange; fiber bundles connect the atrioventricular (AV) node to Purkinje fibers. The blue shading indicates the location of the fibrous cardiac skeleton. (Bottom) A superior view of the heart shows the cardiac skeleton with the atria and major vessels removed. **B.** Section of the atrium. Horizontal lines indicate the boundaries of the myocardium (Myo) with the endocardium (End) and the epicardium (Epi). Gaps in the cardiac muscle of the myocardium are artifacts often produced during preparation. **C.** Section of the ventricle. The boundaries between the myocardium (Myo) and the endocardium (End) above and the epicardium (Epi) below are indicated. This section is from a thin portion of the left ventricle. (Azan stain) **D.** Section of the endocardial surface of the ventricle. The arrow indicates the nucleus of an endothelial cell lining the surface of the endocardium (End). Branched cardiac myocytes of the myocardium (Myo) become separated during preparation. (Azan stain) **E.** Section of the epicardial surface of the ventricle. The epicardium (Epi) contains a large amount of adipose tissue and blood vessels. The lumen of a coronary artery is indicated by an asterisk (*). (Azan stain) **F.** Section of the endocardial surface of the ventricle with Purkinje fibers. The arrow indicates the nucleus of an endothelial cell, which sits on a thin layer of loose connective tissue. Large Purkinje fibers are found in the sub-endocardial space below this connective tissue.

BLOOD VESSELS

A series of systemic **arteries** conducts blood pumped from the left ventricle throughout the body. These arteries supply **capillaries** and then **postcapillary venules**, where the exchange of solutes between the blood and tissues occurs. Larger venules and **veins** then return blood back to the heart. Blood is transported to the lungs from the right ventricle in pulmonary arteries and returned to the left atrium in pulmonary veins.

GENERAL FEATURES OF BLOOD VESSELS

The conducting vessels, the arteries and veins, are constructed from three layers, called **tunics** (Figure 7-2A). Where large vessels communicate with the heart, these tunics are continuous with the three cardiac layers. Capillaries and postcapillary venules consist of a single layer of endothelial cells and occasionally periendothelial cells, or pericytes, all enclosed by a single basal lamina (Figure 7-2B). These exchangers are continuous with the tunica intima of adjacent vessels. Arteries and veins share the three tissue layers, from deep to superficial: tunica intima, tunica media, and tunica adventitia.

TUNICA INTIMA

Endothelial cells are the key feature of the thin layer, the tunica intima. These cells line the lumens of all vessels and serve many functions beyond providing an antithrombotic surface. Endothelial cells sit on a thin layer of loose connective tissue, which may be visible as a distinct **subendothelial** layer in large vessels.

ENDOTHELIAL CELLS The importance of endothelial cells seems inversely proportional to their diminutive size. Two of the well-known roles of endothelial cells include providing an antithrombotic surface for blood and facilitating solute exchange in capillaries and venules. These cells have many other activities; for example,

- Endothelial cells secrete several signaling molecules, such as **nitric oxide** and **endothelin**, which regulate the contraction of smooth muscle in nearby vessels and, therefore, influence blood pressure and the distribution of blood flow (Figure 7-2C). This activity is critical for normal physiologic function, and its perturbation may play a role in various pathologies, such as high blood pressure and the development of cancer.

- Endothelial cells express cell adhesion molecules, such as **selectins**, which cause many white blood cells to slowly roll along the surface of vessels as the blood flows. Signals generated by an infection can then recruit these loosely adherent immune system cells to exit small blood vessels, a process called **diapedesis** or **extravasation**, and enter tissues (Figure 7-2D).

PERICYTES Pericytes are cells that are occasionally found immediately outside the endothelium of capillaries and venules and are embedded in the basal lamina surrounding these vessels (Figure 7-2B). Pericytes have been implicated in many activities, including maintenance of the basal lamina, production of factors involved in hemostasis, contraction to control the diameter of the microvessels, and serving as stem cells for regeneration of blood vessels and other tissues.

TUNICA MEDIA

The tunica media is the middle layer of conducting vessels (especially arteries) and contains helically arranged bands of smooth muscle whose continuous, graded contractions control the diameter of the lumen. Sympathetic innervation of the tunica adventitia and diffusible chemical mediators, such as those secreted by endothelial cells, influence contractions of the smooth muscle. The smooth muscle cells in this layer also produce a reticular connective tissue containing **type III collagen** and **elastic fibers**.

▼ The perturbation of elastic elements produced by fibrillin mutations seen in **Marfan syndrome** can result in rupture of this layer in large vessels, such as the aorta, and can be fatal. ▼

TUNICA ADVENTITIA

The outer layer of conducting vessels is referred to as either the tunica adventitia, or tunica externa. The tunica adventitia consists of connective tissue populated by fibroblasts and smooth muscle cells and contains **type I collagen**. This layer frequently blends with the connective tissue associated with the organs and structures through which the vessels pass. For example, as an artery courses through a muscle, the tunica adventitia of the artery would blend with the epimysium of the muscle. The tunica adventitia of large veins is often thicker than the tunica media, and it may contain a significant amount of smooth muscle involved in controlling the diameter of these vessels.

VASA VASORUM

Cells in the tunica adventitia and tunica media of large vessels may require their own blood supply because they are too far from the blood adjacent to the tunica intima (Figure 7-2A). Blood vessels that supply these portions of large vessels are called vasa vasorum.

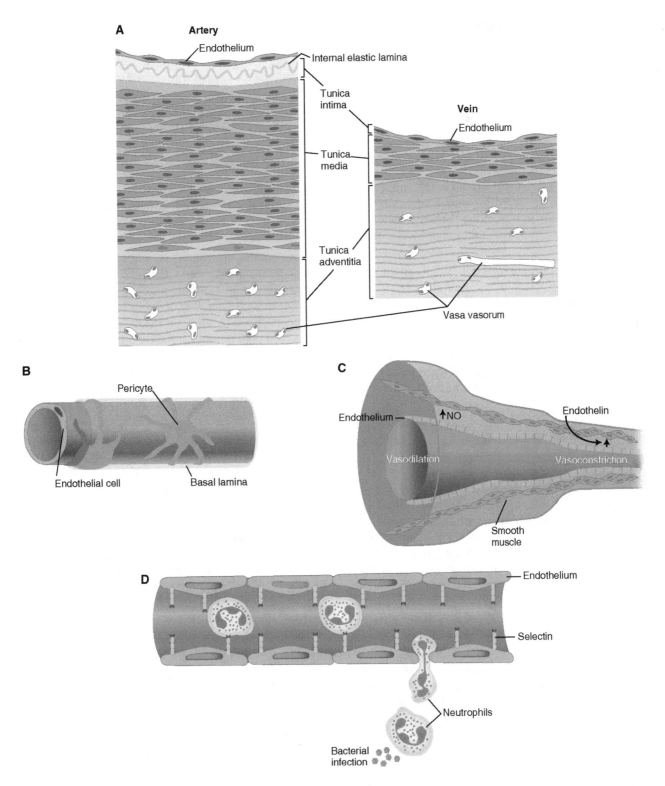

Figure 7-2: Blood vessels and endothelial cells. **A.** Comparison of the three layers of a large artery and vein. The tunica intima of arteries contains a prominent internal elastic lamina that is absent in veins. The tunica media of both vessels contains type III collagen and elastic fibers; this layer is much thicker in the artery. The tunica adventitia contains type I collagen and nerve fibers. In large vessels, the tunica adventitia contains blood vessels, the vasa vasorum. **B.** Capillaries and postcapillary venules are exchange vessels. Each consists of a tube formed from endothelial cells, with pericytes found at irregular intervals on the outside. Both cells are embedded in a single basal lamina. Pericytes help control vessel diameter and maintain the basal lamina. **C.** Signaling to smooth muscle by endothelial cells. Nitric oxide (NO) relaxes vascular smooth muscle and dilates vessels, whereas endothelin causes vasoconstriction. The chemical environment sensed by the endothelial cells, including neurotransmitters, influences the release of these factors. **D.** Adhesion of white blood cells to endothelial cells. Endothelial cells normally express selectins that cause white blood cells, such as polymorphonuclear neutrophils, to loosely adhere to the vessel walls. Signals released from an infection cause neutrophils to push between the endothelial cells (which is called diapedesis or extravasation), enter the tissue, migrate to the site of the infection, and attack the bacteria.

ARTERIES AND ARTERIOLES

Arteries transport blood away from the heart, regulate pressure, and control the distribution of blood to all parts of the body. The arterial system normally contains about 20% of the blood volume. All three tunics are present in arteries and arterioles. This section will discuss special features of the major types of these vessels.

ELASTIC ARTERIES

Elastic arteries are found near the heart, and serve to smooth the extreme pressure waves produced by ventricular contraction and to conduct blood to major regions of the body. These large diameter vessels include the aorta and the pulmonary, common carotid, subclavian, and renal arteries.

- The tunica media contains layers of elastic fibers produced by smooth muscle cells. These layers are arranged in sheets, called **elastic lamellae** or **elastic membranes**.

- The tunica media of the aorta contains 50–60 layers of elastic fibers (Figure 7-3A). Small gaps in the layers, called **fenestrae**, facilitate diffusion to support metabolism of the smooth muscle cells. Elastic recoil of the elastic fibers serves to transform the abrupt pressure pulses from the heart into a more continuous, high-pressure flow.

- Vasa vasorum are found in the thin tunica adventitia and in the outer layers of the tunica media.

SYSTEMIC ARTERIES

A set of systemic arteries of decreasing caliber, from approximately 10 cm to 0.1 mm, distributes blood from elastic arteries and supplies to tissues in individual organs. Compared to elastic arteries, the walls of systemic arteries are thick relative to their diameter, and their mechanical properties are dominated by the state of contraction of the smooth muscle in the tunica media. The larger systemic arteries are often called **muscular arteries**. The pulmonary arteries carry deoxygenated blood to the lungs and, therefore, are not considered systemic arteries by anatomists. However, the histology of pulmonary arteries is the same as that of systemic arteries.

- Many systemic arteries that supply the deep regions of the body are found close to the veins serving the same area and are associated with the peripheral nerves, forming **neurovascular bundles**. For example, the brachial artery, vein, and nerve all course in the same fascial sheath in the arm. In such cases, the vein typically has a less regular and often larger lumen compared to the corresponding artery (Figure 7-3B).

- Large systemic arteries are characterized by a prominent ring of elastic fibers, the **internal elastic lamina**, which is found at the borders of the tunica intima and tunica media. A thinner **external elastic lamina** is located at the border between the media and the adventitia (Figure 7-3C).

- The tunica media of a muscular artery may contain as many as 40 layers of smooth muscle in the largest vessels. Elastic fibers are present in this layer, but they are less dense than in

elastic arteries. Five or six layers of smooth muscle cells are found in the tunica media of the smallest systemic arteries. As the arterial diameter decreases, the internal elastic lamina thins and the external lamina becomes scant.

- The tunica adventitia is a relatively thicker layer in the systemic arteries than in the elastic arteries.

- Arterioles are the smallest elements of the arterial system (Figure 7-3D). Arterioles will be discussed in the following section, along with the exchange system.

▽ **Developmental and pathologic changes** normally occur over the lifespan of an individual, and many of these changes can have a detrimental effect as the individual ages. The previous descriptions are typical of arteries in young adults.

- The amount of extracellular elastic material increases with age. For instance, at birth, the tunica media of elastic arteries contains a significant amount of elastin, but it takes many months for the number of elastic lamellae to approach that found in adults. After the age of 25 years, the volume of smooth muscle in many arteries tends to decrease, while that of elastic fibers increases. The tunica intima tends to thicken in the larger arteries. These changes result in a stiffening of the arteries.

- **Pathologic processes** such as the deposition of calcium salts in the walls of arteries are also common with increasing age. **Arteriosclerosis** ("hardening of arteries") produced, for instance, by calcification or excessive elastin deposition associated with aging, results in increased blood pressure. Deposition of lipid on the tunica intima frequently results in inflammation, damage, clot formation, occlusion, and vessel stiffness in the complex process of **atherosclerosis**.

- The **increases in blood pressure and reduced blood flow** to organs that result from stiffened and damaged arteries may contribute to the degenerative processes of aging. ▽

ARTERIAL BRANCHING

In many tissues, small arteries branch and reconnect (**anastomose**), which provides several pathways for blood to reach capillary beds (Figure 7-3E). Anastomoses are important in many sites, including at joints such as the shoulder or knee where normal movements "pinch off" some arterial branches.

- In **end arteries**, these anastomoses are absent or inadequate so that if the main vessel is blocked (e.g., by a blood clot), the tissue supplied by that vessel dies.

- Some coronary arteries are functional end arteries because interconnections made by the coronary artery are limited. A clot in the lumen of a particular coronary artery can result in a **myocardial infarction**; the heart muscle supplied by the artery is damaged, and the patient suffers chest pain and other symptoms of a heart attack.

- Areas of the kidney are entirely dependent on blood from a single segmental artery branch, which are **structural end arteries**. A clot in a segmental branch results in the death of a portion of the kidney.

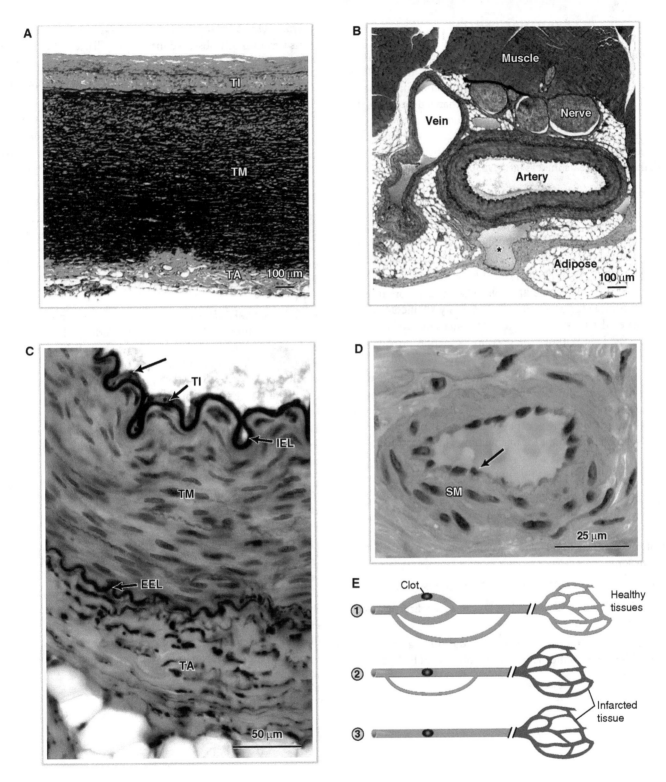

Figure 7-3: Arteries. **A.** Section of the aorta. The elastic stain demonstrates the large number of elastic laminae in the tunica media (TM). The tunica intima (TI) is relatively thick, indicating that this section is from a middle-aged or older person. (Verhoeff stain) **B.** Cross-section of a neurovascular bundle. A small artery and corresponding vein are seen in association with nerves, skeletal muscle, and a lymphatic vessel (*). The tunica intimae of the two blood vessels merge with adipose tissue and with each other. **C.** Section of the wall of a small artery. A higher power view of a portion of the artery seen in part B shows the prominent internal elastic lamina (IEL) at the border of the tunica intima (TI) and the tunica media (TM). The arrow indicates the nucleus of an endothelial cell. The external elastic lamina (EEL) sits at the border with the tunica adventitia (TA). (H&E and elastic stain) **D.** Cross-section of an arteriole. The arrow indicates the nucleus of an endothelial cell. The nuclei of a few layers of smooth muscle cells (SM) constitute the tunica media of this vessel. The tunica adventitia is continuous with the dense irregular connective tissue surrounding the arteriole. **E.** Patterns of arterial branching. (1) Arterial anastomoses can be sufficiently robust so that a clot in one arterial branch does not lead to the death of tissues downstream. (2) In functional end arteries, the blockage of the dominant vessel produces an infarct. (3) Anatomic end arteries lack collateral branching.

EXCHANGE SYSTEM

Arterioles control the blood flow into capillaries and, therefore, will be discussed in this section. Capillaries and postcapillary venules normally contain about 10% of the blood in the body. These are the sites where material is exchanged between the blood and tissues, which is the essential business of the circulatory system. Arterioles, capillaries, and postcapillary venules are all less than 0.1 mm in diameter, too small to be seen by the eye and, therefore, are defined as **microvasculature**.

ARTERIOLES

Arterioles are terminal portions of the arterial system. The three tissue layers in arterioles are thin; the tunica media consists of one or two layers of smooth muscle, and the internal and external elastic laminae also are less prominent.

- An arteriole ends in a **metarteriole**, where the smooth muscle of the tunica media becomes a **precapillary sphincter**, which regulates the flow of blood into the adjacent capillary bed (Figure 7-4A). Without the pressure reduction provided by arterioles and metarterioles, capillaries would rupture.

- The rate of opening and closing of the sphincters is influenced by sympathetic innervation and signaling factors, many of which are released from nearby endothelial cells. Because the total volume available in capillary beds is large, regulation of the sphincters is critical for maintaining blood pressure.

EXCHANGERS

Capillaries and postcapillary venules are the critical sites of exchange of material between blood and tissues. These vessels are tubes constructed from endothelial cells and use a common set of mechanisms to facilitate exchange, with specialized features in some organs.

MECHANISMS OF EXCHANGE ACROSS THE ENDOTHELIUM (FIGURE 7-4B)

- Simple **diffusion** of gases such as O_2 can occur directly across endothelial membranes.

- Solute-specific membrane transport **carriers** can facilitate the diffusion of solutes across the membranes of the endothelial cells.

- **Transcytosis** of large molecules via endocytosis and exocytosis is commonly observed in many capillaries.

- **Leaks** between tight junctions linking endothelial cells are produced by a variety of signaling molecules such as histamine, serotonin, and bradykinin.

- Except for diffusion, endothelial cells have the potential to regulate the last three processes. The **blood–brain barrier** is a property of the endothelia in the central nervous system, where the extent of transcytosis and leaks is tightly restricted.

CONTINUOUS CAPILLARIES Continuous capillaries are the most common type of capillaries and are found, for instance, in brain, muscle, and connective tissue (Figure 7-4C). The endothelial cells are joined together by tight junctions and form a continuous tube. The diameter of continuous capillaries is often less than that of an erythrocyte (7.4 μm).

FENESTRATED CAPILLARIES Fenestrated capillaries are more permeable structures and are found in the kidney, choroid plexus, endocrine organs, and the gut. The endothelial walls contain transcellular windows, or **fenestrae**, that are covered with a thin diaphragm (Figure 7-4B). The basal lamina is continuous and usually thicker in fenestrated capillaries.

SINUSOIDS Sinusoids are highly permeable vessels found in the liver, bone marrow, and spleen. The endothelium has larger openings than are found in fenestrated capillaries and lack diaphragms (Figure 7-4B). The basal lamina becomes thin and incomplete over the openings in the endothelium, further enhancing permeability.

POSTCAPILLARY VENULES Capillaries merge with postcapillary (pericytic) venules, which resemble continuous capillaries but are wider structures with diameters as large as 50 μm (Figure 7-4D). These vessels are a significant site of vascular exchange because the basal lamina is scant and the tight junctions between endothelial cells are often less extensive than in continuous capillaries. Vasoactive compounds, such as histamine, can cause significant fluid loss from these venules. Pericytes are associated with this first element of the venous system.

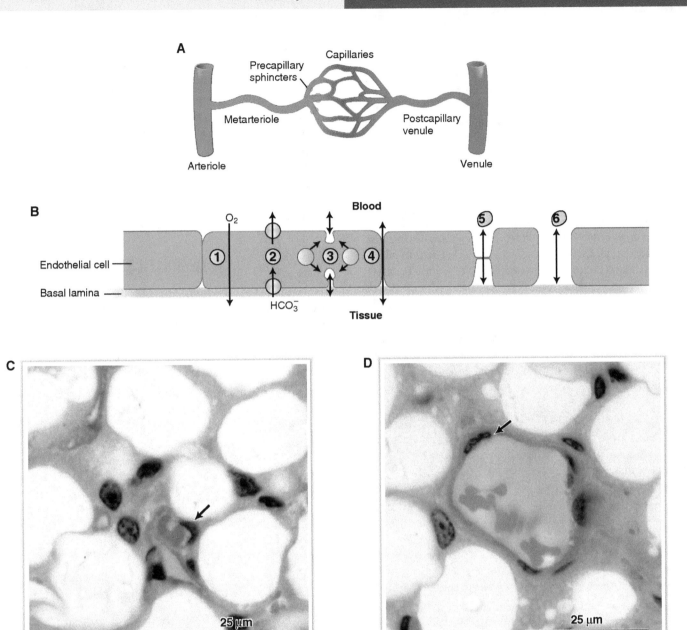

Figure 7-4: Exchangers: capillaries and postcapillary venules. **A.** Microvascular components of a capillary bed. Metarterioles are the terminal portions of arterioles where smooth muscle of the tunica media serves as precapillary sphincters, which control the flow of blood into the adjacent capillary. Blood flow through a capillary is not continuous because the sphincters regularly open and close. Chemical mediators regulate the rate of sphincter activity and, therefore, the blood flow. **B.** Modes of exchange across endothelial cells. (1) Lipid-soluble molecules, such as O_2, can pass freely across endothelial cells. (2) Integral membrane carriers facilitate the movement of ions, such as HCO_3^-, across both surfaces of endothelial cells. (3) Transcytosis can move large molecules by the rapid shuttling of endocytic and exocytic vesicles. (4) Leaks between tight junctions can be a major source of solute movement. (5) Diaphragm-covered fenestrations are found in glomerular capillaries of the kidney and serve as local sites of increased exchange. (6) Direct openings in the wall of endothelial cells are present in sinusoids in the liver and bone marrow. **C.** Cross-section of a capillary in adipose tissue. The arrow indicates the nucleus of an endothelial cell lining a capillary. The diameter of the capillary at this point is about 6 μm, and the red blood cell in the lumen must fold to allow its passage. **D.** Cross-section of a venule in adipose tissue. An arrow indicates the nucleus of one of the endothelial cells lining a venule. The wall appears to be the thickness of one single endothelial cell in most sites.

VENULES AND VEINS

After passing through the exchangers, blood is returned to the heart in the venous system. Veins of increasing caliber are the principal component of this low-pressure system and typically contain about 70% of the blood. The walls of veins are thinner, the lumens are larger, and the smooth muscle is more loosely arranged than in their corresponding arteries.

VENULES

The pericytes of postcapillary venules are eventually replaced with one or two layers of smooth muscle in the tunica media, forming **large** or **muscular venules**, although this organization is far looser than in arterioles (Figure 7-5A). The intimal and adventitial layers become somewhat thicker, because these vessels are now conducting vessels rather than exchange vessels. Venules are considered microvasculature and thus are smaller than 0.1 mm in diameter.

- A specialized type of **high endothelial venule** is found in some lymphatic organs (e.g., lymph nodes and the thymus). These venules are lined with cuboidal endothelial cells that allow lymphocytes to move freely between blood and tissue.

SYSTEMIC VEINS

Veins range from 0.1 mm to more than 10 mm in diameter. All three tunics are present, and are thinner than those in a corresponding small artery (Figure 7-5B; see Figure 7-3B).

- **Medium veins** are from 2 mm to 10 mm in diameter and may contain **valves**, which are formed by outfoldings of the tunica intima to prevent backflow of fluid. The valves are commonly found in veins of the lower leg. The tunica media is usually thinner than the tunica adventitia in these vessels. Bundles of smooth muscle cells interspersed between type I collagen fibers are found in the tunica adventitia.

- **Large veins** have diameters greater than 10 mm (Figure 7-5C). They include the vena cava and the portal, splenic, superior mesenteric, renal, external iliac, and azygos veins. The borders of the tunica media of large veins are often difficult to identify, whereas the tunica adventitia is a thicker layer containing some smooth muscle, bundles of collagen fibers, and a significant amount of elastic fibers.

SPECIALIZED VASCULAR CONNECTIONS

In some cases, the connection between the arterial system and the venous system involves components other than a single capillary bed (Figure 7-5D).

- **Arteriovenous anastomosis.** This anastomosis involves a direct connection of an arteriole to a venule, which bypasses a capillary bed. Arteriovenous anastomoses are found in the skin, where they can prevent blood flow to the surface in order to conserve heat, and in erectile tissue.

- **Portal system.** This system consists of two distinct capillary beds linked in series by a connecting vessel. This arrangement is found in the circulation of blood from the small intestines to the liver in the abdomen and from the hypothalamus to the pituitary in the brain.

- **Glomerular system.** This system consists of two distinct capillary beds linked in series by an efferent arteriole in the blood filtration system in the kidney.

LYMPHATICS

Tissues are bathed in **lymph**, the fluid that escapes from capillaries and venules. Lymph is returned to the circulatory system via the lymphatic system, which eventually joins large veins near the heart. A lymphatic vessel can be seen in the neurovascular bundle shown in Figure 7-3B. Unlike the cardiovascular system, the lymphatic system lacks a pump.

- Lymphatic capillaries are lined by endothelial cells that lack tight junctions and are extremely porous.

- The walls of larger lymphatic ducts are supported by smooth muscle and connective tissue but lack the clear-cut layering of blood vessels. These ducts contain many valves that prevent fluid from flowing backward in this low-pressure system. Body movements that compress lymphatic vessels are largely responsible for energizing lymph flow. For example, the return of lymph from the lower limbs against the force of gravity depends upon the contraction of muscles adjacent to lymphatic vessels.

- Lymphatic vessels are interrupted by **lymph nodes**, an important part of the immune system that will be discussed in Chapter 9.

ANGIOGENESIS

The formation of new blood vessels from existing ones is known as angiogenesis and is initiated by endothelial cells, which can divide, grow, and form tubes. Angiogenesis occurs normally during embryonic development, postnatal growth, and throughout life in wound repair. The process is influenced by dozens of signaling molecules, called **angiogenesis factors**.

▽ Solid tumors secrete angiogenesis factors to stimulate their vascularization. This is a key step in malignant growth. As a result, angiogenesis is an important potential target of cancer chemotherapy. ▽

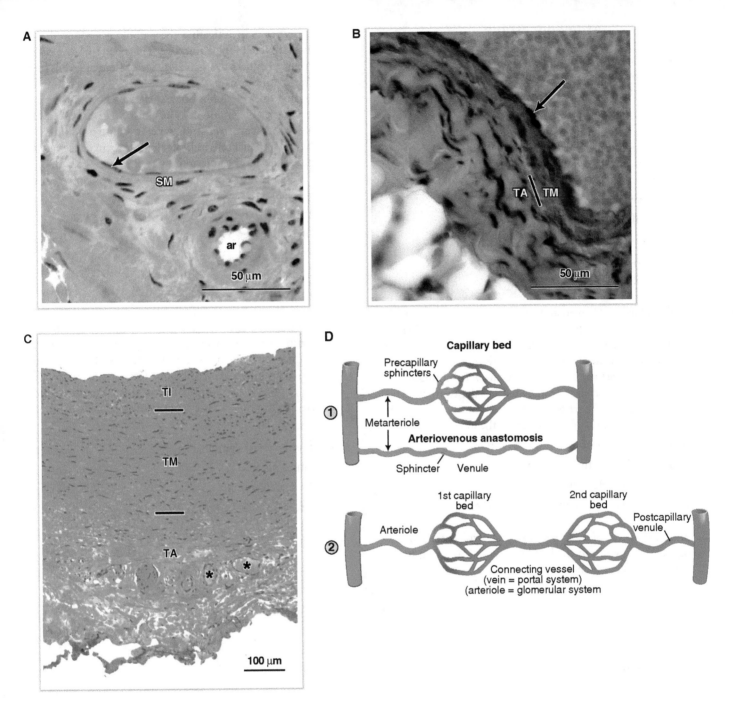

Figure 7-5: Venules, veins, and special vascular systems. **A.** Cross-section of a large venule. The arrow indicates the nucleus of an endothelial cell in a large or muscular venule that is packed with red blood cells. The wall contains two or three layers of nuclei, some of which have the corkscrew shape that is characteristic of smooth muscle cells (SM). An arteriole (ar) is present for comparison. **B.** Section of the wall of a small vein. This is a higher power view of a portion of the vein seen in Figure 7-3B. The arrow indicates an endothelial cell nucleus, which is all that is visible of the tunica intima. The tunica media (TM) contains two or three layers of smooth muscle cells. The tunica adventitia (TA) merges with adjacent adipose tissue. **C.** Section of the wall of the vena cava. The approximate locations of the borders of the tunica media (TM) are indicated by horizontal lines; a precise boundary is not visible. The tunica intima (TI) is relatively thick, and the extensive tunica adventitia (TA) contains several profiles of small blood vessels, the vasa vasorum, two of which are indicated by an asterisk (*). **D.** Specialized vascular connections. (1) Arteriovenous anastomoses can bypass adjacent capillary beds. (2) Two separate capillary beds are connected in series in some organs. In the case of the portal system, the connecting vessel is a vein that delivers blood from the small intestine to the liver. The connective vessel is an arteriole in the glomerular system (kidney) or hypothalamus to the pituitary gland (brain).

STUDY QUESTIONS

Directions: Each of the numbered items or incomplete statements is followed by lettered options. Select the **one** lettered option that is **best** in each case.

1. A researcher opens the chest of an anesthetized mouse and microinjects a small amount of a fluorescent dye into a single cardiac muscle cell at the base of the left ventricle. Over time, the dye spreads toward the left atrium. Which structure listed below is most likely responsible for facilitating this spread?

   A. Endomysium

   B. Intercalated discs

   C. Parasympathetic fibers

   D. Sarcoplasmic reticulum

   E. T tubules

2. While hiking in Canyonlands National Park, a 24-year-old medical student experienced a puncture wound on her leg from a cactus needle. Several days later, she noticed that the wound was swollen and tender, indicating an infection. Because of a class she had taken, she knew that immune system cells are attracted from the blood to the connective tissue involved. What barrier did these cells most likely have to move past to exit the blood and enter the infected tissue?

   A. Collagen fibers of the tunica adventitia

   B. Elastic fibers of the tunica intima

   C. Smooth muscle of the tunica media

   D. Tight junctions between endothelial cells

3. Endothelial cells of a capillary bed can regulate the amount of blood flowing into the bed by releasing signaling molecules that affect which nearby structure?

   A. End arteries

   B. Lymphatics

   C. Metarterioles

   D. Pericytic venules

   E. Vasa vasorum

4. A researcher is investigating the pathology produced by a newly discovered jungle microbe. This bacterium releases a toxin that produces dramatic leakage of fluid and cells from capillaries and postcapillary venules. The researcher determines that this reaction is not associated with an increase in the release of histamine or other known vasoactive molecules. To her surprise, studies using cultures of purified human endothelial cells indicate that the toxin does not bind to these cells and has no noticeable effect on their physiology. Which of the following would be the researcher's next logical cellular target?

   A. Atrial myocytes

   B. Fibroblasts

   C. Pericytes

   D. Smooth muscle

   E. Sympathetic neurons

5. Ehlers–Danlos syndrome is a group of inherited disorders. One form of this syndrome apparently involves the inability to produce type III collagen. In this case, death often results from massive internal bleeding due to spontaneous rupture of the aorta. Which portion of the aorta should be most affected by this form of the syndrome?

   A. Tunica intima

   B. Internal elastic lamina

   C. Tunica media

   D. Tunica adventitia

   E. External elastic lamina

ANSWERS

1—B: Cardiac myocytes form a functional syncytium because of the presence of intercalated discs that link cells mechanically and electrochemically. Gap junctions in the intercalated discs will allow the diffusion of the dye from cell to cell. The other structures listed, endomysium, parasympathetic fibers, sarcoplasmic reticulum, and T tubules, while present in the heart, would not be involved in facilitating this process.

2—D: Cells will gain entry to connective tissue by the process of diapedesis (also called extravasation), crawling between endothelial cells of capillaries and postcapillary venules. This indicates that tight junctions between endothelial cells are the barriers that they must pass to gain access to the connective tissue.

3—C: Smooth muscle sphincters of metarterioles regulate the flow of blood into the capillary beds they supply. The smooth muscle cells are sensitive to chemical signals, including those released by endothelial cells.

4—C: Pericytes are the only cells listed that are routinely associated with capillaries and postcapillary venules. These cells provide many important activities, including the production of factors involved in maintaining the integrity of microvessels.

5—C: The smooth muscle cells of the tunica media produce the type III collagen fibers and elastic fibers found in this layer. Recall that fibrillin mutations in Marfan syndrome, which perturb elastic fiber assembly, can also result in the rupture of large vessels.

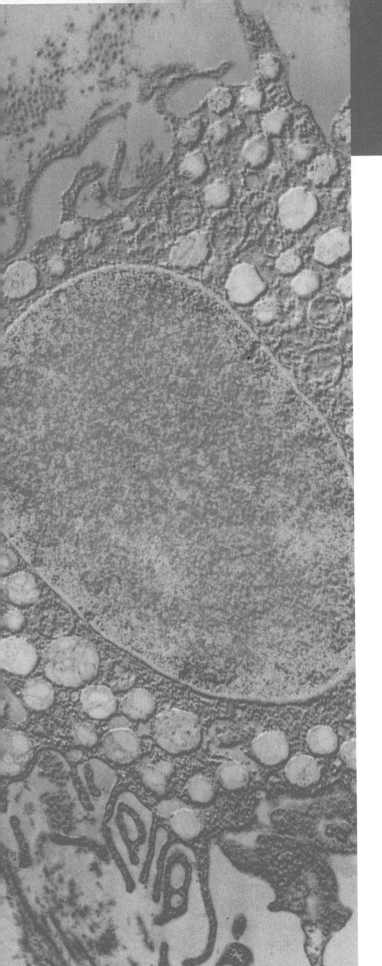

CHAPTER 8

BLOOD AND HEMATOPOIESIS

OVERVIEW

Blood is a complex mixture of cells and fluid transported throughout the body by the cardiovascular system. It can be considered a specialized connective tissue, one that provides functional instead of structural connections between all the body's organs. Blood consists of approximately equal volumes of **plasma** and **cells**.

▨ Plasma is a liquid extracellular matrix that supplies body tissues with material necessary for metabolism, removes wastes, and serves as a dynamic reservoir for maintaining the proper composition of extracellular fluid in the body. Plasma proteins are primarily synthesized in the liver; the kidneys regulate the levels of water and ions.

▨ Blood cells, also called the **formed elements** of blood, consist of anucleate erythrocytes (known as red blood cells, or RBCs), nucleated white blood cells (known as white blood cells, or WBCs), and platelets, which are cell fragments. These formed elements are generated in the **bone marrow** and thymus continuously throughout life by the division of hematopoietic stem cells and progenitor cells. The mitotic activity of these cells is controlled and integrated by a homeostatic network of growth factors produced throughout the body.

▨ Blood cells transport O_2 and CO_2 throughout the body (RBCs), provide immune defense (WBCs), and maintain the integrity of blood vessels and aid in blood clotting (platelets).

Blood flows to every organ and returns traces of pathology that may be present anywhere in the body. Laboratory analyses of plasma components, blood cell numbers, and blood cell production provide significant diagnostic information for all branches of clinical medicine.

OVERALL COMPOSITION OF BLOOD

Blood is described either as a fluid or as a specialized connective tissue. When considered as a tissue, blood is comprised of cells suspended in an extracellular matrix, **blood plasma**, which is normally fluid, although it has the potential of becoming semisolid through the process of **clotting**. Centrifugation of fluid blood provides a simple method for separating these two components in a small tube, with blood cells found at the bottom and plasma on top (Figure 8-1A). Blood is sometimes described as a **discontinuous tissue** because its cells are produced in the bone marrow and the thymus, but they function elsewhere, either in blood or in other tissues throughout the body. Similarly, blood cells do not produce their matrix—several other organs are responsible for producing and regulating plasma. Blood is a substantial tissue; the combination of blood and the principal blood cell producing tissue, **hematopoietic bone marrow**, constitute more than 10% of the body's weight.

PLASMA

The fluid portion of blood consists of water, low–molecular-weight solutes, and proteins (Figure 8-1A). Important constituents of plasma include the following:

- **Water.** Comprises about 90% of plasma volume.
- **Low–molecular-weight components.** Na^+, K^+, Ca^{2+}, HCO_3^-, and glucose are major solutes. Their concentrations, as well as the pH of blood, are precisely regulated by the action of the kidneys and the lungs.
- **Proteins.** Plasma contains thousands of proteins and peptides in concentrations that range over 10 orders of magnitude. These include trace levels of hormones produced throughout the body and substantial concentrations of immunoglobulins secreted by plasma cells in connective tissue. The liver is responsible for synthesizing numerous plasma proteins, of which the following three types are especially abundant:
  - **Albumin.** The most abundant plasma protein, normally present at 3–5 g/100 mL. Albumin contributes significantly to the osmolarity of blood and binds and transports a variety of hydrophobic molecules, such as fatty acids, hormones, and drugs.
  - **Fibrinogen.** An abundant protein that is converted by proteolysis to **fibrin**, which is then cross-linked to form a blood clot. When blood is removed and allowed to clot, the clot contracts and the fluid remaining outside the clot is called **serum**.
  - **α and β globulins.** These proteins serve diverse functions, such as transporting metal ions, inhibiting proteases, and participating in inflammatory reactions.

▽ Blood studies, referred to as **biochemical** or **metabolic panels**, are used to determine the levels of clinically relevant plasma components and enzyme activities, and provide important information about the physiologic state of an individual. One comprehensive metabolic panel currently used employs 20 chemical tests that analyze kidney and liver functions and can also detect more general pathology, such as tissue damage. ▽

FORMED ELEMENTS OR CELLS

Three classes of cells are found in blood (Figure 8-1B).

- **Erythrocytes,** or red blood cells (RBCs).
- **Leukocytes,** or nucleated white blood cells (WBCs), which consist of **granulocytes**, **monocytes**, and **lymphocytes**.
- **Platelets,** or thrombocytes.

Blood cells are often referred to as "**formed elements**" to recognize that RBCs and platelets are not typical cells. Human RBCs lack all organelles, including a nucleus, and platelets are small membrane-bound fragments shed from precursor cells that remain in bone marrow.

Blood cells are analyzed by drawing a sample of blood from a vein in the arm and treating the sample to prevent clotting. If the sample is subjected to centrifugation, the RBCs are found at the bottom; their packed cell volume is defined as the **hematocrit** (Figure 8-1A). The WBCs and platelets form a tan or **buffy coat** on top of the RBCs and occupy less than 1% of the volume. For clinical analyses, this centrifugation procedure has been replaced by automated scanning devices that provide a detailed **complete blood count**, or **CBC**, which is the cellular counterpart of the metabolic panel. Centrifugation may be used to provide a concentrated source of WBCs for molecular analyses. The CBC provides information about the number, volume, and hemoglobin content of RBCs as well as an accounting of the proportions of the different types of WBCs present, the **WBC differential**. The ranges of typical values are provided in Figure 8-1B. This differential count, along with the results for RBCs and platelets, provides important information for diagnosing and monitoring various diseases.

▽ Direct observation of stained blood and bone marrow cells by microscopy is frequently performed both to verify abnormal values and to monitor pathologic conditions. The reagents typically employed for blood and bone marrow slides, Wright and Giemsa stains, contain a mixture of acid and basic dyes. These reagents produce **four characteristic colors or color ranges**, depending on the chemistry of the cellular structures with which they react. The reagents are useful in identifying blood cells and their precursors, although it takes practice to calibrate the eye to account for the sometimes wide natural variation encountered on these slides. For example,

- Eosinophilic or acidophilic staining—red to orange
- Basophilic staining—purple to black
- Neutrophilic staining—pink to tan to nearly invisible
- Polychromatophilic staining—blue to gray ▽

A Blood composition

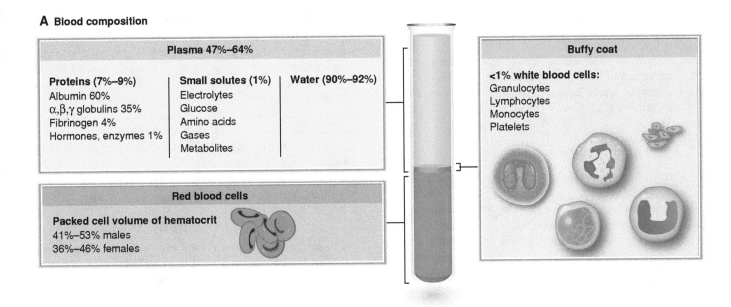

Plasma 47%–64%		
Proteins (7%–9%) Albumin 60% α, β, γ globulins 35% Fibrinogen 4% Hormones, enzymes 1%	**Small solutes (1%)** Electrolytes Glucose Amino acids Gases Metabolites	**Water (90%–92%)**

Red blood cells

Packed cell volume of hematocrit
41%–53% males
36%–46% females

Buffy coat

<1% white blood cells:
Granulocytes
Lymphocytes
Monocytes
Platelets

B Formed elements

NAME		DESCRIPTION	CELLS/μL
Erythrocytes		No nucleus, biconcave disc, thin in center, eosinophilic cytoplasm, 7 μm diameter.	3.5–6 million
White blood cells		Generally spherical, nucleated cells	4,500–11,000
Granulocytes • Neutrophil		3–5 lobed nucleus, neutrophilic cytoplasm, 12–15 μm diameter	2,400–6,800
• Eosinophil		Bilobed nucleus, eosinophilic granules, 12–15 μm diameter	45–300
• Basophil		2–3 lobed nucleus, basophilic granules, 12–15 μm diameter	0–80
Agranulocytes • Lymphocyte		Spherical nucleus, thin, rim of basophilic cytoplasm, two sizes: small 6–8 μm diameter, large 10–20 μm diameter	1,100–3,600
• Monocyte		Large kidney- or U-shaped, nucleus, cytoplasm slightly basophilic, 12–20 μm diameter	135–770
Platelets		No nucleus, basophilic cytoplasm, individual fragments 2–4 μm diameter often appear in clumps	150,000–400,000

Figure 8-1: Composition and formed elements of blood. **A.** Separation of blood into plasma and cells by centrifugation. The percent values listed provide a typical range of volumes, with the major variation due to individual differences in hematocrit (volume of packed erythrocytes). **B.** Overview of blood cells. The range of cell numbers considered normal can vary among clinical laboratories. Neutrophils typically account for 60–70% of leukocytes, 25–30% lymphocytes, 3–7% monocytes, 1–3% eosinophils, and 0–1% basophils.

RED BLOOD CELLS

RBCs are the only formed elements that normally remain in the blood for their entire lifespan. They are essentially membrane-bound containers of hemoglobin, specialized to carry O_2 to tissues and traverse small diameter capillaries. RBCs account for 40–50% of the total volume of blood and 99% of total blood cells.

STRUCTURE OF RED BLOOD CELLS

The average human RBC is about 7 μm in diameter (usual range is 6–8 μm) and 2-μm thick. On blood smears, RBCs serve as useful size markers. The cells adopt a **biconcave disc** shape—thick on the edge and thin in the middle. This shape results in lighter staining of the middle of the cells, called **central pallor**, when cells are viewed on slides (Figure 8-2A). Pathologic conditions that result in the loss of biconcavity can be recognized on slides by the presence of many RBCs with uniform staining density. The biconcave geometry provides two important benefits: a high surface to volume ratio, which facilitates diffusion across the membrane, and the ability to bend and deform passively. Unlike the nucleated WBCs, RBCs are immotile and cannot actively change their shape to crawl through small passages, so easy deformability is required to traverse capillaries with diameters less than 7 μm.

COMPONENTS

RBCs lack all organelles, and their cytoplasm consists primarily of hemoglobin, a heme-containing protein. Because mitochondria are absent, the energy required for a limited number of metabolic functions, such as keeping hemoglobin in a reduced state, is derived from the pentose-phosphate pathway and glycolysis. The other components of the RBC, the membrane and associated cytoskeleton, have been intensively studied because they can be easily isolated by lysing the cells and washing away cytoplasmic contents. Analyses of purified RBC membranes reveal a specialized cytoskeleton based on **actin** filaments and tetramers of the protein **spectrin**, which associate with integral membrane protein assemblies that include **band 3** and **glycophorin** by the linker proteins **ankyrin** and **4.1** (Figure 8-2B). It is the interactions of the cytoskeleton and these membrane proteins that determine the biconcave disc shape of these cells.

FUNCTION OF RED BLOOD CELLS

RBCs mediate the efficient transport of O_2 and CO_2 between lungs and tissues. When RBCs enter capillaries in the lungs, O_2, present at high levels, binds to hemoglobin. When RBCs enter capillaries in tissues, the conditions favor the release of O_2, which diffuses across the membrane to supply nearby cells. CO_2 generated by metabolism in the tissues is present at high concentrations and diffuses into RBCs, where it is largely converted by the enzyme carbonic anhydrase into carbonic acid, which quickly dissociates into HCO_3^-. The release of the HCO_3^- anion from RBCs to blood is facilitated by the membrane protein, band 3, which also is an anion transporter. The bulk of the CO_2 from tissues is carried back to the lungs in the form of HCO_3^- ions. In the lungs, CO_2 diffuses from blood to air, which favors formation of CO_2 from HCO_3^-, which is then exhaled. This reverse reaction is also efficiently catalyzed by carbonic anhydrase (Figure 8-2C).

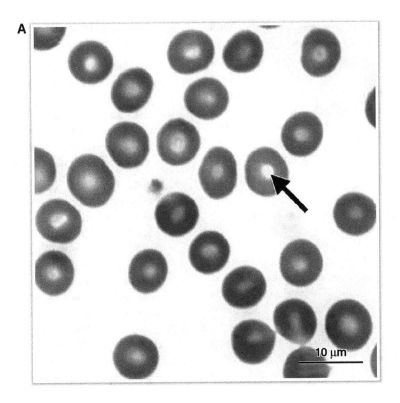

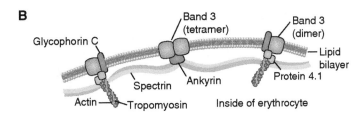

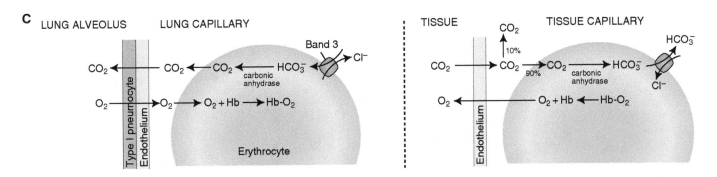

Figure 8-2: Erythrocyte (RBC) structure, activities, and hereditary alterations. **A.** RBCs (normal blood smear). The biconcave disc shape normally results in a lighter density in the middle of the cells, known as central pallor (*arrow*). **B.** RBC membrane. Multiprotein, integral membrane complexes that include band 3 and glycophorin serve as attachment sites for a specialized cytoskeleton, which consists of short actin fibers stabilized by tropomyosin, which interacts with tetramers of spectrin. These membrane-cytoskeletal interactions stabilize the RBC membrane and are responsible for maintaining the biconcave disc shape of these cells. **C.** RBCs and gas transport. In the lungs, the high concentration of O_2 and the relatively high pH favor O_2 entering the blood and binding to hemoglobin, whereas the low CO_2 concentration favors the production of CO_2 from carbonate by carbonic anhydrase, its diffusion into alveoli, and exhalation (*left*). In the tissues, the reduced pH and O_2 levels favor the release of O_2 and the uptake of CO_2 (*right*). (*continued on page 109*)

A variety of **clinical conditions are associated with RBC abnormalities**, including the following:

Anemia is usually defined as a reduction in the amount of functional hemoglobin in blood. This can be caused by a decrease in the number or size of RBCs or the amount of functional hemoglobin carried per cell. Anemia results in reduced O_2 delivery to tissues and produces various symptoms, such as weakness and fatigue.

Polycythemia is an increase in hematocrit, which can result from low O_2 stress in tissues caused by lung disease or time spent at high altitude.

Sickle cell disease results from an altered hemoglobin molecule, HbS, produced by a point mutation. HbS is less soluble when deoxygenated, and can associate into polymers that bind to the membrane and also make the cytoplasm more viscous. This results in cells adopting a sickle-like shape and becoming less flexible (Figure 8-2D). These cells can occlude capillaries and small vessels, causing infarcts and tissue damage. Sickle cells are also more fragile; their lifetime may be reduced to less than 20 days, producing a chronic hemolytic anemia. The mutation is found most frequently in populations originating from tropical and subtropical areas and, when present in heterozygotes, appears to reduce the severity of malarial infections.

Thalassemia is a set of inherited diseases that cause a relative overproduction of one of the two chains of hemoglobin. The excess chains can form precipitates that bind to the membrane, causing a local stiffness. Such cells are recognized in the spleen, where the stiff portions or entire cells are removed, resulting in anemia. These diseases were first described in populations living near the Mediterranean Basin.

Hereditary spherocytosis is a disorder that affects the function of the membrane and cytoskeletal components responsible for maintaining the shape and flexibility of RBCs, such as spectrin, ankyrin, or band 3. Hereditary spherocytosis affects about 1 in 2,000 individuals of northern European descent. As these RBCs travel through the spleen, macrophages detect their altered membrane properties and remove portions of some cells and destroy others entirely. The processed RBCs adopt a spherical shape and lack central pallor on smears (Figure 8-2E). These spherically shaped cells are inflexible and will be frequently destroyed on subsequent passes through the spleen, resulting in a chronic anemia. Splenectomy is a common therapy for patients with hereditary spherocytosis. ▼

D

E

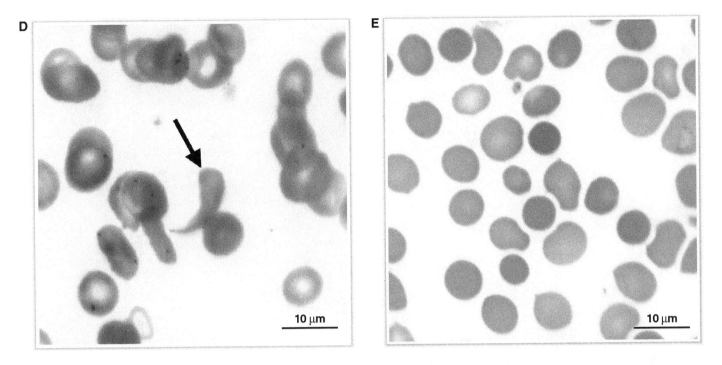

Figure 8-2: (*continued*) **D.** RBCs from a patient with sickle cell disease. A distorted sickle-shaped cell is indicated by the arrow. **E.** RBCs from a patient with hereditary spherocytosis. Note the many small cells that lack central pallor. Samples D and E courtesy of Dr. Mohamed Salama, University of Utah School of Medicine, Salt Lake City, Utah. (A, D, and E Wright stain)

WHITE BLOOD CELLS

WBCs, or nucleated WBCs, consist of **granulocytes**, **monocytes**, and **lymphocytes**. These cells travel for variable lengths of time in blood before eventually exiting a vessel and entering another tissue or undergoing apoptosis. They are all elements of the immune system, and were first introduced in Chapter 2 as residents of connective tissues. Some of the functions of WBCs will be further considered in this chapter, and additional information will be presented in Chapter 9.

GRANULOCYTES

The three nucleated blood cells called granulocytes (neutrophils, eosinophils, and basophils) are members of the **innate immune system**. When viewed on blood smears, all three appear as spheres, ranging from 12 μm to 15 μm in diameter and have a lobed nucleus. Granulocytes are easily distinguished by their specific membrane-bound cytoplasmic granules (hence their name), which play a key role in their function and stain with characteristic colors associated with their names.

NEUTROPHILS Neutrophils are the most numerous WBCs, accounting for about 60% of the total, and contain granules that appear neutrophilic, or nearly invisible, on a Wright-stained preparation. They have a short lifespan, spending less than 1 week in the bloodstream, and typically survive a day or so if recruited into connective tissue.

- **Structure and components.** Neutrophils are readily identified by their multilobed nuclei; the number of lobes increases from two to five as the cells age. They are frequently called polymorphonuclear neutrophils, or PMNs (Figure 8-3A). Neutrophils contain several types of membrane-bound granules, usually classified into two groups: **azurophilic** or **primary granules**, which contain defensins and other bactericidal components, and **specific** or **secondary granules**, which contain lysozyme and enzymes used to generate reactive O_2 compounds. Few mitochondria are present. These cells are adapted for survival in anaerobic conditions.

- **Function.** Neutrophils are absent from tissues unless attracted to sites of inflammation. They are specialized for engulfing and killing bacteria and fungi, but also remove uninfected tissue damaged by injury. Their response to inflammatory signals involves the steps illustrated in Figure 8-3B.

  - **Tissue entry.** It was mentioned in Chapter 7 that endothelial cells express **selectin** proteins to which neutrophils bind and release, allowing them to roll slowly along the walls of vessels. Molecules that signal inflammation, such as **chemokines**, are generated at sites of infection and tissue damage and cause nearby endothelial cells to increase the expression of selectins and also to activate **integrins** expressed by neutrophils. As a result, neutrophils bind tightly to vessel walls and then crawl between endothelial cells, a movement called **diapedesis**.

  - **Migration.** Once in connective tissue, the neutrophils migrate toward the source of the inflammatory signals, a process called **chemotaxis**, to arrive at the site of infection or damage.

  - **Phagocytosis and killing.** Cell-surface receptors on neutrophils bind to pathogens or damaged cells and induce their internalization, which is called **phagocytosis**. The

primary and secondary granules of the neutrophils then fuse with the phagosome, releasing the **killing factors** to attack the pathogens. An important component of this attack is the activation of enzymes from the secondary granules that produce **respiratory burst**. This reaction creates several O_2 radicals that can kill bacteria and fungi. Finally, lysosomes fuse with the phagosome to digest the remains. Once released from the bone marrow, neutrophils have a limited synthetic capacity and undergo apoptosis when their supply of granules is exhausted. The debris left from this process is known as **pus**, which is produced at sites of a significant bacterial infection.

▽ **Inherited defects in neutrophil function** result in frequent bacterial and fungal infections that appear early in life. Mutations in components of the respiratory burst result in cells that engulf bacteria, but cannot efficiently kill them. Accumulations of phagocytes result, forming structures called **granulomas**. Mutations that render the integrins of the neutrophil defective, prevent them from efficiently adhering to blood vessels, undergoing diapedesis, and phagocytosing pathogens. When infection occurs, the **leukocyte adhesion deficiency** results in few neutrophils present at affected sites in connective tissue, whereas the number of neutrophils in blood may be increased many times above normal. Infants with these defects require careful monitoring and early antimicrobial therapy. ▽

EOSINOPHILS Eosinophils usually account for only 2% of the WBCs. However, they are easily identified on blood smears because of their bright orange secondary granules. Eosinophils are found in the connective tissues associated with barrier epithelia, such as that lining the respiratory and digestive systems, and are attracted to sites of parasitic infections.

- **Structure and components.** The nuclei of eosinophils are characteristically only bilobed (Figure 8-3C). The secondary granules of these cells contain toxic proteins, principally **major basic protein**, as well as peroxidase and collagenase. Major basic protein is highly concentrated in these granules and is responsible for their intense staining with eosin or other acidic dyes.

- **Function.** Although eosinophils are capable of phagocytosis, they are able to kill pathogens that are too large to engulf, such as parasitic worms, because they release components from their secondary granules by exocytosis (Figure 8-3D). This release normally requires the activation of eosinophils by cytokines and the presence of antibodies bound to the surface of parasites. Major basic protein binds to cell surfaces and is extremely cytotoxic, but the mechanism involved is not understood. The peroxidase released by the eosinophils generates toxic compounds that can damage parasites. Eosinophils also release signaling molecules with a wide variety of inflammatory effects.

▽ An accumulation of eosinophils in airways can result in tissue damage that produces chronic **asthma** due largely to the release of cytotoxic components from secondary granules. The incidence of airway diseases and other autoimmune reactions is increasing in developed countries, and is thought to be related in part to improved hygiene, antibiotic use, and vaccination. With fewer pathogens to attack, immune cells, such as eosinophils, may respond to minor irritations and damage normal tissue. ▽

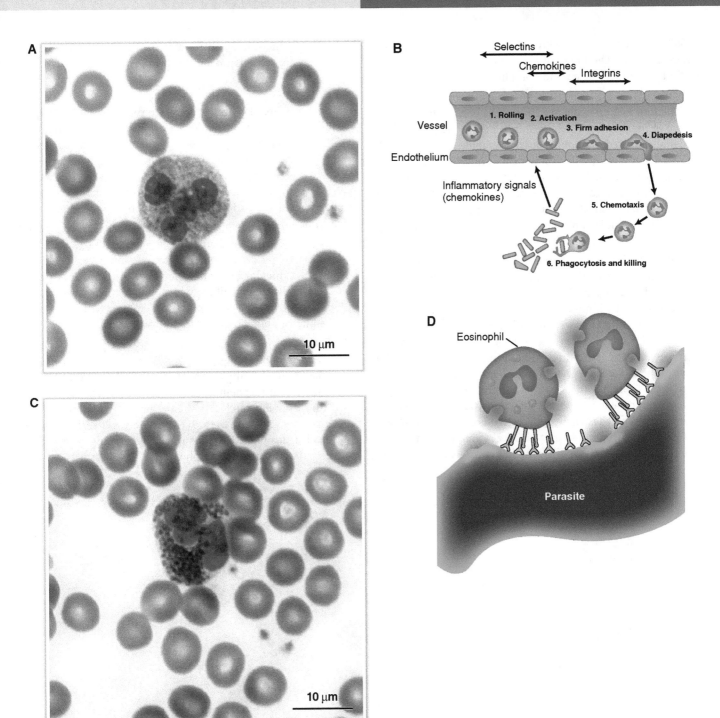

Figure 8-3: Neutrophil and eosinophil structures and activities. **A.** Neutrophil. Four nuclear lobes are visible. The cytoplasm has a faint texture, but the numerous primary and secondary granules are not readily apparent. **B.** Neutrophil responses to infection. Macrophages at the site of an infection secrete chemokines, which activate integrins in neutrophils, resulting in binding and diapedesis. Once in connective tissue, neutrophils follow substances released by bacteria and damaged tissue to their source and then phagocytose and destroy bacteria and damaged cells. **C.** Eosinophil. The eosinophilic secondary granules are readily apparent and do not obscure the nuclear lobes. (A and C, Wright stain) **D.** Eosinophilic attack on a parasitic worm. When eosinophils are stimulated by cytokines, they express high-affinity receptors for the IgE class of antibodies. When an invading parasite is coated with these antibodies, eosinophils will bind and release the cytotoxic contents of their granules without the requirement of phagocytosis. This extracellular release allows eosinophils to kill large invaders.

BASOPHILS Basophils normally constitute less than 1% of circulating WBCs. The rarity of basophils makes them difficult to find on blood smears and complicates the study of their biology.

- **Structure and components.** The nuclei of basophils contain two or more lobes, which are difficult to see because of their densely stained, basophilic specific granules (Figure 8-4A). These granules contain components similar to those found in **mast cells**, such as histamine and heparin, chemokines that attract neutrophils and eosinophils, and proteases that can degrade connective tissue fibers. Heparin, an anticoagulant, is an acidic molecule responsible for the basophilic staining of the granules.

- **Function.** The factors present in the specific granules of basophils, and in the far more abundant mast cells, are released when IgE antibodies attached to the cell surface are cross-linked by binding to an antigen (Figure 8-4B). These factors produce powerful inflammatory effects upon release, including increasing vascular permeability by histamine and attracting and modulating other cells of the immune system by cytokines, chemokines, and leukotrienes. These activities help orchestrate the immune system's response to a variety of infections, but they also create unpleasant and destructive effects that occur during **allergic reactions.** Basophils are not identical to mast cells; the specific granules of basophils contain cytokines that promote the production of antibodies by the acquired immune system in response to an infection.

MONOCYTES

Monocytes are the immediate precursors of **macrophages**, which are members of the innate immune system, were introduced in Chapter 2 as permanent residents of connective tissue. Monocytes normally constitute about 5% of circulating leukocytes. After leaving the bone marrow, monocytes circulate for 1–2 days, enter a tissue, and mature. Mature monocytes, macrophages, can live for months and are capable of dividing.

- **Structure and components.** Monocytes are the largest blood cells, 12–20 μm in diameter (Figure 8-4C). The monocyte nucleus typically has an indentation, and it can be shaped like a kidney or "U," or it may have a lumpy appearance. These cells contain all the typical cellular organelles, including lysosomes and enzymes, for generating O_2 radicals via the respiratory burst required to destroy pathogens.

- **Function.** Monocytes and macrophages are phagocytes capable of engulfing and destroying pathogens. Macrophages are a key source of inflammatory signals—they release chemokines that attract neutrophils to the site of bacterial infection. After the neutrophil response has waned, additional macrophages migrate into the area to phagocytose debris, including dead neutrophils. Macrophages are also antigen-presenting cells that can activate lymphocytes.

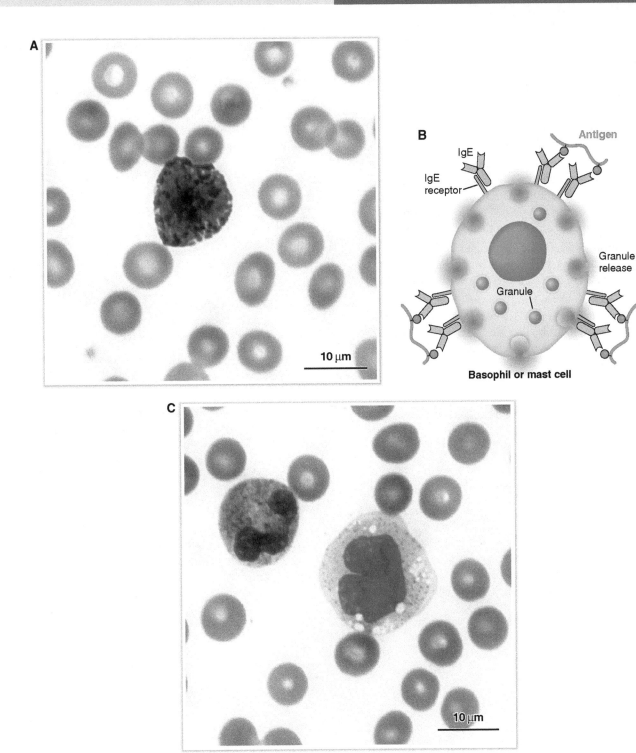

Figure 8-4: Basophils, monocytes, lymphocytes, and platelets. **A.** Basophil. These cells are rare; their dark granules facilitate identification, although they obscure observation of the nucleus, which contains two or more lobes. (Wright stain) **B.** Antigen-mediated degranulation of basophils. Basophils, mast cells, and stimulated eosinophils express a high affinity receptor for the IgE class of antibodies. The granules of basophils and mast cells contain a variety of inflammatory mediators, which are released when an antigen cross-links the receptors. **C.** A monocyte and a neutrophil. The monocyte (*center*) can be identified by its larger size, less condensed nucleus with a small cleft or fold visible, and the presence of a few vacuoles. (Wright stain) (*continued on page 115*)

LYMPHOCYTES

Lymphocytes are spherically shaped cells with round, dark-staining nuclei that occupy most of the cell's volume. Small lymphocytes, 6–8 μm in diameter, constitute 85–95% of these cells; the remainder, large lymphocytes, are 10–20 μm in diameter. Together, lymphocytes comprise about 30% of the circulating WBCs. The lifespan of lymphocytes varies considerably, from weeks to years.

- **Structure and components.** Small lymphocytes are components of the adaptive immune system and consist of B cells (5–25%) and T cells (65–75%), but specific antibody reagents are required to distinguish them (Figure 8-4D).

- Large lymphocytes (not shown) consist of **natural killer cells**, which are part of the innate immune system, and lymphocytes that have been activated by exposure to antigen.

- **Function.** Lymphocytes, similar to the other WBCs, function outside the blood in tissues. B cells are the precursors of plasma cells, which secrete antibodies (see Figure 8-7B). T cells are the precursors of helper T lymphocytes, cytotoxic T lymphocytes, and T regulatory cells.

PLATELETS

Platelets are cell fragments involved in hemostasis—the prevention of blood loss from vessels.

STRUCTURE AND COMPONENTS OF PLATELETS

Platelets are irregularly shaped cell fragments about 2–3 μm in diameter. They lack a nucleus but contain other cellular organelles such as mitochondria, microtubules, endoplasmic reticulum, lysosomes, and an actomyosin-based cytoskeleton. They also contain several types of membrane-bound granules that store signaling molecules, such as **serotonin** and **platelet-derived growth factor**, as well as factors involved in blood clotting. Platelets are often found in clumps on blood smears (Figure 8-4E).

FUNCTION OF PLATELETS

Complex interactions between endothelial cells and platelets normally keep platelets inactive. When a vessel is ruptured, however, platelets come in contact with collagen and bind avidly to it. This contact, as well as exposure to other factors in the matrix outside vessels, activates the platelets. The activated platelets move and aggregate, forming plugs that can seal small breaks. They interact with the fibrin clot that forms at the lesion, stimulating further clot formation and later acting to contract the clot, helping to close ruptures. Platelets also release factors from their granules that stimulate fibroblasts and other cells to migrate to the area and repair the damage. It is estimated that about 10% of the platelets are used to maintain vascular integrity by repairing routine damage to small vessels, and the remainder serve as a reserve to protect against larger wounds.

▽ Common **clinical conditions involving platelets** are recognized by long-term alterations in their circulating levels, as follows:

- **Thrombocytopenia** is an abnormally low platelet count, and is associated with frequent bruising and bleeding. It can be caused by decreased production of platelets in the bone marrow due to cancer or severe infection, by increased destruction, which might result from autoimmune reactions against platelet components, or by drugs such as chemotherapeutics.

- **Thrombocytosis** is an abnormally high platelet count. It may result from an infection that stimulates platelet production, or it may be related to disorders in bone marrow, such as the leukemias. Individuals with high platelet counts may be asymptomatic, but are assumed to be at increased risk of excessive clotting and thrombosis. ▼

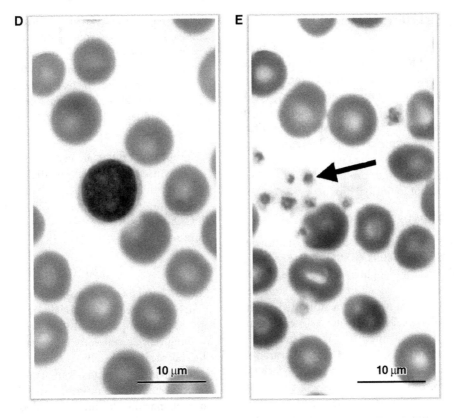

Figure 8-4: (*continued*) **D.** A small lymphocyte. Small lymphocytes are approximately the size of erythrocytes; they contain a dense, uniform nucleus surrounded by a thin rim of cytoplasm. **E.** A group of platelets. These small fragments (*arrow*), typically 2–4 μm, adhere readily to slides and may begin to spread out before being fixed, so they have a somewhat irregular outline. (Wright stain)

HEMATOPOIESIS

Hematopoietic marrow normally accounts for 5% of body weight. It provides a favorable environment for the progenitor cells that produce mature blood cells. Growth factors regulate the activities of these progenitors to maintain the proper balance of blood cells in health and to regulate production in response to disease or injury. Diagnosis of diseases that affect the blood may require biopsy of marrow. Histologic analysis of marrow samples is used to assess the overall health of this tissue and to determine if the lineages that produce RBCs, granulocytes, and platelets are present in normal proportions.

STEM AND PROGENITOR CELLS

All blood cells are ultimately derived from a population of **pluripotent hematopoietic stem cells (HSCs)**. These cells can divide "asymmetrically" so the daughter cells can be of two types; that is, either more HSCs, which maintain the stem cell population, or progenitor cells, which are committed to the eventual production of blood cells. Progenitor cells can maintain their population through asymmetric division, but are continuously replenished from the HSC population. Researchers have identified many of these progenitors as **"colony-forming units" (CFUs)**, which are capable of producing a small colony of blood cells in the spleen of a mouse or in a culture dish. The results of numerous experiments are summarized in Figure 8-5A. This figure provides a generally accepted scheme by which sets of progressively more restricted progenitor cells eventually produce all the cells found in blood. Not all details for every linage have been fully established, and there may be some developmental differences between mice (where much of the experimental genetic research has been done) and humans. The terminology employed refers both to the types of cells produced and their appearance. For example,

- The least specialized CFUs appear to be a **lymphoid progenitor**, committed to producing B and T lymphocytes, and a **myeloid progenitor**, also called the **GEMM** because it produces **g**ranulocytes, **e**rythrocytes, **m**onocytes, and platelets (from **m**egakaryocytes).
- The HSCs and progenitor cells resemble medium-sized lymphocytes, and cannot be distinguished on slides treated with typical stains. However, the final stages of this process, beginning with cells called **blasts**, are associated with distinctive morphologies that can be recognized on Wright-stained preparations.

GROWTH FACTOR CONTROL

Both the rate at which hematopoietic cells divide and the choice of what kind of progenitor cells will be produced are influenced by a large number of hormone-like molecules or **growth factors**, some of which are indicated in Figure 8-5A. These factors are produced both locally at the site of hematopoiesis, or in tissues throughout the body, and circulate widely. Several of these factors were found to favor the formation of particular CFUs in culture and are called colony-stimulating factors (**CSFs**). A variety of cells release **GM-CSF** in response to an infection, which accelerates the production of granulocytes and monocytes. After a major hemorrhage, the resulting lack of adequate O_2 carried by blood results in the release of **erythropoietin (EPO)** from the kidneys, which acts on the CFU for RBCs to stimulate production.

▽ **Growth factors** are used clinically. For instance, GM-CSF is used to stimulate neutrophil and monocyte production after cancer chemotherapy, which often inhibits blood cell production, to help protect against opportunistic infections. EPO is used to treat patients with chronic anemia resulting from severe kidney disease. ▼

HEMATOPOIETIC NICHES

Low levels of HSCs and progenitor cells are found in the circulation, but they require a special microenvironment of cytokines and physical support, a **niche**, to function properly. Blood-forming cells appear early in human embryogenesis, arising in the yolk sac in week 2. By week 6, several types of blood cells are produced in the liver, an organ that provides an important niche before birth. During months 2 to 3, hematopoietic niches arise in bone marrow. Production of blood in marrow becomes significant by month 5; after birth, this medullary hematopoiesis, or myelopoiesis, eventually becomes dominant. Fibroblast-like **stromal cells** provide the niche in bone marrow. Hematopoietic marrow consists of aggregates of stromal and hematopoietic cells, called **cords**, adjacent to highly permeable **sinusoidal capillaries**. Over time, hematopoietic marrow, or red marrow, becomes limited to bones near the center of the body, principally in the pelvis and spine, whereas in other bones, the stromal cells transform into adipocytes and the marrow appears yellow. If hematopoietic marrow is damaged, niches can reappear in yellow marrow and even spread to organs, such as the spleen.

▽ Many pathologic conditions, such as infections, anemia, or cancer, are associated with or produced by significant alterations in circulating blood cells. These changes either result from alterations in the production of cells in the marrow or utilization and destruction of cells in the rest of the body. In some cases, a complete diagnosis will depend on **an assessment of activities in the marrow**, which requires direct sampling, an invasive procedure. Hematopoietic marrow samples are usually obtained at the iliac crest and may involve both the insertion of a syringe needle to obtain an aspirate sample by suction as well as the use of a larger bore trephine needle to remove an intact core sample of the marrow. Smears of an aspirate sample spread the hematopoietic cells, allowing detailed cytologic analysis. Samples are stained and analyzed similar to blood smears, as discussed in the following section. Core samples are fixed and sectioned to provide a picture of the overall organization and cellularity of the marrow (Figure 8-5B). ▼

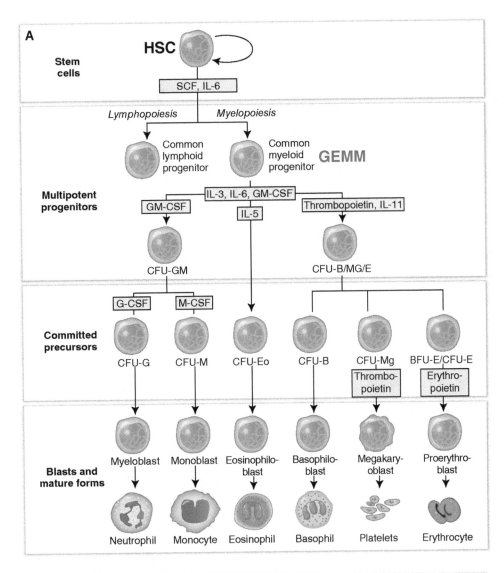

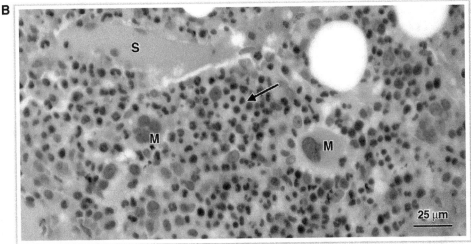

Figure 8-5: Overview of hematopoiesis. **A.** Derivation of blood cells. All blood cells are derived from a pluripotent hematopoietic stem cell (HSC). This gives rise to a series of progenitor cells of increasingly fixed potential, and finally "blast" cells committed to producing one lineage found in blood. Some of the growth factors that influence mitotic rates and cell fate choices are indicated in boxes. CSF, colony-stimulating factor; IL, interleukin; SCF, stem cell factor. **B.** Bone marrow core biopsy sample. Cords of hematopoietic tissue lie between large diameter sinusoids (S). Clusters of cells involved in erythropoiesis can be identified by their round nuclei (*arrow*). Cells with tube-like, elongated nuclei are involved in granulocyte development (granulopoiesis). Megakaryocytes (M) can be identified by their large size and complex nuclear outlines. The large clear areas are sites occupied by adipocytes. The amount of adipose tissue in hematopoietic marrow increases with age; this section was taken from a young person. (H&E stain)

ANALYSIS OF BONE MARROW ASPIRATES

Cells of the terminal hematopoietic lineages that produce RBCs, granulocytes, and platelets are plentiful and morphologically distinct and, therefore, are the cells typically analyzed in marrow aspirate preparations to assess the overall state of this tissue and aid clinical diagnoses. Cytologic analyses of lymphocytes and monocytes are performed less commonly. There are few monocytes and their precursors, monoblasts and promonoblasts, in normal marrow. Small numbers of lymphocytes and their precursors are present, but without the use of special reagents these cells are indistinguishable from several types of progenitor cells. Plasma cells derived from activated B cells, however, are easily identified (Figure 8-7B).

Distinguishing the erythroid from the granulocyte lineages and recognizing the stages in each is the main challenge presented by marrow aspirates. Identification is complicated by the large number of cells that are randomly mixed during preparation. The key differences are summarized in the diagrams of Figure 8-6A and C and described below. Both cytoplasmic and nuclear features are useful for identification, and it is important to recognize that nucleoli appear as light circles surrounded by a dark rim by Wright staining. Granulocyte precursors typically outnumber RBC precursors by a ratio of 2:1.

THE ERYTHROBLAST SERIES

The cells responsible for **erythropoiesis** are typically round, and both they and their nuclei become smaller as development progresses (Figure 8-6B). Development begins with the synthesis of large amounts of mRNAs required for the production of hemoglobin and the elaboration of the RBC cytoskeleton and membrane. Toward the end of the series, organelles disintegrate and the nucleus is finally extruded. Four cell divisions can occur up to the polychromatophilic stage, and the normal process requires about 7 days for cells to enter the circulation.

- **Proerythroblast.** The nucleus contains finely textured chromatin and nucleoli; the cytoplasm is usually dark blue due to the high concentration of mRNA. Proerythroblasts contain erythropoietin receptors, and this hormone may play a role in stimulating the division or differentiation these cells.

- **Basophilic erythroblast.** The nucleus lacks nucleoli, and the chromatin begins to condense and clump; the cytoplasm remains dark blue.

- **Polychromatophilic erythroblast.** The nucleus has shrunken considerably, and the chromatin is highly condensed. The cytoplasm is usually a blue-gray color due to the increased presence of hemoglobin and the decreased amount of mRNA.

- **Orthochromatophilic erythroblast.** The nucleus is usually a featureless disc. The cytoplasm is typically not the true color of an RBC, as the name suggests, but tends to be grayer. Orthochromatophilic erythroblasts are no longer capable of mitosis.

- **Reticulocyte.** When the nucleus is extruded, the cell is called a reticulocyte because it contains residual ribosomes that form a reticulum, or network, which is only visible if the cells are treated with special stains (not shown). Reticulocytes are considered immature, but they can exit the bone marrow. They normally account for 0.5–1.5% of the circulating RBCs, and quickly become mature RBCs as the ribosomes disintegrate.

THE GRANULOCYTE SERIES

The cells responsible for **granulopoiesis** tend to be more oval than round. Their nuclei undergo a variety of shape changes, and granules appear in two stages. Four to five cell divisions can occur up to the myelocyte stage, and the complete process requires 10–11 days. Mature and nearly mature cells remain in the bone marrow for several days before entering the circulation.

- **Myeloblast.** These cells are characterized by scant cytoplasm, which is less basophilic than that of erythroblasts, and a large nucleus that contains lacy chromatin and nucleoli (Figure 8-6D). By definition, myeloblasts are said to lack granules. However, it is not possible to distinguish, with certainty, myeloblasts from other immature cells such as lymphoblasts or megakaryoblasts. Therefore, the use of reagents, such as specific antibodies, is required if unambiguous identification is required.

- **Promyelocyte.** The nucleus of a promyelocyte is usually elongated and contains nucleoli. Dark staining **primary granules** appear in these cells, which provides a key feature and makes these cells the first granulocyte precursor that can be identified on standard preparations. The granules can obscure the nucleoli in some cells. Promyelocytes contain relatively more cytoplasm than myeloblasts (Figure 8-6E).

- **Myelocyte.** Nucleoli are no longer found in this stage and the primary granules, although still present, lose components responsible for their dark staining. **Secondary granules** appear at this stage, so neutrophilic, eosinophilic, and basophilic myelocytes can be distinguished. The expanded Golgi apparatus involved in the production of secondary granules often can be seen as a lighter area, referred to as a "dawn," near the nucleus (Figure 8-6F). As would be expected from the numbers present in blood, cells involved in neutrophil production are by far the most numerous from this stage onward.

- **Metamyelocyte.** An indentation appears in the nucleus at this stage (Figure 8-6G). Metamyelocytes are no longer capable of dividing.

- **Band form.** The narrowing that begins in the metamyelocyte eventually continues to both ends, converting the nucleus into a tube shaped like a "U" or a "J"; the cells with this feature are called band forms or stab forms (Figure 8-6H).

- **Segmented form.** Alteration of the band form nucleus then continues to produce the polymorphonuclear segments that are typical of the mature forms of the three types of granulocytes.

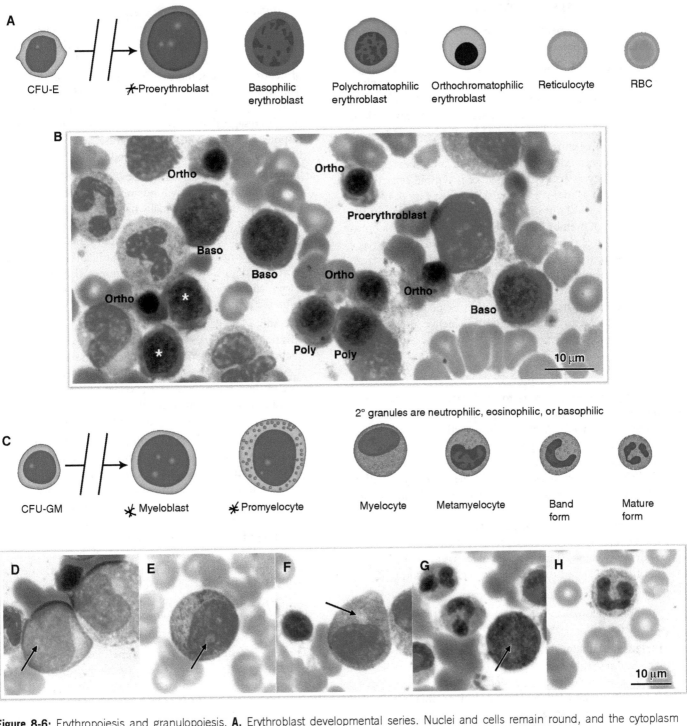

Figure 8-6: Erythropoiesis and granulopoiesis. **A.** Erythroblast developmental series. Nuclei and cells remain round, and the cytoplasm becomes more eosinophilic as development proceeds. Nucleoli are only present in the proerythroblast stage. CFUE, colony-forming unit-erythrocyte. **B.** Bone marrow aspirate of the erythrocyte (RBC) series. A proerythroblast is identified by its large nucleus and fine chromatin against which nucleoli are visible as light areas. Several basophilic erythroblasts (Baso), polychromatophilic erythroblasts (Poly), and orthochromatophilic erythroblasts (Ortho) are indicated. Two cells, intermediate in size and color present between the basophilic and polychromatophilic stage, are marked with asterisks (*). (Wright stain) **C.** Granulocyte developmental series. Nuclei tend to be oval at the beginning of development and undergo dramatic shape changes. Primary granules appear in the promyelocyte stage and lose their color at the myelocyte stage, when secondary granules appear. GFU-GM, colony-forming unit-granulocyte-macrophage. **D–H.** A montage of the granulocyte developmental series from a bone marrow aspirate. **D.** The arrow indicates a faint nucleolus in a myeloblast or myeloblast-like cell. **E.** A promyelocyte with visible nucleoli (*arrow*) contains many primary cytoplasmic granules. **F.** A neutrophilic myelocyte contains secondary granules; the Golgi apparatus is visible (*arrow*). **G.** An eosinophilic metamyelocyte is identified by its secondary granules and the indentation in the nucleus (*arrow*). Two segmented neutrophils are also present in this panel. **H.** The band form neutrophil nucleus appears as an unsegmented tube. (D–H, Wright stain)

PLATELET PRODUCTION

Platelets are cell fragments shed from large cells that reside and remain in hematopoietic marrow.

Megakaryocytes. These large cells are derived from **megakaryoblasts** through a process that involves **endomitosis**—the synchronous division of nuclei without cell division. As mentioned above, megakaryoblasts resemble myeloblasts and lymphoblasts and cannot be identified unambiguously. However, once endomitosis begins, expansion of the cytoplasm accompanies nuclear division and the resulting megakaryocytes are extremely large polyploid cells, which may contain as many as 32 nuclei and are easy to identify (Figure 8-7A). Maturation involves the production in the cytoplasm of membrane-bound granules and the elaboration of long, thin processes that extend through the endothelium into the sinusoids. Platelets are shed from these extensions into blood and survive for 10 days; megakaryocytes produce several thousand platelets and then undergo apoptosis. The production of platelets requires the growth factor **thrombopoietin**, which is produced in the liver and kidney.

RESPONSES TO STRESS

The ability of growth factors to modulate mitotic rates and developmental choices, tunes the process of hematopoiesis to suit the body's needs. A change in growth factor levels can regulate both the rate of production of blood cells and the timing of their release from marrow. The numbers of circulating WBCs considered to be normal in the human population varies over a twofold range, which likely results from individual differences in the interaction of growth-factor signaling and cellular production in response to everyday stresses, such as minor infections. A stress of sufficient strength and duration, however, will result in increased signaling that produces a significant change in the profile of cells circulating in the blood.

▽ Some features related to common **stress responses** include:

Increased reticulocyte ratio. The fraction of reticulocytes in circulation may be elevated above 1.5% after a severe hemorrhage. Until production has replaced lost cells, the hematocrit will remain below normal, but an increase in the fraction of reticulocytes represents the accelerated production and release of RBCs, which is an adaptive response driven by higher levels of EPO. On the other hand, a low reticulocyte fraction in the presence of a long-standing anemia indicates a failure of hematopoiesis, which warrants further clinical investigation.

Neutrophilia. A noticeable increase in the number of circulating neutrophils, called neutrophilia, can result from a bacterial infection. Cytokines released from the site of infection stimulate both production and accelerate the rate of release of neutrophils from marrow, which will then be available to deal with the pathogens. A large number of segmented and band form cells are normally present in the marrow and constitute a reserve that can be rapidly released into the blood in response to inflammatory signals. This reserve pool can be 10 times the number present in the blood.

Demargination. As explained above, neutrophils that leave the marrow continuously bind and release from vessel walls throughout the body; normally half of the cells are bound at any time and constitute another reserve population. A variety of signals can release these **marginated cells** to provide an increased number in the fluid phase available to respond to a local infection. Corticosteroid hormones also promote this release and can produce neutrophilia, which may be misinterpreted as an indication of infection.

Left shift. The neutrophilia produced as a consequence of infection is often accompanied by an increase in the fraction of circulating band form neutrophils and, in severe cases, even the release of metamyelocytes and myelocytes. The appearance of these less mature cells is called a "**left shift**," because diagrams of hematopoietic developmental sequences are usually presented as progressing from left to right (Figure 8-7C).

Toxic changes. A severe infection may stimulate the production and release of neutrophils with abnormal cytoplasmic features. The appearance of vacuoles and dark granules in neutrophils, called **toxic changes**, are considered signs of a strong response to a significant inflammatory situation (Figure 8-7D and E). The "toxic" granules are thought to be primary granules that have maintained some of the staining characteristic of the promyelocyte stage. Basophilic cytoplasmic inclusions, called Döhle bodies, are also observed as part of these toxic changes. These represent aggregates of RNA and ribosomes, assumed to accumulate due to accelerated cellular production, and may be indicative of imperfect cellular maturation. ▽

The responses described here are sometimes called **reactive** to indicate that they are triggered by a pathologic stimulus and provide an adaptive reaction that returns conditions to normal. For instance, neutrophilia is a typical reactive response to a bacterial infection. However, there can be other nonadaptive causes of neutrophilia, such as leukemia, and it is often important to distinguish reactive from nonreactive changes. This can require a bone marrow biopsy and the use of molecular tools, such as flow cytometry, to determine the identity of the cells involved. Thus, the term "reactive" denotes changes that are nonneoplastic.

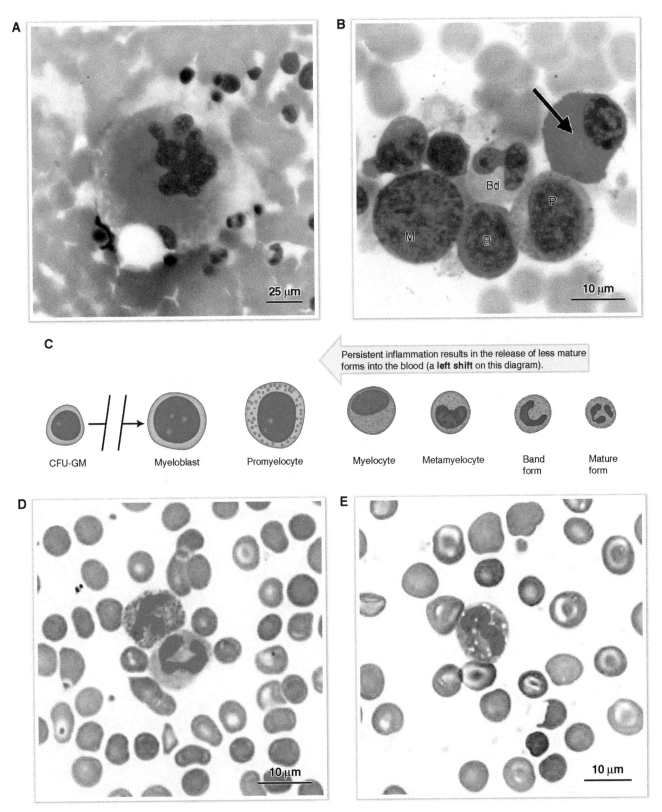

Figure 8-7: Megakaryocytes, plasma cells, and adaptive changes. **A.** Megakaryocyte in a bone marrow aspirate smear. Both the immense size of these cells and their large accumulation of nuclei are apparent. **B.** Plasma cell and other cells in a bone marrow aspirate smear. The nucleus of a plasma cell is usually eccentric and the light area in the cytoplasm (*arrow*) represents the Golgi apparatus. Other cells indicated are the eosinophilic myelocyte (M), band form neutrophil (Bd), late basophilic erythroblast (B), and a late promyelocyte with few primary granules visible (P). **C.** Left shift of neutrophils. The growth factor signaling produced by a severe infection or other inflammatory condition will cause the premature release of neutrophil precursors. Band form neutrophils normally account for as many as 5% of leukocytes; if a higher percentage is observed, a clinically important reaction may be indicated. CFU-GM, colony-forming unit-granulocyte, macrophage. **D.** Blood smear of neutrophil with toxic granules. The speckled appearance of the upper neutrophil is consistent with a long-term infection or another inflammatory condition. **E.** Blood smear of neutrophil with vacuoles. The vacuoles present in this neutrophil are consistent with a persistent infection, as is a left shift, resulting in an increased number of circulating band forms. Samples D and E courtesy of Dr. Mohamed Salama, University of Utah School of Medicine, Salt Lake City, Utah. (A, B, D, and E, Wright stain)

STUDY QUESTIONS

Directions: Each of the numbered items or incomplete statements is followed by lettered options. Select the **one** lettered option that is **best** in each case.

Questions 1–4

Refer to the micrograph below to answer the following questions.

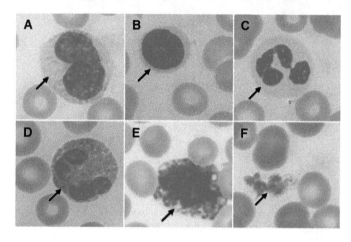

1. A 5-year-old child is brought to the clinic by her parents because she has a fever and malaise. An increase in which type of formed blood element would be most likely consistent with the presence of a bacterial infection in the child?

2. A 25-year-old Peace Corps worker who recently returned from Africa to the United States is seen in the clinic because of a complaint of acute onset of abdominal pain and fever. The physician suspects schistosomiasis, a disease caused by a round worm of the *Schistosoma* genus, which is endemic in the country where the patient has been serving. An increase in which type of formed blood element would be most likely consistent with this diagnosis?

3. A 10-year-old boy with chronic kidney disease is seen in the clinic because he has recently experienced nosebleeds. Laboratory studies show a high BUN (blood urea nitrogen) and abnormal electrolyte levels, a constellation of symptoms known as uremia and associated with declining renal function. Clotting issues are associated with uremia and may be the likely cause for the boy's nosebleeds. In this case, which formed blood element was most likely affected by the boy's uremia?

4. Which blood cell contains histamine and heparin?

5. The iron in the heme group of hemoglobin must be kept in the reduced ferrous (Fe^{2+}) state so that the hemoglobin molecule can bind O_2. The enzyme responsible for this reduction reaction is methemoglobin reductase. The energy required for this reduction is supplied by which of the following?

   A. Band 3

   B. Electron transport reactions

   C. Glycolysis

   D. Lipid oxidation

   E. Oxidative phosphorylation

6. Infectious mononucleosis is caused by a viral infection. One notable laboratory finding is an increased number of circulating white blood cells that contain an unlobed nucleus. In older literature, these cells were called "mononuclear leukocytes," the name from which "mononucleosis" derives. This is not a very useful term because cells with unlobed nuclei could be either monocytes or another white blood cell. Which other white blood cell listed below would be a cell with an unlobed nucleus?

   A. Basophil

   B. Eosinophil

   C. Lymphocyte

   D. Neutrophil

   E. Platelet

ANSWERS

1—C: The polymorphonuclear neutrophil (PMN) is the principal innate immune system cell involved in fighting bacterial infections.

2—D: The eosinophil is the principal innate immune system cell involved in fighting parasitic worm infections. The acute onset of schistosomiasis includes symptoms experienced by this patient; an elevated eosinophil count also would be expected. Stool samples also are analyzed microscopically to determine the presence of the parasite's eggs to confirm the diagnosis.

3—F: Platelets are the formed element involved in blood clotting.

4—E: Basophils are characterized by large, dark-staining granules that contain histamine and heparin, as well as several cytokines.

5—C: Red blood cells contain no organelles; therefore, they derive energy from cytoplasmic enzymes, principally from glycolysis. The pentose phosphate pathway is also present in these cells.

6—C: Monocytes and lymphocytes both contain unilobular nuclei. The virus that causes mononucleosis, the Epstein–Barr virus, infects B cells. The infected B cells are larger than normal and resemble monocytes, which accounts for the original misnaming of the disease.

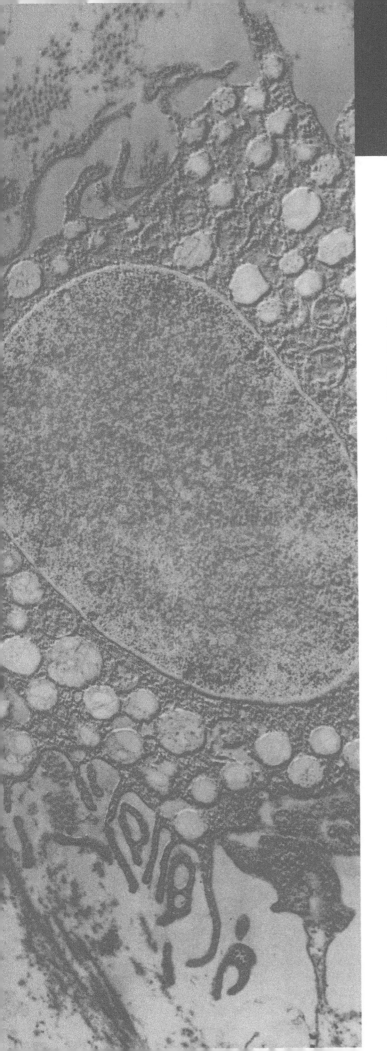

IMMUNE SYSTEM

OVERVIEW

The immune system protects the body against invasion by micro-organisms, removes tissues damaged by trauma, and eliminates malignant growths. Recognizing these problems requires detecting foreign or unexpected molecules, called **antigens**, against the back-ground of self-molecules. This recognition is accomplished by membrane-bound **antigen receptors** on immune system cells. When antigen receptors bind an antigen, these cells initiate **effector** responses to manage the problems.

The immune system provides **innate** and **acquired** responses, which differ in several ways but most notably by the nature of the antigen receptors on the two types of cells. Innate responses are present at birth and provide protection from common patho-gens. Acquired responses develop over time, which is why many illnesses are common early in life. Once acquired responses develop, however, they provide long-lasting immunity to many diseases. Innate and acquired elements are interdependent and work in tandem.

This chapter briefly reviews features of the innate immune system, which were originally introduced in Chapter 2 and 8, and then describes in more detail the features of the acquired system and **lymphoid tissue**.

RESPONSES OF THE IMMUNE SYSTEM

The defense against infection involves all cells in the body. The cells and activities that play major roles in this defense are classified as providing either innate or acquired immune responses. The **innate immune** responses are generally rapid, stereotyped, and short-term reactions directed against common pathogens and damaged tissue. The **acquired immune** responses are slower to develop, but are extremely flexible and can provide long-term protection. The activities of innate and acquired elements are highly complex and intricately interwoven.

COMMON FEATURES OF IMMUNE RESPONSES

A normal immune response by innate or acquired cells involves antigen receptors to recognize a problem, such as infection caused by pathogens or tissue damage, inflammatory signals to alert other cells, and effector functions to attack the problem (Figure 9-1A).

- **Antigen receptors.** Immune cells must recognize the presence of pathogens or altered host cells to initiate a response. This recognition is accomplished by specialized receptors that bind foreign, or nonself, molecules with high affinity and initiate intracellular signaling events. Molecules that stimulate an immune response are called **antigens**, and most antigens are proteins. To avoid reactions against "self," antigen receptors should not recognize molecules normally present in the body.

- **Inflammatory signals.** Once an antigen or tissue damage has been detected, both immune and nonimmune system cells broadcast inflammatory signals to recruit and activate other cells. Signaling includes the release of **cytokines**, which can act at long distances, as well as changes in the cell-surface components on infected and damaged cells.

- **Effector functions.** The destruction of pathogens and the removal of damaged or altered host tissue are major activities of immune system cells. It is critical to direct these activities at appropriate targets or autoimmune diseases result.

DIFFERENCES BETWEEN INNATE AND ACQUIRED ANTIGEN RECEPTORS

The genes coding for antigen receptors on acquired immune cells are constructed by an unusual mechanism that distinguishes the adaptive from the innate immune system.

- **Innate receptor genes.** The genes coding for innate antigen receptors are present in the germline DNA. These receptors have been shaped by billions of years of evolution. For example, the important family of **toll-like receptors (TLRs)** for antigens was first discovered in fruit flies (Figure 9-1B). Innate receptors are specific for components of pathogens, such as lipopolysaccharides produced by bacteria, which are not present in humans. These receptors, therefore, do not bind or respond to self-antigens.

- **Acquired receptor genes.** The genes coding for acquired antigen receptors are not present in the germline DNA. Rather, they are created by molecular mechanisms that **rearrange the receptor genes** as acquired immune cells are born and develop and before antigens are encountered. The rearrangements are highly variable and unpredictable, and it is likely that each acquired immune system cell arises with a different antigen receptor.

These two strategies of gene assembly result in different response times. Innate immune system cells are born with a limited set of antigen receptors and are able to respond immediately when stimulated. Acquired immune system cells with receptors specific for a particular antigen are initially present in small numbers. The production of a sufficient population of cells capable of a significant response to an antigen requires cell division and, therefore, time. In many cases, the innate response rapidly limits an infection, which the acquired response eventually eliminates and then prevents from recurring. The acquired strategy carries a price: the rearrangement of receptor genes can produce cells with receptors that bind to self-molecules, which must be eliminated to avoid autoimmune reactions. The acquired immune system is also responsible for allergic reactions and the rejection of transplanted organs.

PROPERTIES OF THE INNATE IMMUNE SYSTEM

The innate immune system consists of a diverse group of specialized cells as well as soluble proteins and activities common to many cells.

CELLS

Innate immune system cells include **macrophages** and **granulocytes**, described in previous chapters, and **natural killer (NK) cells** and **antigen-presenting cells (APCs)**, which will be discussed in this chapter. Nonimmune system cells can also furnish protective innate responses; recall the **defensins** produced by epithelial cells. The production of a class of cytokines, called **interferons**, provides another important example, shown in Figure 9-1C. When a cell is infected by a virus, transcription factors that stimulate the production and secretion of type 1 interferons are activated. These bind to neighboring cells and initiate a set of reactions that reduce their susceptibility to viral infection, helping to limit the proliferation of the virus. Interferons also recruit immune system cells, such as NK cells, which can destroy virally infected cells.

COMPLEMENT

Complement is a set of serum proteins produced by the liver, and serves numerous important innate functions. A fragment of one of the proteins, known as C3b, can be covalently attached to the surfaces of pathogens and serves as part of an enzyme that stimulates the addition of more C3b. Cell-surface C3b and its products are **opsonins**, molecules that mark targets for immune reactions. These reactions include phagocytosis by macrophages and neutrophils, which have receptors that bind C3b, and attack by other complement proteins, known as the membrane attack complex (**MAC**), which can lyse cells (Figure 9-1D).

▼ **Deficiencies in C3b** deposition are associated with an increased risk of a wide variety of bacterial infections, whereas defects in MAC increase the frequency of infections caused by *Neisseria* bacteria. Deficiencies in the protective activities that remove C3b from host cells result in autoimmune attacks. For example, red blood cells (RBCs) are lysed by MAC in the syndrome of **paroxysmal nocturnal hemoglobinuria**. ▼

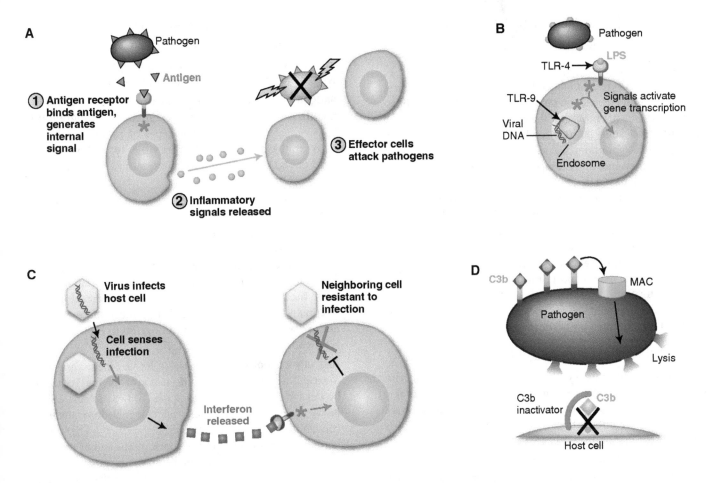

Figure 9-1: Some general features of immune responses. **A.** Steps from recognition to action. (1) Binding antigens to antigen receptors generates internal signals. (2) Antigen-stimulated cells release inflammatory signals that activate other cells. (3) Effector cells attack the source of antigens with toxic chemicals and engulf debris. **B.** Toll-like receptors (TLRs). Shown are examples of antigen receptors present on innate immune cells that recognize molecules specific to microbes and viruses. TLR-4 on the outside of the cells binds lipopolysaccharides (LPS) produced by gram-negative bacteria, whereas TLR-9 in endosomes recognizes DNA from bacteria and viruses. Receptor binding results in the activation of genes for inflammatory cytokines. **C.** Interferons protect against viral infection. Any cell infected with a virus responds by producing interferons. These cytokines bind to neighboring cells and initiate reactions that help resist viral infection. **D.** Complement proteins identify cells for attack. The attachment of complement protein C3b to the surface of cells induces several responses, including the assembly of the membrane-attack complex (MAC), that pokes a hole in the cell membrane, resulting in cell lysis. Host cells protect themselves by expressing membrane proteins that inactivate C3b.

PROPERTIES OF THE ACQUIRED IMMUNE SYSTEM

The acquired immune response is provided entirely by two types of lymphocytes, **B cells** and **T cells** (Figure 9-2A). These cells are born and develop in **central (or primary) lymphoid tissues**, which, in mammals, is the **thymus** for T cells and the **hematopoietic bone marrow** for B cells (Figure 9-2B). After exiting central lymphoid tissue, lymphocytes are considered **naïve** until they encounter antigens in **peripheral (or secondary) lymphoid tissue**, such as in **lymph nodes** or the **spleen**. Naïve lymphocytes activated by antigen undergo mitosis and then function throughout the body. Most lymphocytes are found in lymphatic tissues, but they continuously **recirculate** from site to site in the body via the blood and lymphatic vessels. The development, activation, and functioning of lymphocytes is a continuum that may be most easily understood by first outlining the activation and subsequent activities of lymphocytes. Birth and development will then be considered when describing the structure and function of the thymus.

CLONAL SELECTION, DELETION, AND ANERGY

After the birth and initial development of naïve B and T cells, which involve receptor gene rearrangements, each cell contains many **identical copies** of a unique, type-specific surface-antigen receptor. Interactions of the receptor with antigens determine the fate of these cells (Figure 9-2C).

Initial antigen-binding effects. When a naïve cell exits the central lymphoid tissue and initially encounters an antigen to which its antigen receptor can bind with high binding affinity, intracellular signals are generated. These signals result in either mitotic activation of the lymphocyte, a "Go" signal, or a "Stop" signal is produced that causes the cell to undergo apoptosis or to become inert to future encounters with the antigen, a condition called **anergy**. The mitotic activation is called **clonal selection** because the antigen binding results in the production of a clone of cells bearing the antigen receptor. In this case, apoptosis is called **clonal deletion** because a potential clone of cells is eliminated. It is the **context** in which antigen is bound that determines if clonal selection or if clonal deletion or anergy occurs. The context involves antigen presentation (see below).

Effector and memory cells. After several days to 1 week or more, some of the cells of a clone stimulated to divide by antigen binding become **effector cells**, which produce an immune response directed at the antigen. This is called a **primary response** because it arises during the initial development of a clone. Other cells of the growing clone become temporarily inert and are able to persist for a long period as **memory cells**, which constitute a set of preselected cells, all with the same antigen receptor. The later appearance of the antigen can trigger the activation of these memory cells and result in the rapid production of a much larger population of effector cells and a strong **secondary response**. Secondary responses persist and provide long-term protection against many diseases. **Immunologic memory** is an important feature of the acquired immune system, which the innate immune system lacks.

PROPERTIES OF B CELLS

B cells normally constitute 5–25% of lymphocytes circulating in blood and live for months. The antigen receptors on B cells are converted into the secreted effector agents, **antibodies**.

Antigen receptors. Surface-antigen receptors on B cells are membrane-anchored proteins called **immunoglobulins** (Figure 9-2D). There are five types of immunoglobulins; however, the antigen receptors are either type M or D, called sIgM or sIgD, respectively, to denote surface immunoglobulin M or D.

Effector cells. Cells from successfully activated B-cell clones differentiate into **plasma cells**, which reside in connective tissue and are the effector elements derived from clonally activated B cells. Plasma cells secrete **soluble** forms of immunoglobulins, called **antibodies**; the four different types are IgM, IgA, IgE, or IgG. IgD is not secreted. Secretion occurs when the gene coding for the surface-antigen receptor is differentially spliced in plasma cells to remove the membrane anchor, which permits the immunoglobulin molecule to be released from the cell by exocytosis.

Antibody functions. Antibody molecules have the antigen-binding specificity of the surface-antigen receptor on the B cell. Antibodies and the receptors are composed of two proteins, called a **heavy** chain and a **light** chain, which assemble into **monomers** containing two copies of each chain (Figure 9-2D). The antigen-binding region on antibodies is found at one end of the molecule, called the **variable region**, because it is transcribed from the portion of the heavy and light chain genes that have undergone genetic rearrangement. The other end of the immunoglobulin is called the **constant region**, which contains sequences that activate complement and serve as receptors that stimulate phagocytosis by macrophages and neutrophils. Antibodies bound to the surface of a virus or bacteria can prevent the pathogen from binding to cells as well as target the pathogen for attack by complement or phagocytosis by macrophages and neutrophils. The protection provided by antibodies is called **humoral immunity**, because these proteins are found in the blood, one of the four classic humors of the body.

Antibody isotypes. IgM is the first antibody secreted during a primary response. It is released as a large, **pentameric** molecule of five IgM monomers linked by disulfide bonds and a **J (joining) chain** protein (Figure 9-2E). IgM is very effective in attracting complement when bound to the surface of a pathogen, but it is not very soluble. A secondary antibody response involves a switch in isotype from IgM to one of the other molecular types. **IgA** is produced by plasma cells in the lymphatic tissue near the digestive tract and transported into the lumen of the gut as a **dimer** to bind to organisms present in the gut. **IgE** monomers bind to mast and basophil cells to provide antigen specificity for those cells. **IgG** monomers are extremely soluble and are found in plasma and extracellular fluid at high concentrations. IgG provides protection against pathogens that cross a barrier epithelium. The exchange in isotype from IgM to one of the other isotypes involves DNA rearrangements that pair the antigen-binding variable region sequence on the heavy chain gene with a sequence coding for a constant region of IgA, IgE, or IgG.

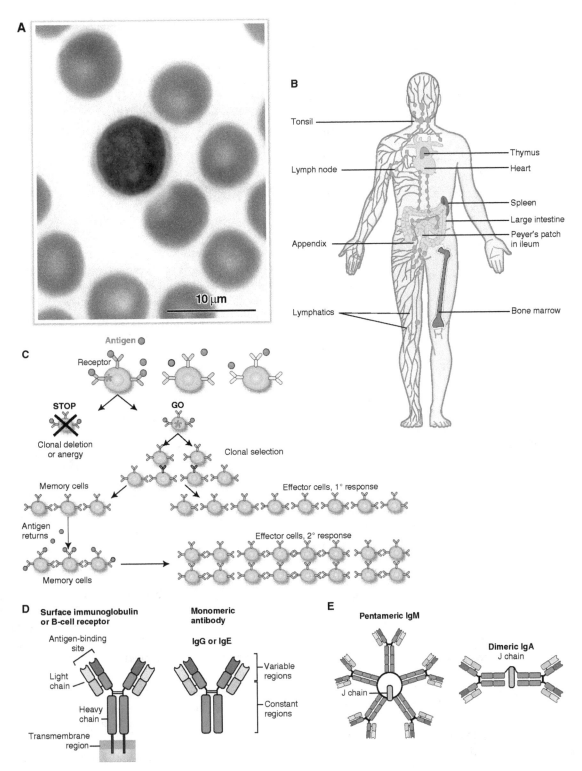

Figure 9-2: Lymphoid tissue, lymphocyte activation, and B-cell activities. **A.** Lymphocyte in a blood smear. It is not possible to differentiate between B cells and T cells on standard blood smears. (Wright stain) **B.** Lymphoid tissues. In humans, the thymus and hematopoietic bone marrow are the central lymphoid tissues where naïve lymphocytes develop. Lymphocytes are activated in many peripheral lymphoid organs and recirculate between lymphatics and the blood stream. **C.** Clonal selection, deletion, and anergy. Each naïve lymphocyte has a unique antigen receptor. Encounter with an antigen that binds tightly to the receptor results in either cell division and clonal selection or the death or inactivation of the cell. Effector and memory cells are produced by clonal selection. Memory cells persist and provide strong secondary responses to the reappearance of antigen. **D.** B-cell antigen receptors and antibodies. The B-cell receptor is an immunoglobulin molecule anchored to the cell by a transmembrane region (left panel). Effector cells, called plasma cells, secrete the immunoglobulin, called an antibody, by an alternative splicing event that removes the transmembrane region (right panel). IgG and IgE type antibodies are considered monomeric antibodies that contain two antigen-binding sites. **E.** Multimeric antibodies. The first type of antibody secreted during the primary response to antigen is IgM, which consists of pentameric aggregates of five monomers held together by sulfhydryl bonds and a joining chain (J chain). IgM efficiently stimulates complement binding. IgA is assembled into dimers by the addition of a J chain, and is secreted by epithelial cells into the lumens of the digestive and respiratory tracts to bind to microbes.

PROPERTIES OF T CELLS

T cells normally account for 65–75% of circulating lymphocytes and can live for years. The "T" denotes that these cells require the thymus for their development.

Antigen receptors. The surface-antigen receptor on T cells is called the T-cell receptor, or **TCR**. TCRs consist of membrane-anchored **dimers** and associate with several other proteins involved with signaling and antigen recognition. Key accessory molecules involved with antigen recognition are the coreceptors, known as **CD4** and **CD8**. The TCR plus coreceptor is restricted to binding antigen presented on the **major histocompatibility complex (MHC)** proteins of other cells, described as follows (Figure 9-3A):

Effector cells. Unlike B cells, T cells do not secrete their antigen receptor; rather, the effector functions of T cells involve the secretion of either cytokines or toxins.

Helper T cells (T_h). The effector function of this subset of CD4 T cells is cytokine secretion. T_h cells provide cytokines that are critical for the normal activities of many other immune system cells, thus the term "helper cells." For instance, cytokines from T_h cells are required for the differentiation of B cells into plasma cells.

Cytotoxic T cells (T_c). The effector function of these CD8 T cells is the secretion of factors that kill cells. These factors are stored in membrane-bound granules and include **granulysin** and **perforin**, which produce pores in the membranes of cells targeted for attack by T_c cells, as well as **granzymes**, the proteases that enter the target cells through the pores and induce apoptosis. T_c cells can also express the **Fas ligand**, which communicates an apoptotic signal to cells carrying the **Fas receptor**. This protective function is called **cellular immunity** because it involves close contact between the T_c cell and the target cell.

Regulatory T cells (T_{reg}). The effector function of this subset of CD4 T cells is to suppress the effector activities of T_h and T_c cells. The suppressive mechanisms are not completely understood, but they appear to involve the release of immunosuppressive cytokines and direct cell–cell contacts between the T_{reg} cells and other effector cells.

ANTIGEN PRESENTATION TO NAÏVE LYMPHOCYTES

The clonal selection of naïve T lymphocytes requires the presentation of antigens on a membrane protein assembly, called the **MHC**, expressed on the surface of specialized **APCs**. APCs provide additional signals and cytokines required to activate T cells. B lymphocytes are less dependent on antigen presentation, but their activation is facilitated by APCs.

Antigen presentation to B cells. Although B cells can interact with antigen on their own, they often encounter antigen trapped on the surface of **follicular dendritic cells (FDCs)** in peripheral lymphoid tissues (Figure 9-3B). FDCs are fibroblast-like cells bearing long processes that can trap and maintain antigens on their surface for long periods. This trapping and presentation to B cells does not involve MHC molecules. These cells also supply cytokines that promote B-cell survival.

The initial interaction with antigen activates a B cell, but additional interactions with T_h cells are necessary for the cell to divide and produce a clone.

Antigen presentation to T_h cells by APCs. The TCR–CD4 combination on T_h cells can only bind to antigen that is presented on the **MHC class II** molecules, whose expression is limited to **macrophages**, **dendritic cells**, and **B cells**. These cells are called **professional APCs** and are derived from the pluripotential hematopoietic stem cell; dendritic cells are closely related to macrophages. APCs present antigens to T_h cells that arise outside the cells. The antigens are internalized by phagocytosis, and proteins are degraded in endocytic vesicles to peptides. The peptides bind to MHC class II molecules, which are also present in the vesicles, and these complexes then are transported to the cell surface (Figure 9-3C, left side). When a naïve T_h cell encounters an APC presenting an antigenic peptide on an MHC class II molecule to which its TCR can bind, a protein kinase associated with CD4 is activated. This kinase then initiates the first of two signals required for the expression of genes required for the T_h cell to begin cell division (Figure 9-3A).

Antigen presentation to T_c cells by APCs. The TCR–CD8 combination on T_c cells is restricted to binding antigen that is present on **MHC class I** molecules, which are found on most cells in the body, except RBCs. The activation of naïve T_c cells, however, only occurs on professional APCs, principally **dendritic cells**. The antigens presented at the cell surface on MHC class I molecules are peptides cut from proteins in the **cytoplasm** of dendritic cells by proteolytic cleavage in **proteasomes**. These peptides are transported into the endoplasmic reticulum, bound to MHC class I proteins, and then sent to the cell surface (Figure 9-3C, right side). There are two ways that dendritic cells can acquire and present a pathogenic protein (e.g., from a virus) to a naïve T_c cell. First, the virus can infect the dendritic cell. Second, the dendritic cell can phagocytose debris from other infected cells and release some of the internalized material from endosomes into its cytoplasm. This process, called **cross-presentation**, is not well understood. In either case, antigen binding of the TCR of a naïve T_c cell generates the first signal required for mitotic activation (Figure 9-3A).

Costimulation by APCs. The professional APCs can also express a cell-surface protein, called **B7**, which binds to the **CD28** protein on naïve T_h or T_c cells. The interaction of B7 with CD28 and the simultaneous binding of the T cell's TCR to a specific antigen presented by the APC on an MHC molecule produce a second activating signal in the T cell. This second signal results in the production of the cytokine **interleukin-2 (IL-2)**. Stimulation of the T cell by its own IL-2 is necessary for the cell to divide (Figure 9-3D). Professional APCs only express B7 when they are in an inflammatory condition, such as when their TLRs have signaled the presence of pathogenic molecules. In the absence of costimulation by B7, the signal generated by the TCR responding to a specific antigen on the MHC of an APC causes the naïve T cell to become **anergic**. Costimulation helps to ensure that the induction of an adaptive immune response is associated with an infection.

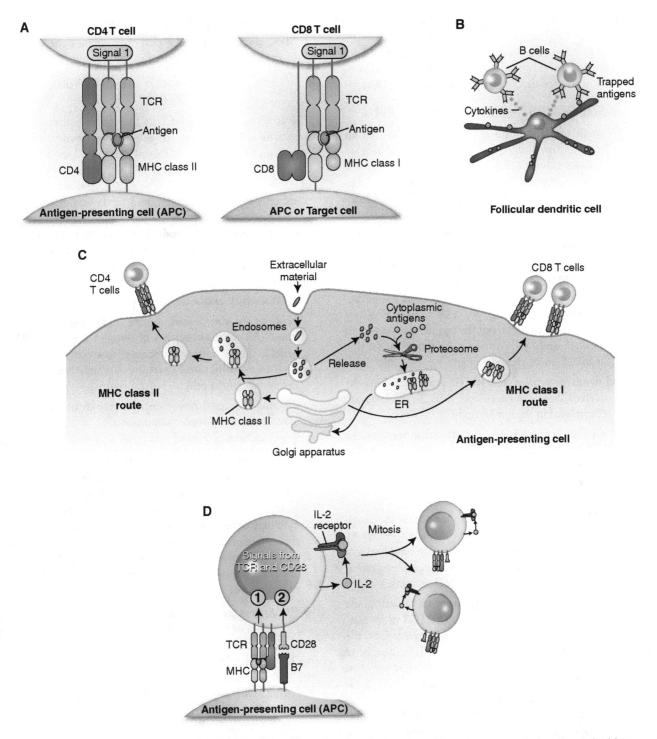

Figure 9-3: T-cell antigen receptors and antigen presentation to lymphocytes. **A.** T cells bind to antigen presented on the major histocompat-ibility complex (MHC). The T-cell receptor (TCR) requires a coreceptor protein: CD4 on helper T (T_h) cells and CD8 on cytotoxic T (T_c) cells. T_h cells are restricted to scanning for antigen on MHC class II molecules expressed on specialized antigen-presenting cells and T_c cells to antigens expressed on MHC class I molecules, which are present on all cells except RBCs. Binding of an antigen generates the first signal required to activate a naïve T cell. **B.** Follicular dendritic cells (FDCs) interact with B cells. These cells reside in peripheral lymphoid tissues and secrete cytokines that attract naïve B cells and promote their survival. In addition, FDCs trap antigens and maintain them on their sur-faces. Naïve B cells may interact with these antigens. **C.** Antigen presentation to naïve T cells. Antigen-presenting cells (APCs) express both MHC class I and class II proteins. As shown on the right side, class I proteins are loaded with peptide antigens in the endoplasmic reticulum (ER). These peptides are from proteins produced in the cytoplasm of APCs or from extracellular proteins released from endocytic vesicles. The MHC class I proteins plus peptides are then displayed on the cell surface for naïve T_c cells. Class II proteins (left side) are loaded in endosomes with peptide antigens derived solely from extracellular material, such as bacteria and dead cells. **D.** Costimulation of naïve T cells. Binding of antigen by the TCR provides the first signal (1), and binding of B7 by the CD28 protein of a naïve T cell provides the second signal (2) required to activate the cell. It then expresses the interleukin-2 (IL-2) receptor and secretes IL-2, which promotes cell division.

EXAMPLES OF ADAPTIVE IMMUNE RESPONSES

Outlining the immune responses of two typical infections demonstrates the intricate and effective coordination of the cells and processes that have been described previously in this chapter.

HUMORAL IMMUNITY—ANTIBODY PRODUCTION

The development of plasma cells secreting antibodies that are reactive against a protein antigen normally requires clonal selection in lymph nodes of both B and T_h cells that recognize features of the protein. As an example, consider a small wound on a fingertip that breaks through the epidermis and allows bacteria access to the underlying dermis. The rich extracellular fluid in this connective tissue will support rapid bacterial growth. If the innate immune system fails to eliminate the invaders quickly, acquired immune system cells will be activated in lymph nodes and antibodies will be produced that help eliminate the bacteria (Figure 9-4A).

- **Dendritic cells** in the area detect the presence of the pathogens and become activated (e.g., via TLR signaling). These cells will engulf some bacteria, exit the connective tissue via a lymphatic vessel, and arrive in a nearby **lymph node** that drains near the elbow or shoulder.

- Activated dendritic cells express the B7 protein and display peptides derived from the bacteria on MHC class II molecules to cells present in the lymph node.

- Binding of TCRs on naïve T_h cells to bacterial antigens on the dendritic cells' MHC class II proteins, the CD28–B7 interactions, and the subsequent IL-2 production and signaling stimulate mitosis, which produces clones of T_h cells with TCRs specific for bacterial proteins.

- Bacteria and bacterial proteins also enter the node via the lymph and interact with sIgM and sIgD on naïve B cells, either directly or on the surface of the FDCs. B cells, which can bind bacterial antigens, will internalize them and display peptides derived from bacterial proteins on their MHC class II antigens.

- T_h and B cells activated by the same antigen are present in the node, and some will come into contact. The TCR of an activated T_h cell recognizes antigen peptides presented on the MHC class II antigen of the B cell, and these cells bind together to form **cognate pairs**. The activated T_h cell expresses a protein called the **CD40 ligand**, which binds to **CD40** on its partner B cell. Similar to the B7–CD28 interaction between APCs and T cells, this binding provides a **second** intracellular stimulatory signal to the B cell. The T_h cell then begins secreting cytokines, such as **IL-4**, which drive the proliferation of the B cell and the eventual production of effector plasma cells.

- The plasma cells begin secreting antibodies, which exit the lymph node, enter the circulatory system, and eventually are delivered back to the connective tissue in the finger, where they bind to the bacteria and help eliminate the infection.

- Note that both members of the cognate pair recognize the same antigen, which requires the selections of distinct antigen receptors on naïve B and T_h cells. The specific portion of the antigen recognized by the B-cell receptor, called an **epitope**, is not necessarily identical to that recognized by the T_h cell's TCR. The B-cell receptor and the antibody related to it bind to the epitope of an intact protein. This protein is cut into short peptides inside the B cell and then presented on MHC class II proteins, so that the cognate T_h cell's TCR may bind to a sequence unrelated to the epitope.

CELLULAR IMMUNITY—KILLING VIRALLY INFECTED CELLS

The adaptive responses of cells to viral infections, interferon production, and the NK cell activity described above are not always successful. A major role of cytotoxic T cells is to patrol the body and kill host cells infected with viruses. The activation of naïve T_c cells with TCRs specific for viral antigens by dendritic cells in peripheral lymphoid tissue was described previously. Once activated, these T_c cells exit the lymphoid tissue and circulate throughout the body. They no longer require costimulatory signals, and will kill host cells they encounter that present viral peptides on their MCH class I proteins (Figure 9-4B).

- In all cells, cytoplasmic proteins are processed by **proteasomes**, which cleave them into peptides.

- Peptides are transported into the rough endoplasmic reticulum (RER) by **transporters of antigen peptides** (**TAP proteins**). The peptides become associated with newly synthesized **MHC class I** molecules, which then are routed to the surface of the cells.

- Cytotoxic T cells move continuously through tissues and scan MHC class I molecules on cells.

- If a cell is infected with a virus, peptides from viral proteins are presented on MHC class I molecules.

- When a T_c cell with a TCR specific for a viral peptide encounters a cell presenting the peptide on its MHC class I molecules, it binds with high affinity. This stimulates the release of the **cytotoxic compounds** from granules stored in the T_c cell onto the surface of the infected host cell, which undergoes apoptosis. Viral nucleic acids are cleaved, and the dying cells are engulfed and destroyed by macrophages, preventing the further spread of the virus.

▼ **Viruses** are under intense negative selection by the robust action of cytotoxic T cells; therefore, it is not surprising that some viruses have evolved evasive strategies. For example, the **herpes simplex virus** produces proteins that interfere with the TAP function. As a consequence, MHC class I proteins are not loaded with peptides in the RER and are not transported to the cell surface. This prevents surveillance and killing of infected cells by cytotoxic T cells.

NK cells are large lymphocytes that can counter this viral strategy of reducing MHC expression. NK cells contain the same cytotoxic factors, perforins and granzymes, as do cytotoxic T cells. NK cells also have receptors that bind to MHC class I molecules on other cells. However, these receptors do not monitor antigens presented on MHC and, thus, NK cells are considered elements of the innate immune system. The MHC class I receptor on NK cells inhibits their killing activity. NK cells also contain receptors that monitor proteins expressed on cell surfaces in response to **physiologic stress**. When an NK cell encounters a virally infected cell with low levels of MHC class I on its surface, the inhibitory signals are weak and the excitatory signals generated from the stress receptors cause the release of cytotoxic factors that kill the infected cell (Figure 9-4C). ▼

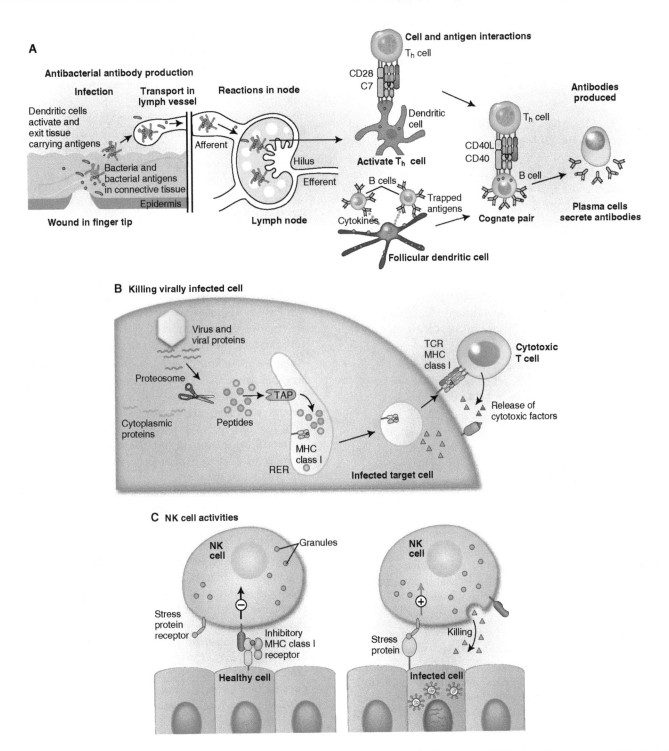

Figure 9-4: Adaptive immune activities and natural killer (NK) cells. **A.** Humoral response to a bacterial infection. The presence of bacteria in the skin activates antigen-presenting dendritic cells, which internalize pathogens and antigens, exit the tissue, and enter a lymphatic vessel. The dendritic cells express B7 and present antigens and activate some naïve T$_h$ cells in a lymph node downstream. Naïve B cells in the node will be independently activated by antigen entering from the infection. Cognate pairs of activated T$_h$ and B cells form, and cytokines provided by the T$_h$ cell drive proliferation of B cells, which produce plasma cells that secrete antibodies reactive against bacterial antigens. **B.** Cellular response to a viral infection. Peptides from viral proteins are produced by proteolytic cleavage in proteasomes. These peptides are transported into the rough endoplasmic reticulum (RER) by the transporter of antigen peptides (TAP protein). Peptides are loaded on major histocompatibility complex (MHC) class I proteins and shuttled to the cell surface. When a T$_c$ cell, which has been previously activated by viral peptides on a dendritic cell in a lymph node encounters the peptide on the infected cell, it releases its cytotoxic factors and kills the cell. **C.** NK cells and the MHC. NK cells express cytotoxic activity that is inhibited by receptors for MHC class I and stimulated by receptors for stress proteins on other cells. If a virally infected cell has downregulated MHC class I expression, it will escape being killed by T$_c$ cells, as described in part B of this figure. However, an NK cell will detect the lack of MHC and the presence of stress proteins and kill the cell.

THYMUS

The thymus is a central lymphoid organ that lies in the mediastinum with its base on the pericardium and the apex in the neck. The thymus produces naïve T cells from precursors born in the bone marrow and is fully functional at birth. The **rearrangement of the TCR genes** occurs in the thymus, followed by two **selections**, which help ensure that the novel TCRs are functional and do not recognize self-molecules. The thymus begins to atrophy in infancy during a process known as **involution**, which accelerates after puberty, with fat replacing functional tissue. The population of T cells produced early in life persists by longevity and self-renewal.

STRUCTURE OF THE THYMUS

The thymus is unusual in that its stroma is composed of epithelial cells, called **epithelial reticular cells**. It is packed with T cells, called **thymocytes**, and also contains macrophages and dendritic cells. The thymus is divided into **cortical** and **medullary** regions (Figure 9-5A and B).

- **Capsule.** A thin, connective tissue capsule covers the thymus. Penetrations of the capsule, called **septa**, divide the organ into lobes. Shorter elements of **septa** enter the organ and terminate at the medulla.

- **Cortex.** The cortex is located immediately beneath the capsule and is characterized by an extremely high density of thymocytes, which obscure the epithelial cells that form a cellular reticulum. There are no extracellular fibers in the cortex. Dendritic cells and macrophages are also present.

- **Medulla.** The medulla can be identified by its lower density of thymocytes. Dendritic cells and macrophages thus are relatively more common but cannot be distinguished by conventional staining. Epithelial cells in the medulla occasionally wrap around themselves to form **Hassall's corpuscles**, a distinctive feature of the thymus. Hassall's corpuscles enlarge as a person ages (Figure 9-5C).

- **Blood vessels.** The thymus is supplied by branches of the subclavian artery. Small arteries and arterioles course in the septa and enter the parenchyma at the junction of the cortex and medulla. Processes of the epithelial cells surround capillaries in the cortex to form a **blood–thymus barrier**, which prevents antigens from entering the area. **High endothelial venules (HEVs)** are found in the medulla, and are sites where progenitor cells from the bone marrow enter the organ and naïve T cells exit (Figure 9-5D).

T-CELL DEVELOPMENT IN THE THYMUS

Lymphocyte progenitor cells are born in hematopoietic bone marrow and are not committed to becoming B or T cells. Some lymphocyte progenitor cells exit the marrow and enter the thymus via HEVs in the medulla. Interaction of these cells with signals from epithelial reticular cells induces them to commit to the process of T-cell development. The cells divide for a time and then migrate into the cortex and undergo a series of steps that result in naïve CD4 and CD8 T cells that have rearranged the genes coding for a TCR (Figure 9-5E).

- **TCR gene rearrangement.** As thymocytes enter the cortex, they use **DNA recombinase** enzymes to rearrange the genes coding for the TCR proteins. Successful expression of these proteins generates a signal that allows the cell to live. Cells in which the rearrangements fail to produce useful receptors undergo apoptosis and are removed by macrophages in the cortex. During this process, cells expressing the α and β receptor genes also express both the CD4 and CD8 coreceptors, and are called **double-positive thymocytes**.

- **Positive selection.** The epithelial reticular cells in the cortex express both MHC class I and class II molecules, which present self-peptides. Double-positive thymocytes contact these cells; those cells that can bind an MHC with **moderate affinity** receive a **positive selective signal** and continue to develop. The successful cells become either a **single-positive** CD4 or CD8 T cell, depending on the class of MHC that resulted in the positive selection. Thymocytes that fail to bind an MHC molecule within several days undergo apoptosis. Positive selection ensures that the T-cell receptor is able to recognize the host MHC molecules.

- **Negative selection.** Thymocytes that bind with **high affinity** to either MHC class I or class II molecules expressed on epithelial cells, macrophages, or dendritic cells receive **negative selective signals**, such as the signal delivered by the **Fas ligand**, and undergo apoptosis. Because self-peptides are being presented on these cells, negative selection removes a large number of cells with receptors that would recognize self-antigens when they leave the thymus. Negative selection is extended to proteins created outside the thymus by the **autoimmune regulator (AIRE)** transcription factor, which is expressed in some epithelial cells and results in the expression of organ-specific genes (e.g., insulin) in these cells.

- **Naïve T cells and peripheral tolerance.** Only 2–5% of thymocytes rearrange their TCR genes, successfully express receptors, receive positive selection, and avoid negative selection. These cells, known as **naïve T cells**, have returned to the medulla and are ready to leave the thymus via HEVs and encounter antigen in peripheral lymphoid tissue. The removal of self-reactive cells by negative selection is called **central tolerance**. As explained above, self-reactive cells that escape this selection are often inactivated by anergy if they encounter self-antigens without costimulatory signals produced by inflammation, a process called **peripheral tolerance**.

HEMATOPOIETIC BONE MARROW AND B-CELL DEVELOPMENT

The development of B cells in marrow is largely analogous to that of the development of T cells in the thymus. **Stromal cells** in the marrow provide the functions of epithelial reticular cells that are found in the thymus. The B-cell receptor genes are recombined using the same DNA recombinases used for the TCR genes. Positive selection can be considered the successful expression of a surface immunoglobulin on a B cell. Negative selection is limited, and because a large number of B cells with self-reactive receptors leave the marrow and enter the circulation, **peripheral tolerance** is especially important to inactivate self-reactive B cells.

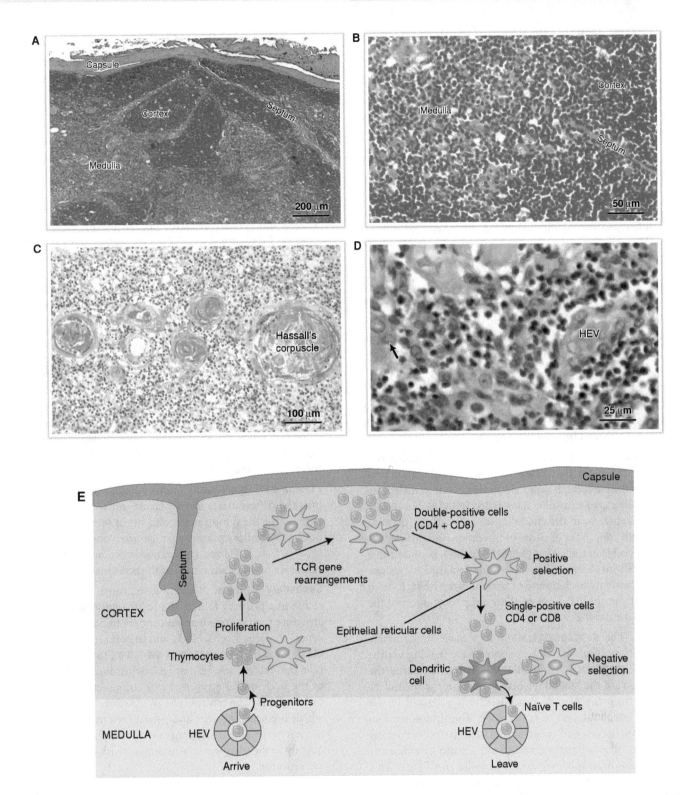

Figure 9-5: The thymus. **A.** Section of a fetal thymus. The cortex, adjacent to the capsule, is densely packed with small thymocytes; medullary areas are far less cellular. **B.** Section of a fetal thymus. A septum, a short extension of the capsule, ends at the boundary between the cortex (right) and the medulla. **C.** Involution in an adult thymus. The overall cell density in this organ is low, and there are no obvious cortical and medullary areas. Epithelial reticular cells wrap around themselves and form large structures called Hassall's corpuscles, which facilitate identification of the organ in sections obtained by biopsy from an older person. **D.** High endothelial venule (HEV) in the medulla of a fetal thymus. The nearly cuboidal endothelial cells are diagnostic. A portion of a small Hassall's corpuscle is indicated (*arrow*). **E.** Development of T cells in the thymus. Progenitor cells from hematopoietic bone marrow enter the thymic medulla by exiting HEVs. Progenitor cells come in contact with epithelial reticular cells, become committed to T-cell development and proliferate, and the daughter cells move toward the capsule. T cells rearrange their receptor genes, express coreceptor molecules, and undergo positive selection by contact with epithelial cells. Negative selection to remove self-reactive cells occurs by contact with epithelial cells, macrophages, and dendritic cells. Ninety-eight percent of the cells die and are removed by macrophages; the surviving cells exit the thymus through HEVs.

LYMPH NODES AND NONENCAPSULATED LYMPHATIC TISSUE

Lymph is the fraction of extracellular fluid that is collected and flows in lymphatic vessels and returns to the circulatory system. Lymphatic vessels are interrupted by lymph nodes, which receive pathogens and their antigenic products as well as activated dendritic cells arriving from an infection upstream. This section will review information that has been presented previously to emphasize that although they have a rather bland appearance when viewed in sections, peripheral lymphatic tissues are specialized structures that bring together all the cells required to mount an acquired immune response.

STRUCTURE OF THE LYMPH NODE

Lymph nodes are small, round, or bean-shaped, encapsulated organs with a **hilus** at one surface penetrated by blood vessels and a single efferent lymphatic vessel. Several afferent lymphatics are found at other surfaces. Although lymph nodes are not divided into lobes and lobules, they are similar to the thymus in that they contain a cortex and a medulla (Figure 9-6A and B).

Capsule. The lymph node capsule is composed of dense, irregular connective tissue. Penetrations of the capsule, called **trabeculae**, extend into the node for varying distances.

Cortex and medulla. The dense collection of lymphocytes near the capsule is the **cortex**, and the less cellular central region is the **medulla**. B cells tend to be located in the cortex nearest the capsule in collections called **follicles**. T cells are found deeper, near the medulla, in a T cell, or **thymic-dependent** zone. Sites of B-cell proliferation and differentiation are visible in follicles as **germinal centers**, because some of the cells involved are larger and the structure stands out against the dense background of lymphocytes. **HEVs**, located principally in the T-cell zones, allow lymphocytes to exit the blood and enter the node.

Sinuses. The **subcapsular sinus** is a space immediately beneath the capsule in which lymph flows, arriving via the **afferent lymphatics. Trabecular sinuses** then carry the lymph toward the **medulla**, where it flows in the medullary sinuses; from the medullary sinuses, it is collected in the **efferent lymphatic** and exits the node. The sinuses contain extracellular fibers and collections of cells that slow the flow of lymph, allowing pathogens and antigens to be efficiently trapped by macrophages, dendritic cells, and FDCs. This facilitates adaptive immune responses as well as prevents the spread of pathogens into the bloodstream.

CELL ACTIVATION

Lymph nodes are efficient sites for initiating adaptive immune responses because they provide high concentrations of antigens and the APCs required to activate lymphocytes. Naïve B and T cells can enter nodes through the HEVs, or they can "float" in with the lymph from a node upstream. Lymphocytes are the only cells frequently seen in lymph, which is the origin of their name. B cells are specifically recruited into lymph nodes by chemical signals. Important steps in the production of B-cell and T-cell immune responses are as follows:

T-cell activation. Naïve T cells migrate through the cortex of a node and contact dendritic cells displaying antigen peptides on MHC molecules. When a T cell encounters a peptide, to which its TCR binds, mitosis is activated and the growing clone remains in the node. T cells that do not encounter antigen on dendritic cells eventually exit the node in the efferent lymphatic. Naïve T cells continuously recirculate through lymph nodes.

B-cell activation. Naïve B cells are often induced to enter lymph nodes by cytokines that cause their adhesion to HEVs. The naïve B cells then are attracted to primary follicles in the cortex of lymph nodes by signals released by the FDCs. If a B cell encounters its specific antigen in the node, it is likely that there are T_h cells in the same node that have also been activated clonally by the same antigen. This facilitates the formation of cognate pairs of B and T_h cells, resulting in the formation of B-cell clones.

Plasma-cell activation. Some clonally activated B cells quickly differentiate into plasma cells, migrate to the medulla of the lymph node, and begin secreting IgM antibodies. Medullary plasma cells are often observed in groups (Figure 9-6C). Plasma cells can exit the node and take up residence in other sites, such as the bone marrow.

Secondary follicles. In many cases, some activated B cells reside in a primary follicle and form a **germinal center**; the structure then is known as a secondary follicle. The germinal center contains T cells, FDCs, and large B cells, called **centroblasts**, which undergo **isotype switching** to produce plasma cells that secrete IgA, IgE, or IgG, depending on the location of the node and the types of cytokines secreted by T_h cells present there. The centroblasts also produce **memory B cells**, which become quiescent and persist for long periods. These memory cells have undergone isotype switching and respond quickly to the reappearance of antigen with a strong secondary response.

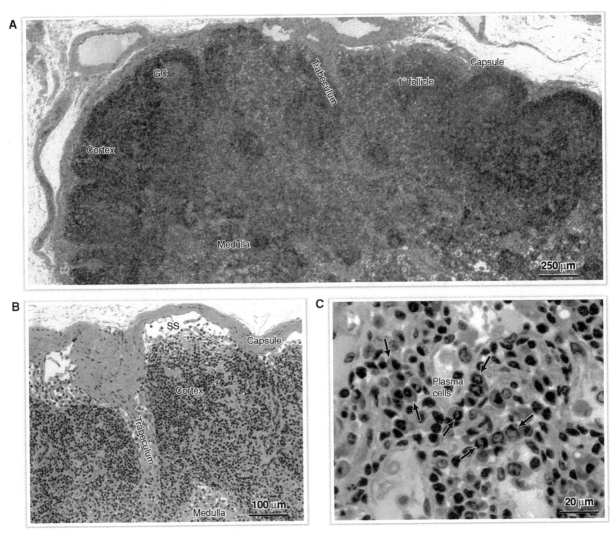

Figure 9-6: Peripheral lymphatic tissues: lymph node, tonsil, Peyer's patch. **A.** Section of a small lymph node. The high cellularity of the cortex under the capsule is apparent. Dark circular areas represent primary follicles, and a germinal center (GC) in a secondary follicle is visible on the left. Extensions of the capsule, called trabeculae, wind irregularly toward the medullary area in the center of the node. **B.** Section of lymph node cortex. Lymph arriving in afferent lymphatics flows in the subcapsular sinus (SS) and then percolates along trabeculae toward the medulla. **C.** Plasma cells in the medulla. Plasma cells differentiated from B cells dividing in the cortex migrate into the medulla and are found in groups called cords. The chromatin in plasma cells becomes clumped around the edge of the nucleus, producing a wheel-like appearance. Three groups of plasma cells are indicated between the arrows. (*continued on page 139*)

NONENCAPSULATED LYMPHATIC TISSUES

Most of the body's peripheral lymphatic tissue consists of aggregates of cells associated with mucosal surfaces, such as that lining the gastrointestinal tract and the respiratory system. These aggregates are collectively referred to as **mucosa-associated lymphoid tissue (MALT)**. They contain the same cells and functions as those of lymph nodes but lack a complete capsule and afferent lymphatics. Antigens gain entry to these structures through the mucosal epithelium. The intestinal epithelium over these aggregates contains specialized cells, called **microfold** or **M cells**, which actively import material from the lumen by transcytosis. The cytokines produced in MALT cause antigen-activated lymphocytes to express a specific set of adhesion molecules, known as **homing receptors**. These homing receptors direct such cells back to mucosal tissues as they recirculate through the body; lymphocytes activated in encapsulated lymph nodes in the presence of other cytokines return to connective tissues near the skin. Similarly, the specific cytokine milieu in MALT induces B cells to produce IgA after isotype switching, whereas that in lymph nodes near the skin results in the production of IgG. Two examples of MALT found in the gastrointestinal tract are as follows:

Tonsils. The oral cavity contains partially encapsulated lymphatic aggregates with one surface covered with mucosal epithelium. **Palatine** and **lingual** tonsils have a stratified squamous epithelium; **pharyngeal** tonsils are covered with respiratory epithelium. Figure 9-6D shows lymphatic follicles in a section from a palatine tonsil.

Peyer's patches. Dome-like lymphatic aggregates in the lamina propria and submucosa, known as Peyer's patches, are the defining features of the ileum (Figure 9-6E). Peyer's patches contain HEVs; M cells are found in the mucosal epithelium. Many of the plasma cells secreting IgA antibodies found in the lamina propria of the gut arise from B cells activated in Peyer's patches because the plasma cells return to mucosal tissues.

▼ Cancer can arise from a cell in any of the stages involved in B-cell development. Plasma-cell tumors are called **multiple myelomas** and often retain the ability to produce and secrete immunoglobulins, either assembled in isotypic forms or as individual chains. Because these tumors are clonal, a patient's serum will contain high levels of the unique proteins produced by the original transformed plasma cell. The abundant monotypic immunoglobulins are easily detected by electrophoresis and serve as a key diagnostic feature for this type of cancer.

The **acquired immune deficiency syndrome (AIDS)** provides dramatic evidence of the importance of the effector activities of T_h cells. AIDS is caused by infection with the **human immunodeficiency virus (HIV)**, which uses the CD4 molecule as a receptor for binding and entering cells. CD4 is expressed on macrophages, dendritic cells, and T_h cells. The virus kills T_h cells and eventually their rate of loss exceeds the rate of their replacement. When the level of T_h cells falls too low, various clinical symptoms appear, including susceptibility to infections caused by common pathogens that are normally kept in check by the adaptive immune system. ▼

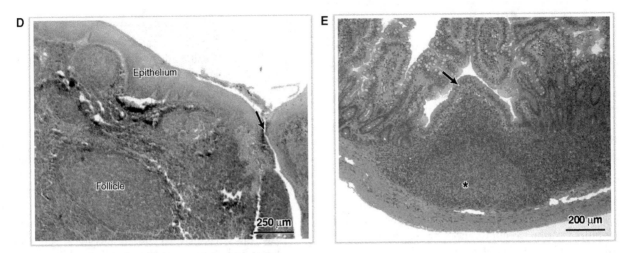

Figure 9-6: (*continued*) **D.** Palatine tonsil. These tonsils are covered by stratified squamous epithelium at surfaces that face the oral cavity. They lack afferent lymphatics, and antigens gain entry through the epithelium. Deep invaginations of the epithelium, called crypts, occur; one is indicated by the arrow. B-cell activation in these organs is common; several follicles are visible in the small portion visible here. **E.** Peyer's patch. These dome-shaped structures (*arrow*) are diagnostic of the ileum and are covered by enterocytes. The epithelium contains microfold cells that transport antigen into the lymphatic tissue underneath. The light area at the bottom of the image (*) may represent a large germinal center.

SPLEEN

The spleen is the largest peripheral lymphoid organ and serves as a biologic filter for blood. It is responsible for removing bloodborne pathogens and for mounting immune responses to antigens present in blood. The spleen is also the site for the turnover of senescent RBCs. The hilus of the spleen is penetrated by the splenic artery and vein, by nerves, and by efferent lymphatics.

STRUCTURE OF THE SPLEEN

The spleen is located in the upper left abdominal quadrant and is organized into areas called **red** and **white pulp**. The terms red and white pulp describe the appearance of a freshly transected organ; red pulp is visibly full of blood in sinusoids and connective tissue areas called cords, whereas white pulp is densely packed with lymphocytes. The microscopic organization of the spleen can be described as follows (Figure 9-7A–C):

Capsule. The spleen is covered with a thick, connective tissue capsule. **Trabeculae** extend into the organ from the capsule and are highly branched. The trabeculae contain small arteries and veins.

Red pulp. This region is filled with blood and accounts for about 80% of the volume of the splenic parenchyma. The blood is found in **splenic sinuses**, porous, thin-walled venules, and **splenic cords**, which consist of a highly cellular, loose connective tissue containing reticular fibers. The spleen is one organ in which blood exits vessels lined with endothelial tissue and circulates through a connective tissue.

White pulp. This region consists of distinct clusters of lymphocytes, each typically organized around an **arteriole** and historically called a **central artery**, which arises from a trabecular artery as it enters the parenchyma. In a newborn before exposure to antigens, these lymphocytes, which are predominantly T cells, surround the artery symmetrically and are called a **periarterial lymphatic sheath (PALS)**. After exposure to antigens and the development of B-cell responses, primary and secondary follicles appear, expanding one side of a PALS so that the central artery becomes eccentric. The boundary between the red and white pulp is known as the **marginal zone**.

Blood flow. Blood enters the parenchyma of the spleen in central arteries, which connect to small **penicillar arterioles** running radially away from the central artery toward the marginal zone (Figure 9-7D). This blood follows one of two routes. In the **closed circulation** route, a penicillar arteriole eventually joins a splenic sinus and, therefore, the blood never leaves an endothelium-lined vessel. In the **open circulation** route, a penicillar arteriole terminates in the marginal zone, releasing its blood into the splenic cords. The plasma and cells flow through the cords and must pass between the loosely adherent endothelial cells of the sinus to regain entry to the circulator system. The sinuses drain into trabecular veins, which supply the splenic vein.

FUNCTIONS OF THE SPLEEN

The trapping of pathogens and antigens and the turnover of RBCs are both facilitated by the open blood circulation that occurs in the spleen.

Immune activities. Macrophages and dendritic cells trap antigens that are released into the marginal zone cords. The macrophages and dendritic cells then migrate back to the white pulp and present the antigens to the large population of T cells present in the PALS. B cells are continuously flowing through the spleen, either fresh from the bone marrow or recirculating after passage through other peripheral immune tissues. The spleen, therefore, functions similar to a lymph node to concentrate antigen, APCs, and lymphocytes together to activate cells efficiently and support clonal selection. Macrophages and neutrophils also engulf and kill pathogens that are released into the cords.

RBC filtration. RBCs accumulate membrane alterations as they age. They become more dense, less flexible, and their surface charge decreases. It is assumed that some of these changes are monitored by macrophages, and this triggers the phagocytosis of RBCs that are older than 120 days as they traverse splenic cords. The complete removal of an entire RBC is called **culling**. Any areas of membrane stiffness slow the movement of RBCs as they squeeze between endothelial cells to re-enter splenic sinuses. Macrophages remove the stiff portions of RBCs in a process called **pitting**. The remainder of the pitted cell can reseal and continue to circulate, although it will be smaller. A fraction of RBCs do not completely extrude the nucleus and the remaining fragments, called **Howell–Jolly** bodies, are pitted when the cells pass through the spleen. The extensive pitting of cells containing defective cytoskeletal proteins is responsible for the spherical appearance of RBCs that occur in **hereditary spherocytosis** (see Chapter 8). Intracellular **malarial parasites** in RBCs retard the entry of the RBCs into sinuses and cause their removal by pitting.

▽ The disruption of the shape and flexibility of RBCs associated with **sickle cell disease** causes significant complications involving the spleen. The spleen in infants and young children who have sickle cell disease is often swollen as the result of trapping numerous sickled cells. Blockage of small vessels results in frequent infarcts in the spleen. The damage accumulates, and by the age of 8 years, most children with sickle cell disease no longer have a functioning spleen.

The loss of the immune activity provided by the spleen renders children with sickle cell disease especially susceptible to bacterial infections caused by *Haemophilus influenzae* and *Streptococcus pneumonia*. These bacteria have thick capsules, and their opsonization by antibodies is important for clearing infections. Surgical removal of the spleen from adult patients also increases the likelihood of such infections. Vaccinations to protect against these bacteria are administered to all patients lacking spleens. Culling of RBCs in asplenic individuals is taken over by macrophages in the liver and bone marrow, but it is typical to observe a higher percentage of old and abnormal RBCs in such cases. ▽

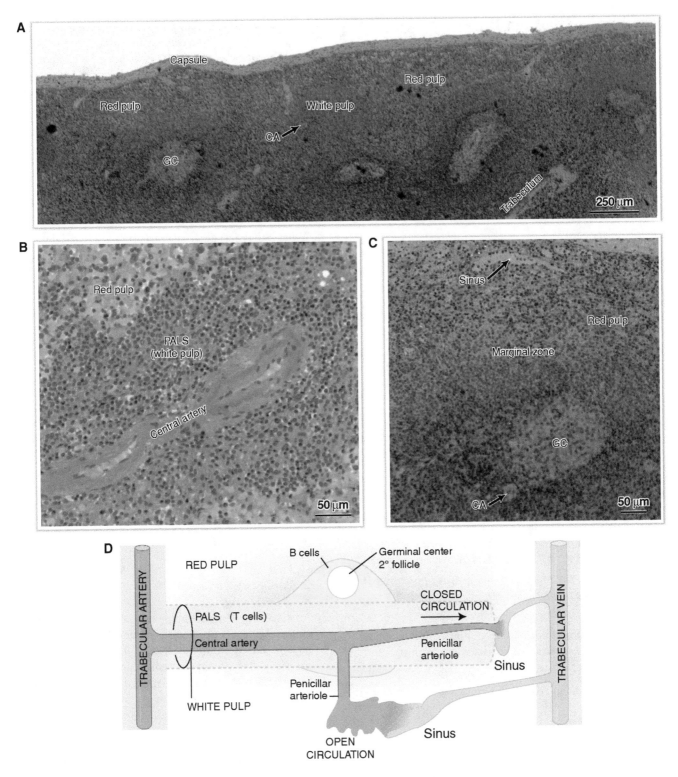

Figure 9-7: The spleen. **A.** Section of spleen. Areas of red and white pulp are intermixed beneath the dense capsule. Central arteries (CA) are associated with the lymphocytes in white pulp, and a germinal center (GC) is seen at the center of a secondary follicle, indicating an advanced B-cell response. Thick trabeculae penetrate the organ and carry vascular elements. **B.** Central artery. The central artery is surrounded by a periarterial lymphatic sheath (PALS), which consists predominantly of T cells. The lower cell density of red pulp is evident. **C.** Red and white pulp. The central artery (CA) is eccentric, in this case due to the proliferation of B cells and the associated germinal center (GC), indicating active production of plasma cells and antibody secretion. By the open theory of circulation, blood from some of the central arteries is released at the marginal zone area and must travel through the red pulp to large splenic sinuses, where it re-enters the circulation. **D.** Blood flow in the spleen. T cells maintain a symmetric relationship in PALS around the central arteries, whereas areas of rapid B-cell expansions in follicles grow away from the artery. Penicillar arterioles branch from the central artery and either connect directly to splenic sinuses (closed circulation) or terminate at the marginal zone, dumping blood into red pulp tissue (open circulation). Red blood cells are processed by macrophages as they traverse red pulp to remove inclusions and destroy old cells.

STUDY QUESTIONS

Directions: Each of the numbered items or incomplete statements is followed by lettered options. Select the **one** lettered option that is **best** in each case.

1. The epithelium that covers portions of the nonencapsulated lymphatic aggregates found along the gastrointestinal tract, such as tonsils and Peyer's patches, is functionally equivalent to which structure of a lymph node?

   A. Afferent lymphatic

   B. Cortex

   C. Efferent lymphatic

   D. Medullary cord

   E. Trabecula

2. Deletions of part of chromosome 22 can produce a variety of developmental defects, such as DiGeorge syndrome, which may include the complete failure of the thymus to develop. What is the most likely effect on the adaptive immune system in children affected with this severe form of DiGeorge syndrome?

   A. B and T cells will be present in the blood and serum antibody levels will be normal

   B. B and T cells will be present in the blood and serum antibody levels will be reduced

   C. B cells will be present in the blood, serum antibody levels will be normal, and T cells will be absent

   D. B cells will be present in the blood, serum antibody levels will be reduced, and T cells will be absent

3. What is the principal reason that some lymphocytes are born with antigen receptors that recognize self-molecules?

   A. Antigen receptors must bind to self-major histocompatability complex molecules

   B. Clonal anergy requires self-reactive receptors

   C. Lymphocyte progenitor cells develop in central lymphoid tissue, which is a specialized, restricted environment

   D. Receptor genes are rearranged during cell development

   E. Self-reactive cells are required for immunologic surveillance

4. A 4-year-old boy is treated in the clinic for a severe ear infection. He has had four similar infections within the past 6 months and was hospitalized for pneumonia about 1 year ago. A blood test indicates IgM well above normal levels but no detectable IgA or IgG. Which activity is most likely defective or missing in this child?

   A. Cytokine secretion

   B. Endocytosis

   C. Isotype switching

   D. Peptide transport into the endoplasmic reticulum

   E. Perforin release

5. What is the most likely fate of a pre-T cell in the thymus that expresses a T-cell receptor that cannot bind to major histocompatibility complex molecules expressed by cells in the thymus?

   A. It will become a T_c cell

   B. It will become a T_h cell

   C. It will become a T_{reg} cell

   D. It will become a memory cell

   E. It will die by apoptosis due to negative selection

   F. It will die by apoptosis due to the lack of positive selection

6. The malarial parasite invades red blood cells and in this location is largely hidden from antibodies and T_c cells. This strategy, however, puts the parasite at risk of which other kind of attack?

   A. Diapedesis

   B. Immunologic surveillance

   C. Negative selection

   D. Peripheral tolerance

   E. Pitting in the spleen

7. Analysis of IgG present in the sera of millions of people vaccinated against the poliovirus reveals that many individuals have produced antibodies that bind to the same regions of the virus. Because an unpredictable process is responsible for producing the variable antigen-binding regions of antibodies, this might seem to be an unlikely situation. Which activity best explains the production of antibodies by different individuals that bind to the same molecular features on the poliovirus?

   A. Anergy

   B. DNA recombination

   C. Clonal selection

   D. Cytokine secretion

   E. Peripheral tolerance

ANSWERS

1—A: The epithelium of many nonencapsulated lymphatic aggregates serves to import antigens. In the case of Peyer's patches, microfold cells in the epithelium introduce antigens by transcytosis. Thus, the epithelium functions as afferent lymphatics for these structures.

2—D: B cells develop in bone marrow, thus naïve B cells will be produced and be present in the blood in the absence of the thymus. T cell development, however, requires the thymus for the first few decades of life, therefore T cells will be absent or severely depleted in the blood. The effective production of an antibody response by activated B cells requires T_h cells, therefore serum antibody levels will be severely reduced due to the absence of T cells in this case.

3—D: The rearrangement of antigen-receptor genes during lymphocyte development in central lymphoid tissues can produce proteins that bind to self-antigens.

4—C: Because the child is able to produce IgM antibodies, but not IgA or IgG, a defect in isotype switching is the most likely cause of the problem. This could be due to the lack of an enzyme required for the DNA recombination that accomplishes the switch from IgM to another isotype, or it could be due to the lack of signaling between T_h and B cells required to initiate the switch.

In either case, these individuals produce high levels of IgM but often experience repeated infections because these antibodies are less effective than other antibody isotypes in combating bacterial infections.

5—F: Thymocytes that are able to rearrange their T-cell receptor genes and express a receptor must be able to bind to major histocompatability complex molecules on epithelial reticular cells with moderate affinity in order to receive positive selective signals. If they do not receive these signals, they undergo apoptosis.

6—E: Macrophages in the spleen will be able to remove the portion of some of the cells containing the parasite because they will retard the re-entry of the red blood cell from the splenic cords back into splenic sinuses. This reaction is known as pitting.

7—C: The process of clonal selection during the development of a successful response to vaccination with poliovirus will ensure the production of clones of B cells with antigen receptors that bind to the virus with high affinity. The virus contains a limited number of epitopes and clonal selection ensures that antibodies will be generated to these features.

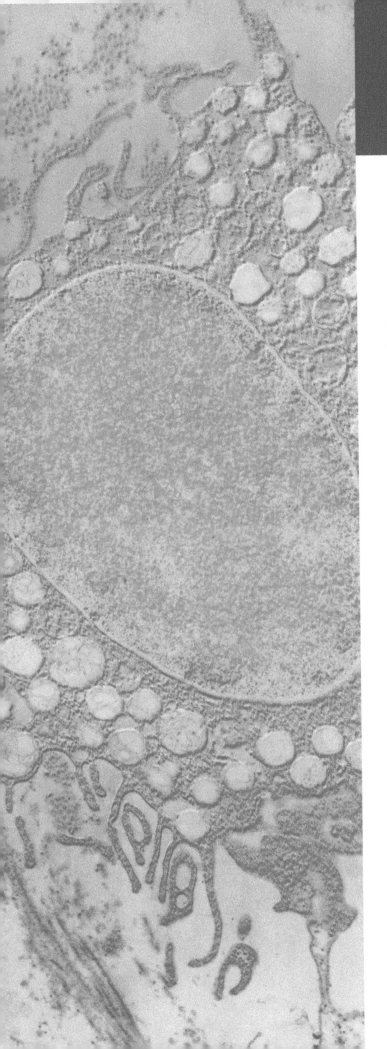

CHAPTER 10

RESPIRATORY SYSTEM

OVERVIEW

The main function of the respiratory system is to supply the body with O_2 from the air and to remove CO_2 from the bloodstream via a system of tubes or airways. Functionally, the respiratory system can be divided into the following **two divisions**, which differ significantly in structure and function:

Conducting division. Consists of a series of airways located both outside and inside the lung. Inspired air is warmed, moistened, and cleaned in the conducting airways before entering the respiratory airways. Conducting airways do not directly participate in gas exchange.

Respiratory division. Consists of airways located entirely within the lungs. Respiratory airways are specialized for gas exchange and are distinguished by the presence of **alveoli**, saclike structures where the exchange of gasses occurs between inhaled air and blood in capillaries.

Airways in the lung branch repeatedly, producing progressively smaller conduits, described collectively as a **bronchial tree.**

The lungs receive blood from two sources: the **pulmonary arteries** deliver deoxygenated blood to capillaries in the alveoli for oxygenation, and the **bronchial arteries** supply nutrients and oxygenated blood to the lung tissue itself.

O_2 from the air is exchanged for CO_2 from the blood across the **blood-air barrier**, a specialized, trilayered region of the alveolus composed of the alveolar and capillary endothelia and their fused basal laminae.

CONDUCTING DIVISION

ORGANIZATION OF CONDUCTING AIRWAYS

Air from the external environment travels through a series of conducting airways (nasal cavity, nasopharynx, larynx, trachea, and extrapulmonary primary bronchi) before entering the lungs. Within the lungs, air continues through still more conducting airways, which branch repeatedly into progressively smaller airways (intrapulmonary bronchi, bronchioles, and terminal bronchioles) (Figure 10-1A). As air leaves the terminal bronchioles, it enters the respiratory airways, where O_2 is exchanged for CO_2. No gas exchange occurs as air is transported through conducting airways, but the inspired air is warmed, moistened, and cleaned (i.e., conditioned) along this journey, which helps to prevent damage to delicate alveoli. This chapter will focus on the functional histology of the trachea and lung.

TRACHEA

The trachea is a flexible tube about 10–12 cm long and 2.5 cm in diameter. It begins at the base of the larynx and ends at the T4 vertebral level, where it bifurcates to form primary bronchi, which then enter the right and left lungs (Figure 10-1A).

MUCOSAL LINING OF THE TRACHEA The trachea and most other large airways in the conducting portion of the respiratory system are lined with **pseudostratified ciliated columnar epithelium (respiratory epithelium)** resting on a lamina propria of loose connective tissue with abundant elastic fibers (Figure 10-1B and C) and immune cells (e.g., lymphocytes, plasma cells, and mast cells). The respiratory epithelium and underlying lamina propria comprise the **mucosa**. All of the respiratory epithelial cells are in contact with the basal lamina, but not all cells extend to the lumen.

There are five types of cells in respiratory epithelium—mucous goblet, ciliated columnar, basal (short), brush, and small granule cells—each with important functions (Figure 10-1B). **Mucous goblet cells** and **ciliated columnar cells** are the most abundant cells, each comprising about 30% of the cells in the epithelium of the upper respiratory tract. These two cells are important in functioning to clean and condition inspired air before it enters the respiratory airways.

- **Mucous goblet cells** secrete mucus onto the luminal surface of the epithelium. The mucus traps inhaled particles (e.g., dust and microorganisms), absorbs water-soluble gases (e.g., SO_2 and ozone), and also helps to moisturize inspired air.
- **Cilia** on **columnar cells** sweep mucus and trapped material, called **sputum**, toward the oral cavity, where it is either swallowed or expectorated. Identifying pathogens present in sputum samples is important in the diagnosis and treatment of respiratory infections.

▽ Normally, cilia move mucus as a sheet. However, in individuals with **cystic fibrosis**, the mucus is unusually viscous, making it difficult to clear the airway, which can result in frequent respiratory infections. This illustrates the important protective function of normal mucus flow. ▽

The other three types of cells in respiratory epithelium are difficult to distinguish in histologic sections, but they can be identified by electron microscopy or immunohistochemistry.

- **Basal (short) cells** are small cells that do not extend to the lumen and comprise about 30% of cells in the epithelium. Basal cells are stem cells that have the ability to divide and replace other cell types in the epithelium.
- **Brush cells** are tall columnar cells that have a tuft of microvilli on their apical surface and afferent nerve endings on their basal surface. Brush cells are thought to function as sensory receptors.
- **Small granule cells** (Kulchitsky cells) are a heterogenous population that is part of the **diffuse neuroendocrine system (DNES)** (see Chapter 14). Their basal cytoplasm is filled with small dense-core granules (100–300 nm in diameter) containing diverse products, which include serotonin, calcitonin, and gastrin-releasing peptide (bombesin) and which are thought to control the function of other cells in respiratory epithelium.

Respiratory epithelium lines most of the larger conducting airways in the upper portions of the respiratory system. However, regions that experience intense direct airflow or other abrasions (e.g., the vocal folds in the larynx and the upper surface of epiglottis) are lined with stratified squamous nonkeratinized epithelium.

▽ In any given region of the respiratory system, the type of epithelium and relative portion of different cell types is not absolutely fixed and can be altered by environmental conditions through the process of **metaplasia** (see Chapter 1). For example, in people who smoke, high concentrations of SO_2 and CO cause an increase in the number of goblet cells relative to the number of ciliated cells. Whereas the increased mucus production helps trap the inhaled pollutants, there are relatively fewer ciliated cells to move the mucus to the oral cavity. As a result, smaller airways can become congested. Respiratory epithelium in a chronic smoker can convert to stratified squamous epithelium, considered **dysplasia**, and is an early step in the initiation of squamous cell carcinoma, the most common type of lung tumor. ▽

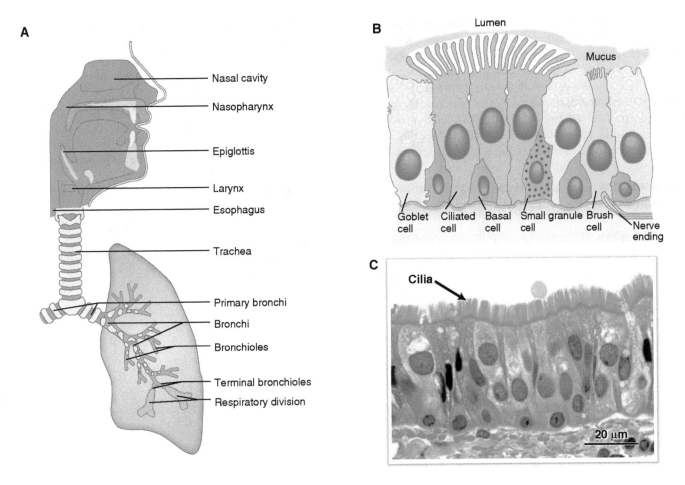

Figure 10-1: Organization of conducting airways and structure of the trachea. **A.** Airways in the conducting division: the nasal cavity, nasopharynx, larynx, trachea, bronchi, and bronchioles. The walls of the trachea and large bronchi are supported by rings or plates of carti-lage, respectively (*blue*). Terminal bronchioles, which branch to produce respiratory bronchioles, are the most distal conducting airways. **B.** Cells in respiratory epithelium: mucous goblet, ciliated columnar, basal (short), brush, and small granule cells. Note that brush cells have afferent nerve endings on their basal surface. **C.** Section through the respiratory epithelium (pseudostratified ciliated columnar epithelium) lining the trachea. Respiratory epithelium is composed of the five types of cells illustrated schematically in part B. All of the cells are in con-tact with the basal lamina, but not all cells extend to the trachea lumen. (*continued on page 147*)

SUBMUCOSA AND ADVENTITIA OF THE TRACHEA Although respiratory epithelium in the trachea and other conducting airways is distinct, there is some disagreement among histologists about the organization of the deeper layers. Some histologists consider that the entire tracheal wall beneath the respiratory epithelium comprises the **lamina propria**, whereas others consider that the regions of the tracheal wall below the epithelium can be divided into lamina propria, **submucosa**, and **adventitia**. Regardless of the nomenclature, the wall of the trachea contains the following structures (Figure 10-1D and E):

Elastic lamina. A dense layer of elastic fibers that enables the trachea and other airways to stretch during inspiration to accommodate inhaled air. The stretched fibers recoil during expiration, which allows air to be expelled and returns airways to their resting state.

Seromucous glands. Numerous branched tubuloalveolar glands whose ducts open onto the epithelial surface. Secretions of the seromucous glands function together with those of goblet cells to humidify inspired air, detoxify soluble gases, and trap inspired foreign particles.

Hyaline cartilage. 15–20 C-shaped rings of hyaline cartilage that line the trachea. The open ends of the rings face posteriorly, toward the esophagus. Cartilage in the trachea and other large conducting airways reinforces the walls, preventing the airways from collapsing.

Trachealis muscle. A band of smooth muscle that joins the open ends of the C-shaped cartilage rings and regulates the tracheal diameter. For example, during the cough reflex, contraction of the trachealis muscle narrows the tracheal lumen, increasing the velocity of expired air to greater than 50 mph, which helps to clear the airway.

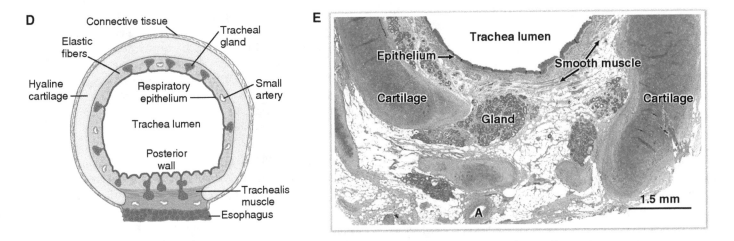

D. Connective tissue, Elastic fibers, Hyaline cartilage, Tracheal gland, Respiratory epithelium, Small artery, Trachea lumen, Posterior wall, Trachealis muscle, Esophagus

E. Trachea lumen, Epithelium, Smooth muscle, Cartilage, Gland, Cartilage, A, 1.5 mm

Figure 10-1: (*continued*) **D.** Histologic organization of the trachea wall. **E.** Section through the posterior region of a human trachea. Bundles of smooth muscle (the trachealis muscle) join the open ends of the cartilage rings. A, small artery. (periodic acid-Schiff reaction with H&E stain)

BRONCHIAL TREE

ORGANIZATION OF THE BRONCHIAL TREE

The trachea bifurcates into the left and right primary bronchi, which supply the left and right lung, respectively. Each lung is organized around its system of airways, referred to metaphorically as the **bronchial tree**, in which the primary bronchus is the trunk and the smaller bronchi, bronchioles, and respiratory airways are its branches. The blood supply to the lungs closely follows the branching pattern of airways in the bronchial tree. Each branch of the bronchial tree ventilates a distinct part of the lung.

The left and right primary bronchi enter the lungs and bifurcate (left) or trifucate (right) into secondary (lobar) bronchi to supply the two lobes of the left and three lobes of the right lung, respectively. Each **secondary bronchus** branches dichotomously (up to 16 times), producing progressively smaller airways. The larger airways are classified as **bronchi**; smaller airways (<5 mm) are called **bronchioles** (Figure 10-2A). The most distal conducting airways are the **terminal bronchioles**. Terminal bronchioles branch to give rise to **respiratory bronchioles**, the initial airways in the respiratory portion of the bronchial tree.

BRONCHI

All conducting airways are built from the same basic components—epithelium, hyaline cartilage, smooth muscle, and elastic fibers—but the histologic appearance gradually changes as you look deeper into the bronchial tree. As each generation of branching produces progressively smaller airways, the walls become thinner and simpler and the organization of the **four basic components** changes, as follows:

- **Epithelium** height and complexity decreases (Figure 10-2C).
- **Cartilage rings** are replaced by isolated blocks or irregular **plates of hyaline cartilage** (Figure 10-2A and B).
- **Trachealis muscle** is replaced with **bundles of smooth muscle** that spiral helically around the airways in a layer between the epithelium and cartilage (Figure 10-2A and B).
- Amount of **cartilage, elastic fibers,** and **smooth muscle,** and the number of **goblet cells** and **glands** decreases. However, elastic fibers and smooth muscle decrease less rapidly than the other components and, therefore, comprise a larger proportion of the wall of smaller conducting airways (Figure 10-2B and D).

Contraction or relaxation of the bands of smooth muscle regulates the diameter of the airways and is under **autonomic control**. The regulation of bronchial and bronchiolar diameter is important in regulating airflow during inspiration and expiration.

- **Parasympathetic stimulation** (via the vagus nerve) causes contraction of smooth muscles and constriction of bronchi and bronchioles.
- **Sympathetic stimulation** causes relaxation of smooth muscles and dilation of bronchi and bronchioles.

 Asthma is a chronic inflammatory disorder characterized by recurrent episodes of airway constriction and increased resistance to airflow (see Chapter 8). It is likely that inflammation causes airways to become hyperreactive to a variety of stimuli, resulting in bronchoconstriction. Because smooth muscle comprises a relatively larger portion of the walls of bronchioles than bronchi, the increased airway resistance in the patient with asthma is primarily due to contraction of bronchiolar smooth muscle. In addition, excessive production of viscous mucus contributes to airway obstruction. ▼

BRONCHUS-ASSOCIATED LYMPHATIC TISSUE The respiratory epithelium and its underlying lamina propria provide the first line of defense against potential airborne pathogens and irritants from the external environment. Not surprisingly, lymphocytes are abundant in the lamina propria of conducting airways. Lymphatic nodules containing T and B lymphocytes (**bronchus-associated lymphatic tissue**, or **BALT**) are common near branch points of the bronchial tree, providing immune protection to the airways.

BRONCHIOLES

After about ten generations of branching, conducting airways lose the distinguishing characteristics of bronchi and acquire features characteristic of bronchioles. **Bronchioles can be distinguished from bronchi by the following features:**

- Cartilage and glands are no longer present (Figure 10-2A and D).
- Pseudostratified columnar epithelium characteristic of larger airways transitions to simple ciliated columnar epithelium, then to simple ciliated cuboidal epithelium in the smallest airways (Figure 10-2C, D, and E).
- Goblet cells are replaced by **Clara cells** (nonciliated bronchiolar secretory cells), tall dome-shaped cells containing secretory granules, which are unique to bronchioles (Figure 10-2E). Clara cells appear to have diverse functions, including secreting proteins and glycosaminoglycans to protect bronchiolar epithelium and reduce the viscosity of surface mucus, degrading airborne toxins, and dividing to regenerate ciliated and nonciliated cells of bronchiolar epithelium.

Bronchioles branch repeatedly, giving rise to **terminal bronchioles**, the smallest airways in the conducting division (Figure 10-3B). Terminal bronchioles are characterized by simple cuboidal epithelium with both ciliated and nonciliated (Clara) cells. Each terminal bronchiole branches, giving rise from two to five **respiratory bronchioles**, the initial airways in the respiratory portion of the respiratory system.

BRONCHIAL AND BRONCHIOLAR BRANCHING There are two obvious functional consequences of the extensive branching of the conducting components—increased gas exchange and decreased air velocity.

- **Gas exchange.** Branching of conducting airways results in increased area for gas exchange in the respiratory airways.
- **Air velocity.** Branching increases the total cross-sectional area of the conducting airways. Thus, due to simple physics, air travels at a lower velocity as it moves deeper in the lungs, which reduces potential damage to the respiratory airways.

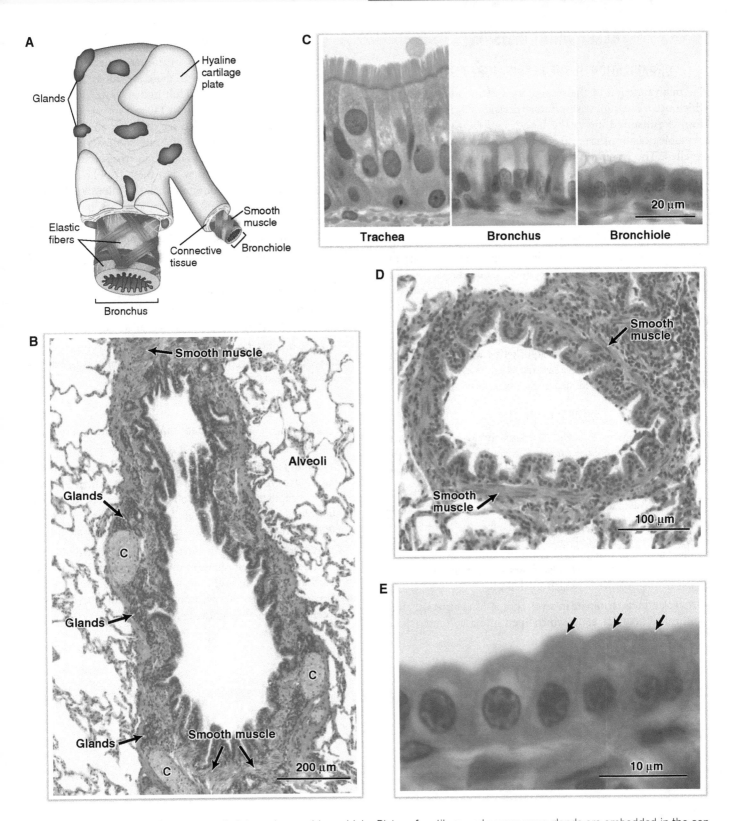

Figure 10-2: Bronchi and bronchioles. **A.** A bronchus and bronchiole. Plates of cartilage and seromucous glands are embedded in the connective tissue walls of bronchi but are absent from bronchioles. Connective tissue is removed in the lower portion of this figure to show the smooth muscle and elastic fibers. Contraction of smooth muscle causes the mucosa to appear folded. **B.** Tangential section through a bronchus. The epithelium of bronchi is lower than respiratory epithelium in the trachea but contains the same types of cells. Bundles of smooth muscle, elastic fibers (not stained), plates of hyaline cartilage, and seromucous glands are present in the bronchial wall. C, hyaline cartilage. **C.** Comparison of epithelium in the trachea, a bronchus and a bronchiole. The height and complexity of the epithelium decreases in smaller airways. **D.** Cross-section through a bronchiole. The airway is lined by cuboidal epithelium composed of ciliated cells and nonciliated Clara cells, but no goblet cells. Note that there are no glands or cartilage in the wall of bronchioles. **E.** Section of a bronchiole showing Clara cells (*arrows*). Clara cells are dome-shaped, nonciliated cells found only in the epithelium of bronchioles.

RESPIRATORY DIVISION

ORGANIZATION OF THE RESPIRATORY AIRWAYS

The main function of the respiratory system is gas exchange, which occurs within its respiratory division. O_2 is transferred from the inspired air to the blood, and CO_2 generated by metabolism is transferred from the blood to the air and then expelled. The respiratory division of the respiratory system consists of all airways and structures in which gas exchange occurs—respiratory bronchioles, alveolar ducts, alveolar sacs, and alveoli (Figure 10-3A).

RESPIRATORY BRONCHIOLES

Respiratory bronchioles are the initial components of the respiratory division of the respiratory system. Similar to the terminal bronchioles, respiratory bronchioles are lined with low cuboidal ciliated epithelium with occasional Clara cells, and are supported by a helical network of smooth muscle cells and elastic fibers. However, the cuboidal epithelium in respiratory bronchioles is interrupted by small sac-like outpocketings, about 200 µm in diameter, called **alveoli**, where gas exchange occurs (Figure 10-3B). Respiratory airways are defined by the presence of alveoli.

ALVEOLAR DUCTS

Respiratory bronchioles branch successively two to three times. As respiratory bronchioles branch, the number of alveoli in the walls increases and the amount of cuboidal epithelium decreases.

Airways completely lined by alveoli, with no cuboidal epithelium, are called **alveolar ducts** (Figure 10-3A, B, and D). The walls of alveolar ducts and individual alveoli are supported by strands of elastic and reticular fibers. The elastic fibers allow alveoli to expand and passively contract with inspiration and expiration; reticular fibers prevent overexpansion and damage of alveoli, which are delicate.

Alveolar ducts terminate in several grape-like clusters of alveoli called **alveolar sacs**, which open into a common space called an **atrium** (Figure 10-3D).

ALVEOLI

Alveoli are the structural and functional units of the lung. Although each alveolus is small, there are about 300 million alveoli tightly packed in each adult human lung, producing a total area for gas exchange of about 140 m².

Each alveolus is a thin-walled sac that is open on one side. The opening is controlled by a delicate sphincter of smooth muscle, which appears as a small knob in cross-section (Figure 10-3C). Gas exchange occurs within these cuplike structures. Adjacent alveoli share a common wall, the **interalveolar septum**, which contains a dense capillary network in which gas exchange occurs. The septa are interrupted by small openings (10–15 µm in diameter), called **pores of Kohn** (Figure 10-3A), which connect the air space of adjacent alveoli. These openings serve to:

Equalize pressure in the alveoli.

Promote collateral circulation of air when small bronchioles are blocked, maximizing the use of available alveoli.

Provide a route for macrophages to travel freely from one alveolus to another; unfortunately, these openings also facilitate the spread of pneumonia and neoplasms.

BLOOD SUPPLY TO THE LUNGS

The lungs receive a **dual blood supply** from the pulmonary arteries and the bronchial arteries (Figure 10-3A).

Pulmonary arteries (functional or pulmonary circulation) deliver deoxygenated blood from the right ventricle of the heart to capillaries in the alveoli for oxygenation. Reoxygenated blood in the alveolar capillaries flows into the **pulmonary venules** and **veins** and is returned to the left atrium.

Bronchial arteries supply nutrients and oxygenated blood to the lung tissue itself, anastomosing with small branches of the pulmonary arteries in respiratory bronchioles. Both sets of arteries, as well as pulmonary lymphatics and nerves, closely follow the branching bronchial tree.

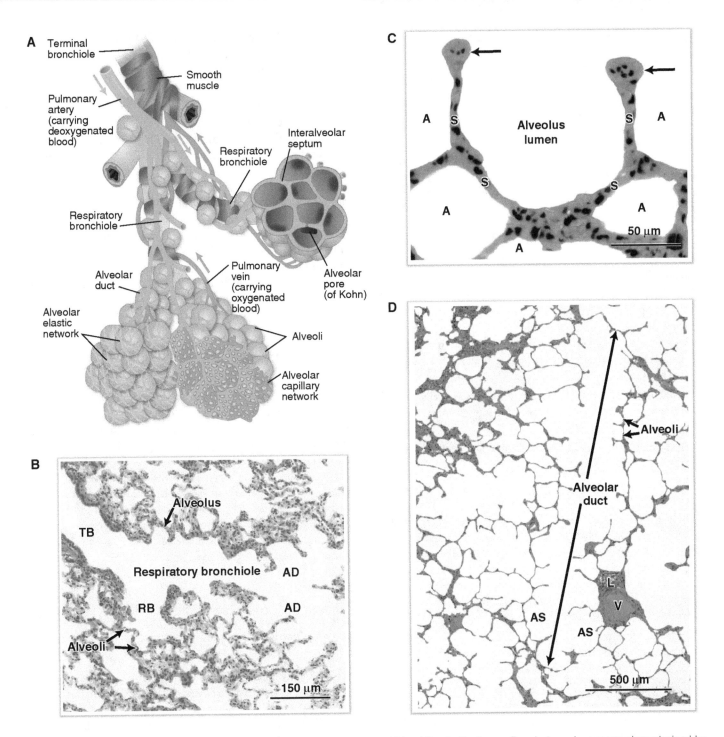

Figure 10-3: Organization of respiratory airways. **A.** Respiratory airways and blood flow in the lungs. Respiratory airways are characterized by outpocketings, called alveoli, where gas exchange occurs. Pulmonary arteries deliver deoxygenated blood (*blue*) to alveolar capillaries, and pulmonary veins return oxygenated blood (*red*) to the left atrium. **B.** Longitudinal section through a respiratory bronchiole. The walls of respiratory bronchioles (RB), the initial airways in the respiratory division, are composed of cuboidal epithelium as in terminal bronchioles (TB), interrupted by alveoli. Respiratory bronchioles branch to produce alveolar ducts (AD), airways composed entirely of alveoli. **C.** Section through several alveoli. Adjacent alveoli (A) share a common wall, the interalveolar septum (S) where gas exchange occurs (see Figure 10-4A). The arrows indicate the smooth muscle cells that control the alveolar opening. **D.** Longitudinal section through an alveolar duct. Alveolar ducts end blindly, with several alveolar sacs (clusters of alveoli) (AS) opening off a common chamber called an atrium (not labeled here). L, lymphatic aggregate; V, blood vessel.

INTERALVEOLAR SEPTA

The walls of alveoli, termed **interalveolar septa**, are composed of the epithelium of two adjacent alveoli, with a capillary-rich interstitium sandwiched between the alveoli (Figure 10-4A). The epithelium that lines alveoli is composed of roughly equal numbers of two types of cells, **type I** and **type II pneumocytes**. In addition, many macrophages are found in loose association with alveoli. The pneumocytes cooperate to promote the critical function of gas exchange between air and tissues, and the macrophages protect against material entering the lungs from outside.

TYPE I PNEUMOCYTES

Type I pneumocytes (type I alveolar cells) are specialized for gas exchange. The following features of type I pneumocytes facilitate gaseous exchange between the alveoli and capillaries:

Cell shape. Type I cells are extremely thin, flattened cells. They are only 25-nm thick, except near the nucleus where all organelles are clustered (Figure 10-4A). Although type I cells represent only 10% of cells in the lung, they cover about 97% of the surfaces.

Fused basal laminae. The basal lamina of type I cells is fused to that of pulmonary capillaries in the interalveolar septa.

Tight junctions. Type I cells are joined to adjacent cells by tight junctions (zonulae occludens), which prevent leakage of fluid into the alveolar space.

TYPE II PNEUMOCYTES

Type II pneumocytes (type II alveolar cells) are large cuboidal cells that secrete **surfactant**, a complex mixture of proteins, phospholipids (primarily dipalmitoyl phosphatidylcholine and phosphatidylglycerol), and glycosaminoglycans. Type II cells also function as stem cells that divide and differentiate into both type I and type II pneumocytes. There is a constant turnover of alveolar epithelial cells, with approximately 1% being replaced daily.

Surfactant released from type II cells spreads over the alveolar epithelium and reduces surface tension in the alveoli, facilitating expansion of the alveoli during inspiration and preventing their collapse during expiration.

Type II cells can be easily identified by electron microscopy due to the presence of numerous membrane-bound **multilamellar bodies** in the cytoplasm, which represent the source of the surfactant. In histologic sections, the cytoplasm appears foamy or vacuolated because lipids in the multilamellar bodies are extracted during preparation of the slides (Figure 10-4B).

Type II cells occupy about 3–5% of the alveolar surface, usually at sites where the walls of the alveoli join. These cells are part of the alveolar epithelium and are joined to type I cells by tight junctions and desmosomes.

▽ During fetal development, surfactant first appears during the last weeks of gestation. Premature infants born before type II pneumocytes are fully developed lack surfactant and have great difficulty breathing, a condition known as **respiratory distress syndrome (RDS)**. Fortunately, symptoms of RSD can be alleviated by the administration of glucocorticoids, which induce synthesis of surfactant. ▼

ALVEOLAR MACROPHAGES

Alveolar macrophages are commonly found in alveoli but are not part of the epithelium per se. They are derived from the monocytes of blood, and are the most numerous cells in the lung. These cells patrol air spaces, phagocytosing and removing debris and bacteria that enter the lungs. Alveolar macrophages are also known as **heart failure cells** because, in the case of congestive heart failure, these cells phagocytose blood that leaks into the alveoli; or they are known as **dust cells**, because they trap inhaled particulates. They are present in interalveolar septa and on the surface of the alveolar lumen (Figure 10-4C). Alveolar macrophages in the lumen eventually migrate or are washed to bronchioles, where they adhere to mucus and are carried up the bronchial tree to the mouth for disposal.

GAS EXCHANGE AND THE BLOOD-AIR BARRIER

Inspired air that enters alveoli must cross the alveolar epithelium and capillary endothelium to reach the blood; CO_2 in the blood must cross these same structures to be expelled (Figure 10-4D). In regions where gas exchange occurs, the epithelium of both type I alveolar cells and capillaries is specialized to facilitate gas exchange. This specialized region of the alveolar wall across which gas exchange occurs is referred to as the **blood-air barrier (blood-gas barrier)**. The blood-air barrier is a trilayered structure consisting of the cytoplasm of the type I alveolar cells (and thin layer of surfactant that covers these cells), the cytoplasm of endothelial cells, and the fused basal laminae of these two cells. The cytoplasm of both the type I alveolar cells and endothelial cells is highly attenuated, such that the total thickness of the wall that separates air from blood is only 0.1–1.5 μm (Figure 10-4A). Because adjacent alveoli share a common interalveolar septum and each capillary contacts epithelium of two alveoli, a similar blood-air barrier is also present on the opposite surface of the septum.

▽ **Emphysema** is a chronic, debilitating disease characterized by destruction of the wall of alveolar ducts, sacs, and individual alveoli. The destruction of alveoli results in a permanent dilation of airspaces, with a reduction in surface area for gas exchange. Emphysema is usually caused by exposure to pollution or by smoking. ▼

REGULATION OF BLOOD pH

Although the main function of the respiratory system is gas exchange, the lungs carry out other important functions; for example, **regulating the pH of blood.**

Most CO_2 generated by metabolism is transported through the blood in the form of bicarbonate ions, HCO_3^- (see Chapter 8). Bicarbonate and CO_2 form a buffer system largely responsible for maintaining the pH of the blood within a narrow range (7.35–7.45). By regulating CO_2 levels, the lungs play a key role in controlling this critical variable.

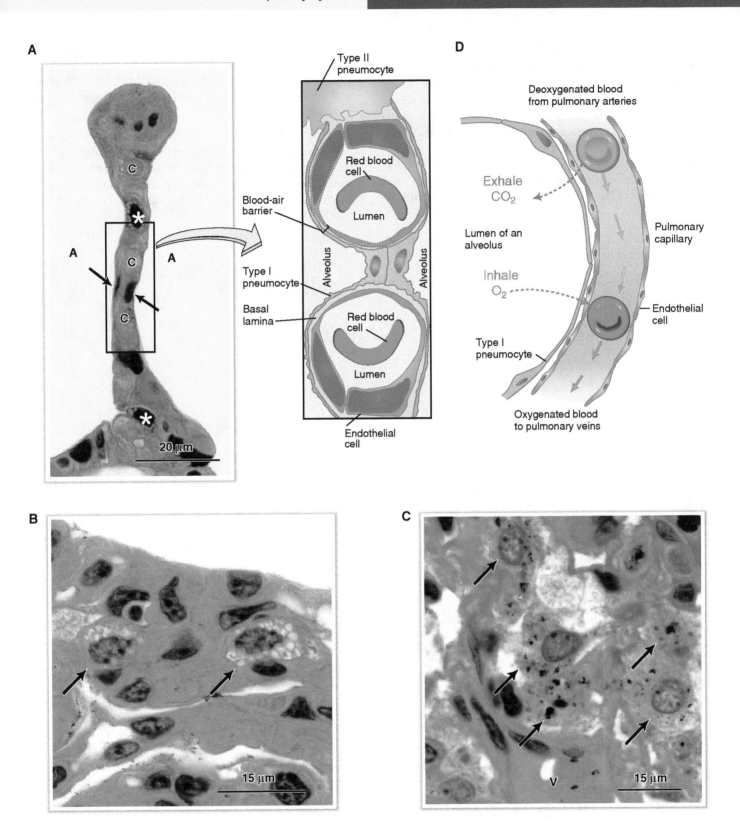

Figure 10-4: Interalveolar septa and alveolar cells. **A.** Section through an interalveolar septum in the lung (left panel). Arrows and asterisks indicate the nuclei of type I and type II pneumocytes, respectively. The boxed area is shown schematically at higher magnification in the right panel. A, alveolar lumen; C, capillary. **B.** Section of lung showing type II pneumocytes. Type II pneumocytes (*arrows*) appear vacuolated due to the extraction of lipids during processing. **C.** Section of lung showing alveolar macrophages in the interstitium. Alveolar macrophages, or dust cells (*arrows*), are easily identified by the debris in the cytoplasm. V, small vein. **D.** Gas exchange in a single alveolar capillary. Deoxygenated blood containing CO_2 generated by metabolism is brought to the lungs via pulmonary arteries. CO_2 is transferred from the blood to air in the alveolus and then expelled. O_2 is transferred from the inspired air to the blood, and the oxygenated blood is then returned to the left atrium via pulmonary veins.

PLEURA

The lungs are covered by a serous membrane, the **visceral pleura**, which is continuous with the **parietal pleura**, which lines the internal surface of the thoracic cavity. The pleura consists of connective tissue containing elastic and collagen fibers covered by a layer of flattened **mesothelial cells**. The mesothelial cells are connected to each other by complex junctions, which prevent air from leaking out of the lungs into the thoracic cavity. The space between the visceral and parietal pleura (the pleural cavity) contains a thin film of lubricating fluid, which permits the lungs to slide smoothly in the thoracic cavity during breathing. The surface tension of the pleural fluid causes the lungs to adhere to the thorax wall and to expand and recoil passively as the thoracic cavity expands and recoils.

▽ **Pneumothorax** is a clinical condition in which air or gas is present in the pleural cavity. For example, a perforating injury to the chest wall can allow air to enter the pleural space, which may cause a collapsed lung due to compression produced by the gas trapped within the pleural cavity. ▼

STUDY QUESTIONS

Directions: Each of the numbered items or incomplete statements is followed by lettered options. Select the **one** lettered option that is **best** in each case.

1. What is the functional unit of the lung?

   A. Alveolus

   B. Bronchopulmonary segment

   C. Intrapulmonary bronchus

   D. Segmental bronchus

   E. Terminal bronchiole

2. A 25-year-old woman delivers a premature infant who develops severe respiratory difficulty. Arterial blood gases show severe hypoxemia. Which of the following cells is most likely too immature to function properly, resulting in these symptoms?

   A. Alveolar macrophage

   B. Basal cell

   C. Goblet cell

   D. Type I pneumocyte

   E. Type II pneumocyte

3. A 60-year-old man makes an appointment with his family physician for a routine check-up. When the physician is taking his history, the man explains that he has smoked two packs of cigarettes a day for almost 45 years. A biopsy of an alveolar septum of his lung shows numerous cells containing black particles. Which of the following cell types most likely contains such particles?

   A. Alveolar macrophages

   B. Ciliated cuboidal cells

   C. Endothelial cells

   D. Goblet cells

   E. Type I pneumocytes

   F. Type II pneumocytes

4. Asthma is a condition in which the mucosal lining of the airways becomes hypersensitive, often resulting in excessive bronchoconstriction of the smooth muscle in the airways. One treatment includes the use of muscarinic receptor antagonists, which cause bronchodilation by blocking the action of

   A. Acetylcholine released from parasympathetic nerves

   B. Acetylcholine released from sympathetic nerves

   C. Norepinephrine released from parasympathetic nerves

   D. Norepinephrine released from sympathetic nerves

5. Drug delivery from the alveolar lumen to the pulmonary circulation is effective due to the thin epithelium and large surface area provided. For example, adrenergic agonists are administered via inhalation for the treatment of asthma. During inhalation, the drug moves down the bronchial tree until it reaches the alveoli. Arrange the following structures in proper sequence for the passage of the drug from an alveolus to an alveolar capillary.

   1 = blood plasma

   2 = fused basal laminae of alveolar and endothelial cells

   3 = surfactant

   4 = cytoplasm of endothelial cell

   5 = cytoplasm of type I pneumocyte

   6 = air in alveolus

   A. $6 - 3 - 4 - 2 - 5 - 1$

   B. $6 - 3 - 5 - 2 - 4 - 1$

   C. $6 - 3 - 5 - 4 - 2 - 1$

   D. $6 - 4 - 2 - 3 - 5 - 1$

   E. $6 - 5 - 3 - 2 - 4 - 1$

6. Which letter on the micrograph below represents the airway in which gas exchange first occurs?

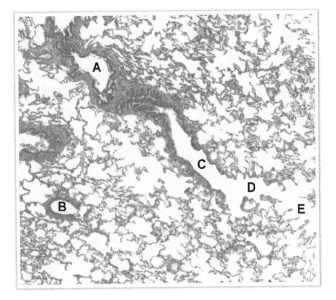

ANSWERS

1—A: The alveolus is the functional unit of the lung because that is where gas exchange occurs.

2—E: Some of the final cells to mature and develop in a fetus are the type II pneumocytes, which are responsible for the production of surfactant. Without surfactant, the alveoli collapse after each breath, resulting in labored breathing by premature infants. Type I pneumocytes form the blood-air barrier.

3—A: Alveolar macrophages (dust cells) are often observed containing black granules of material picked up from the respiratory surfaces (such as dust, pollution, and material from cigarette smoke).

4—A: Parasympathetic nerves are responsible for bronchoconstriction. Smooth muscle lining the airways constricts, resulting in bronchoconstriction. Muscarinic receptor antagonists bind to the acetylcholine receptors on the postsynaptic membrane, thus blocking the action of the parasympathetic acetylcholine neurotransmitters. This would result in reduced or no bronchoconstriction, thus opening up the airways.

5—B: The drug would move from the alveolus, across the surfactant coating on the internal surface of the alveolar sacs, through the cytoplasm of the type I pneumocytes, through the fused basal lamina of the pneumocytes and endothelial cells, through the cytoplasm of the endothelial cell, and into the blood plasma.

6—D: Respiratory bronchioles (D) are the initial components of the respiratory division of the respiratory system, and are the first airway in which gas exchange occurs. Note the presence in airway D of alveoli, where gas exchange occurs. Airways A and B are bronchioles, and airway C is a terminal bronchiole. Airway E is an alveolar duct lined by alveoli, but is distal to D.

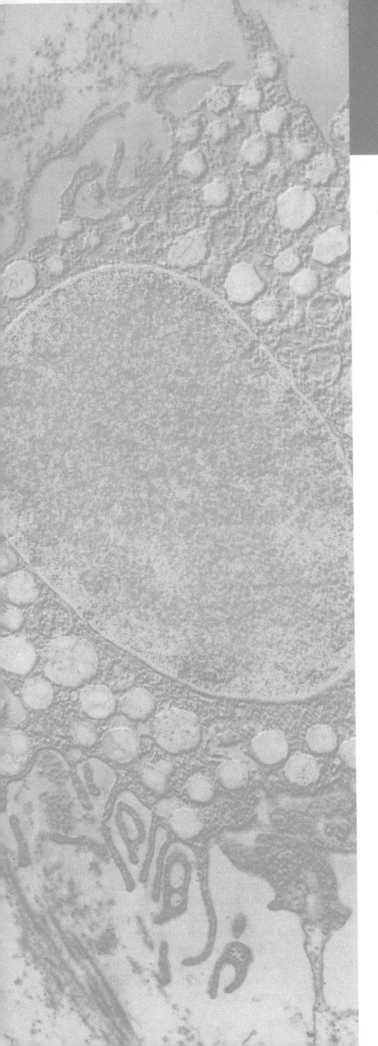

SKIN

OVERVIEW

Skin is the largest organ in the body, constituting approximately 16% of body weight. It forms an essential barrier between the organism and the external environment. A major function of skin, therefore, is **protection** (e.g., against desiccation, bacterial invasion, physical injury, and ultraviolet [UV] radiation). In addition, skin is important for **thermoregulation, synthesis of vitamin D,** and **reception of sensory stimuli.**

The histologic appearance of skin varies from site to site on the body, depending upon the specific function of the skin. However, skin in all regions consists of the following **layers:**

- **Epidermis.** An outer epithelial layer of ectodermal origin that is in contact with the external environment. The epidermis is a continuously renewing tissue composed primarily of cells called **keratinocytes.** In addition, the epidermis contains **melanocytes, Langerhans cells,** and **Merkel cells,** which are specialized for production of melanin, immune protection, and sensory reception, respectively.

- **Dermis.** A deeper layer of connective tissue of mesodermal origin that supports and nourishes the epidermis.

- **Hypodermis.** A subcutaneous layer of loose connective tissue below the dermis that attaches the skin to underlying tissues.

Skin contains various appendages derived from epidermis, including sweat glands, hair follicles and sebaceous glands. Skin is classified as either **thick** or **thin.** Thick skin is found on the palms and the soles. Thin skin is present everywhere else on the body.

▼ Because the skin is readily visible, it is **useful in the diagnosis of genetic and systemic diseases**, which can be confirmed with pathology via minimally invasive techniques. ▼

EPIDERMIS

ORGANIZATION

The epidermis is a **stratified squamous keratinized epithelium** derived from ectoderm, and is composed primarily of a single type of cell called a **keratinocyte**. In addition, three other less abundant types of cells—melanocytes, Langerhans cells, and Merkel cells—are interspersed among keratinocytes.

Skin is classified as either **thick (glabrous)** or **thin (hairy)**, based on the thickness of the epidermis (Figure 11-1A–C). Epidermis in **thick skin**, the type of skin found on the palms, flexor surfaces of the digits, and the soles of feet, is 400 to 600-µm thick. In comparison, epidermis in **thin skin**, which is found everywhere else on the body, is only 75 to150-µm thick.

Throughout the body, the epidermis is specialized to provide a protective barrier between the internal organs and tissues and the external environment. For example, keratinocytes are tightly joined to other keratinocytes by desmosomes and to the basement membrane by hemidesmosomes and integrins. In addition, lipid-rich material extruded from keratinocytes, coupled with an external layer of keratin, creates a nearly impenetrable outer layer of epidermis that keeps microbes out and tissue fluids in.

RENEWAL OF THE EPIDERMIS

The epidermis is subjected to a great deal of wear and tear, and undergoes constant cellular turnover to replace dead or damaged cells. Keratinocytes are produced from stem cells in the innermost layer of the epidermis. They are pushed upward through the epidermis and then die and are sloughed off (**desquamation**) from the outermost layer. The epidermis is replaced about every 3 to 4 weeks, a process that is most rapid in regions that experience the most abrasion, such as the soles of the feet.

The histologic appearance and function of keratinocytes changes dramatically as the keratinocytes differentiate and transit through the epidermis. As keratinocytes mature, they undergo many changes, including enlarging, changing from a cuboidal to a squamous shape, producing increased amounts of keratin protein, synthesizing novel proteins and lipids, losing cytoplasmic organelles, and dying. These changes associated with keratinocyte maturation are manifested in four or five layers in the epidermis of thin and thick skin, respectively.

LAYERS OF THE EPIDERMIS

The five layers of the epidermis, from deep to superficial, are as follows (Figure 11-1C):

1. **Stratum basale (germinativum).** A single basal layer of mitotically active cuboidal keratinocytes, as well as occasional melanocytes and Merkel cells. Scattered stem cells in the stratum basale and lower stratum spinosum produce all of the keratinocytes in the epidermis.

2. **Stratum spinosum or prickle cell layer.** The thickest layer in the epidermis. Keratinocytes in the stratum spinosum range in shape from cuboidal to slightly flattened, but appear spiny in histologic sections due to shrinkage of numerous cytoplasmic projections between cells (Figure 11-1D). Adjacent keratinocytes are joined tightly together by desmosomes at the tips of the projections. Bundles of keratin intermediate filaments converge in the projections and insert into the desmosomes. This network of keratin filaments and desmosomes maintains cohesion among cells and plays an important role in resisting the effects of abrasion.

3. **Stratum granulosum.** Consists of three to five tiers of flattened cells. Keratinocytes in the stratum granulosum produce the following **two types of granules** that contribute essential constituents to the protective barrier function of the epidermis:

   - **Keratohyalin granules.** Irregularly shaped aggregations of proteinaceous material that forms the matrix in which keratin filaments are embedded in the stratum corneum.

   - **Lamellar (membrane-coating) granules.** Membrane-bound organelles that contain a mixture of hydrophobic glycolipids. Lamellar granules fuse with the cell membrane and discharge their contents into the intercellular space between keratinocytes to form a waterproof barrier. This lipid barrier prevents loss of tissue fluids, but also prevents the diffusion of nutrients to more superficial cells, thereby hastening cell death.

     Keratinocytes begin to destroy their cytoplasmic organelles, including the nucleus, in the upper stratum granulosum.

4. **Stratum lucidum.** Consists of several tiers of clear, anucleate, flattened cells and is typically found only in thick skin. The cytoplasm consists almost entirely of keratohyalin protein and tightly packed keratin filaments.

5. **Stratum corneum.** Acellular layer composed of stacks of flattened plates or scales (**squames**) of cross-linked keratin protein and lipids. Although the squames are not living tissue, they perform important functions, including preventing desiccation, invasion by microorganisms, and chemical and physical injury to underlying tissues. The plates are continuously sloughed from the surface of the epidermis and are replaced by differentiation of new cells produced in the stratum basale.

All five layers of the epidermis are present in thick skin. The stratum lucidum is not present in thin skin, and the stratum spinosum, stratum granulosum, and stratum corneum are narrower than in thick skin (Figure 11-1C).

▼ In the epidermis of normal individuals, proliferation and desquamation of keratinocytes are in balance (homeostasis). In patients with **psoriasis**, a common chronic inflammatory skin disease, the proliferation of keratinocytes is greatly accelerated and their maturation is aberrant. This imbalance of proliferation and desquamation results in the accumulation of incompletely differentiated keratinocytes in the stratum corneum, which appear as scaly red plaques, most frequently on the scalp, elbows, and knees. ▼

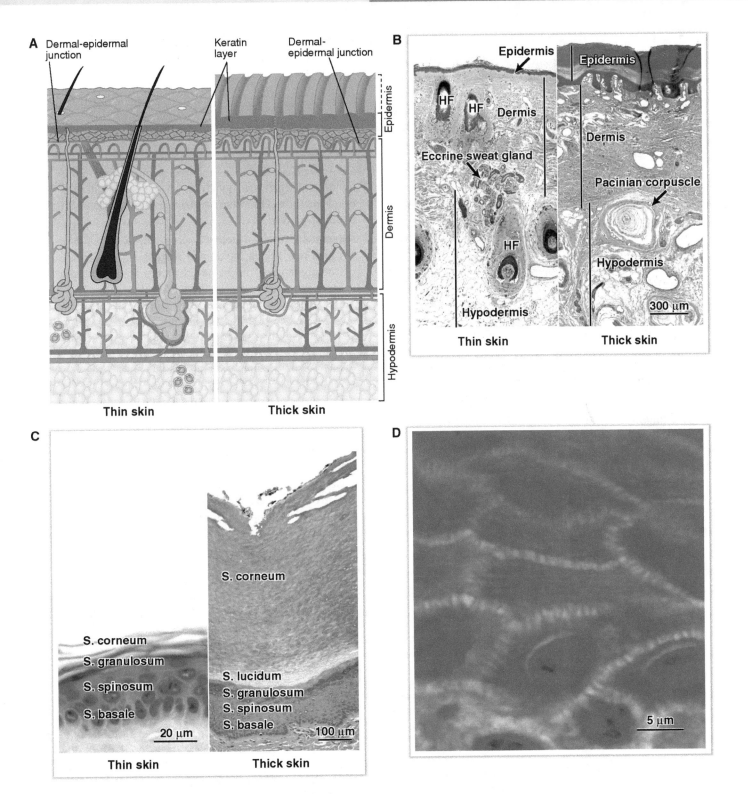

Figure 11-1: Organization of epidermis in thin and thick skin. **A.** Organization and structure of thin and thick skin. **B.** Thin and thick skin from the scalp and fingertip, respectively, of the same individual. HF, hair follicle. **C.** Comparison of the layers of keratinocytes in thin (scalp) and thick (sole of foot) skin. Note that the stratum lucidum is present only in thick skin. Images are at different magnifications. S, stratum. **D.** The stratum spinosum in thick skin (fingertip) shown at high magnification.

MELANOCYTES

Melanocytes are small, pale-staining cells found in the stratum basale and in hair follicles that are derived from neural crest (Figure 11-2A). These cells produce a brown pigment called **melanin**. Melanocytes extend long, thin processes between cells of the stratum spinosum. They do not form desmosomes with neighboring keratinocytes, but are attached to the basement membrane by hemidesmosomes. Melanin is produced from tyrosine in specialized organelles called **melanosomes**, which move into the cellular processes and are transferred to adjacent keratinocytes via **cytocrine secretion**. Thus, nearby keratinocytes usually contain more melanin granules than the melanocytes themselves. Although UV irradiation is essential for the conversion of ergosterol, a derivative of cholesterol, to **vitamin D**, excessive exposure to UV can damage DNA. Melanosomes accumulate over the nucleus of keratinocytes, scattering UV rays and thereby protecting the DNA from UV damage.

The density of melanocytes varies on different parts of the body, ranging from 800/mm² to 2300/mm², but melanocyte density in any particular region is similar among light-skinned and dark-skinned races. Differences in skin color among races are due primarily to variation in the activity of the melanocytes.

▽ **Vitiligo** is a common skin disorder characterized by the partial or complete loss of melanocytes, often by an autoimmune process, which results in patches of unpigmented skin. In contrast, melanocytes are still present in individuals with **albinism**, but they are unable to synthesize melanin due to a lack of or a defect in the enzyme **tyrosinase**. ▽

LANGERHANS CELLS

Langerhans (dendritic) cells are specialized antigen-presenting cells that originate from precursors in the bone marrow, take up residence in the epidermis, and extend long slender cytoplasmic process between keratinocytes, primarily in the stratum spinosum (Figure 11-2A). Langerhans cells serve an important immune function—they phagocytose foreign antigens, migrate to lymph nodes, and present the antigens to T lymphocytes to initiate immune reactions (see Chapter 9).

▽ The number and complexity of Langerhans cells increases in many chronic inflammatory skin diseases, and especially those with an allergic or immune etiology such as **chronic atopic dermatitis**. ▽

MERKEL CELLS

Merkel cells are a population of cells of uncertain functions interspersed among keratinocytes in the stratum basale, primarily in thick skin on the palms and soles (Figure 11-2A). These cells are connected to adjacent keratinocytes by desmosomes, receive afferent nerve terminals on their basal surface, and contain dense-core granules of unknown composition. Merkel cells most likely function as mechanoreceptors, but may be part of the diffuse neuroendocrine system (DNES) (see Chapter 14). They are the source of the rare neoplasm, **Merkel cell carcinoma**.

DERMAL–EPIDERMAL JUNCTION

The epidermis is tightly tethered to the underlying dermis across the basement membrane via a complex structure called the **dermal–epidermal junction**. The dermal–epidermal junction is organized to minimize the risk of separation of the two layers, as follows:

■ **Increased area of contact.** Evaginations of the epidermis, called **rete ridges (epidermal ridges)**, extend downward to interdigitate with finger-like projections from the upper surface of the dermis, called **dermal papillae** (Figure 11-2B and C). These interdigitations greatly increase the area of contact between the epidermis and dermis, and are most prominent in skin regions that experience frequent pressure and friction, such as the fingertips and the soles. In these regions, the interdigitations are visible on the skin surface as the whorls, loops, or arches that comprise a person's fingerprints, and are more accurately called **dermatoglyphs**. In addition, the basal surface of individual keratinocytes and the basement membrane are both convoluted, which further increases the area of contact.

■ **Specialized attachments across the basement membrane.** Cells in the epidermis are firmly joined to the extracellular matrix of the dermis via a complex molecular chain of connections that spans the basement membrane (Figure 11-2D). For example, cells in the stratum basale are attached to the basement membrane by hemidesmosomes and integrins. Hemidesmosomes are linked to the lower layer (the lamina densa) of the basement membrane by **anchoring filaments** composed of **type XVII collagen** and **laminin 5** (laminin 322). The lamina densa is attached to collagen fibers in the upper surface of the dermis by **anchoring fibrils** composed of **type VII collagen** and to elastic fibers in the dermis by **fibrillin microfibrils**.

▽ The dermal–epidermal junction is molecularly complex. Genetic mutations or autoantibodies that disrupt the function of any component of this junction can weaken the attachment of the epidermis to the dermis, resulting in formation of blisters (bullae). The site of the lesion depends on which protein is affected (Figure 11-2D). For example, blisters can occur within the stratum basale (e.g., **epidermolysis bullosa simplex**; keratins 5 and 14, see Chapter 1); between the stratum basale and basement membrane (e.g., **bullous pemphigoid**; hemidesmosome proteins BP230 and BP180); within the basement membrane (e.g., **junctional epidermolysis bullosa**; laminin 5 or type XVII collagen); or between the basement membrane and the dermis (e.g., **epidermolysis bullosa acquisita**; type VII collagen). ▽

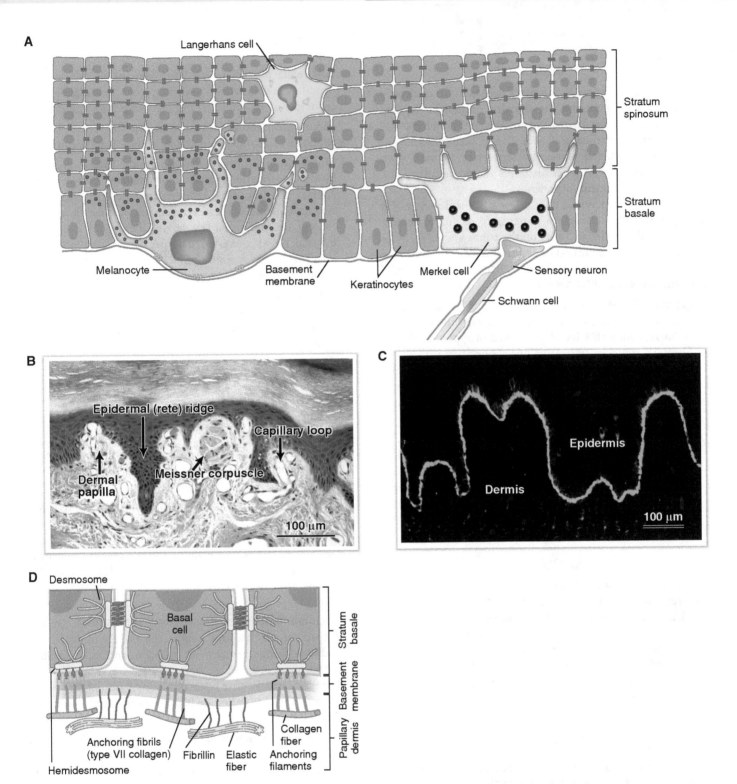

Figure 11-2: Specialized cells in the epidermis and the dermal–epidermal junction. **A.** Location and structure of specialized cells, including a melanocyte, a Langerhans cell, and a Merkel cell, in the epidermis. Note that melanin granules are present in keratinocytes adjacent to the melanocyte. Melanocytes are connected to the basement membrane by hemidesmosomes but do not form desmosomes with keratinocytes. Merkel cells contain dense-core granules, form desmosomes with keratinocytes, and are innervated by afferent (sensory) axons. **B.** Dermal–epidermal junction. Evaginations of the epidermis, called epidermal (rete) ridges, extend downward to interdigitate with projections of the dermis, called dermal papillae. **C.** Dermal–epidermal junction. Fluorescently tagged antibodies (*yellow*) to epitopes in hemidesmosomes label the dermal–epidermal junction in a patient with bullous pemphigoid. Image courtesy of Scott R. Florell, MD, University of Utah School of Medicine, Salt Lake City, Utah. **D.** Dermal–epidermal junction. Keratinocytes in the stratum basale are linked to the extracellular matrix of the dermis via a complex molecular chain of connections that spans the basement membrane. Only some of the constituents of this chain are illustrated here.

DERMIS AND HYPODERMIS

DERMIS

The dermis is the region of skin that lies deep to the epidermis and supports and nourishes the overlying epidermis. The dermis is extensively vascularized and also contains lymphatic vessels, epidermal appendages (sweat glands, hair follicles, and sebaceous glands), and nerves and sensory receptors (Figure 11-3A). In addition, the dermis binds the epidermis to the underlying subcutaneous hypodermis.

Similar to the epidermis, the thickness of the dermis varies among different skin regions, ranging from approximately 0.6 mm on the eyelids to 3 mm on the palms and soles. In all regions, however, the portion of dermis immediately under the basement membrane, the **papillary dermis**, differs histologically from the deeper **reticular dermis** (Figure 11-3B).

- **Papillary dermis.** The outermost portion of the dermis containing the dermal papillae is the papillary dermis. It consists of **loose connective tissue** composed of type III collagen fibers (i.e., reticular fibers) and fine elastic fibers with abundant fibroblasts and other connective cells, including macrophages and mast cells. In addition, capillary loops, free nerve endings, and sensory receptors (Meissner corpuscles and Krause end bulbs) are present in dermal papillae.

- **Reticular dermis.** Below the papillary dermis is a region of **dense irregular connective tissue**, termed the reticular dermis. Cells are sparse in the reticular dermis, but thick bundles of type I collagen fibers and elastic fibers are abundant. In addition, sensory receptors (pacinian corpuscles and Ruffini endings) and epidermal appendages, including sweat glands, sebaceous glands, and hair follicles, are present in reticular dermis.

HYPODERMIS

The undersurface of the dermis is attached to deeper tissues by the **hypodermis** (referred to as **superficial fascia** in gross anatomy), a subcutaneous layer of loose connective tissue containing variable amounts of fat (Figure 11-3A). The loose nature of the hypodermis allows the skin to move over the underlying tissues. In addition, the hypodermis may also contain hair follicles, sweat glands, and sensory receptors (pacinian corpuscles and Ruffini endings).

VASCULAR SUPPLY

The rich vascular supply of the dermis provides nutrients to the epidermis, which is avascular. **Blood vessels form the following two plexuses** in the dermis (Figure 11-3A):

- Subpapillary plexus. Superficially located between the papillary and the reticular dermis, with capillaries looping into dermal papillae to nourish the epidermis (Figure 11-3A and C).

- **Cutaneous plexus.** Located between the dermis and the hypodermis with blood vessels supplying sweat glands, sebaceous glands, hair follicles, and subcutaneous adipose tissue.

Vasculature in the skin also plays an important role in **thermoregulation** by controlling blood flow to superficial dermis.

- Blood flow into superficial vessels and capillaries is regulated by numerous **arteriovenous anastomoses** between deep arteries and veins, and between superficial arterioles and venules. Opening of these shunts in hot weather allows blood to flow into superficial vessels, which promotes cooling. In the cold, blood is diverted to deeper dermis, preventing heat loss. These arteriovenous shunts also play a role in **regulating blood pressure**.

- In the dermis, vascular smooth muscle relaxes in the heat and constricts in the cold. This regulates heat exchange between blood and the environment, which is important for maintaining constant body temperature.

INNERVATION

The skin is richly innervated by both afferent (sensory) and efferent (motor) axons.

- **Free nerve endings.** Terminate between keratinocytes and in the papillary dermis and are important for perception of pain (nociception), itch, and temperature (Figure 11-3A).

- **Meissner corpuscles (tactile corpuscles).** Sensory receptors in dermal papillae, primarily in thick skin of the fingers and palms. Meissner corpuscles consist of several nerve terminals surrounded by stacks of flattened support cells and a connective tissue capsule (Figure 11-3A and C). These corpuscles respond to light touch and are responsible for tactile discrimination.

- **Pacinian corpuscles (lamellated corpuscles).** Onion-shaped, encapsulated mechanoreceptors deep in the dermis and hypodermis, primarily in the palms and soles. Pacinian corpuscles consist of an unmyelinated nerve ending surrounded by concentric layers of modified fibroblasts (Figure 11-3D). These receptors are highly sensitive to vibration and deep pressure.

- In addition, **Merkel disks, Ruffini endings**, and **Krause end bulbs** are specialized sensory receptors in skin that are thought to respond to touch, stretch, and cold, respectively.

- **Autonomic sympathetic innervation.** Efferent (motor) axons control contraction of arrector pili muscles (see below), secretion of sweat glands, and vascular blood flow.

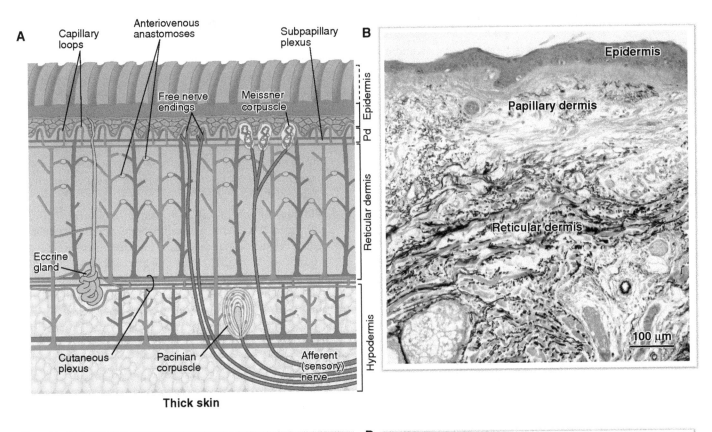

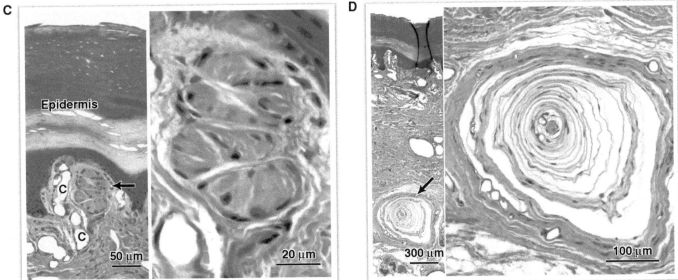

Figure 11-3: Dermis, hypodermis, and sensory receptors. **A.** Thick skin. Pd, Papillary dermis. **B.** Epidermis and dermis in thin skin (scalp). Collagen fibers (*pink*) and elastic fibers (*black*) are thinner in the upper portion of the dermis, termed the papillary dermis, than in the deeper dermis, termed the reticular dermis. (Elastic stain) **C.** Meissner corpuscle. The left panel shows the location of a Meissner corpuscle (*arrow*) in a dermal papilla, which is shown at higher magnification in the right panel. C, capillary loop. **D.** Pacinian corpuscle. The left panel shows the location of a pacinian corpuscle deep in the dermis (*arrow*) which is shown at higher magnification in the right panel.

EPIDERMAL APPENDAGES

Eccrine and apocrine sweat glands, sebaceous glands, and hair follicles are found in the dermis and hypodermis. Sebaceous glands are usually associated with a hair follicle and a bundle of smooth muscle (an arrector pili muscle) in a structure termed a pilosebaceous unit (Figure 11-4). The glands and hair follicles are produced by the proliferation, down-growth, and differentiation of cells in the stratum basale of the epidermis and, therefore, are considered **epidermal appendages**. The three types of glands differ in location, structure, and function.

ECCRINE SWEAT GLANDS

Eccrine sweat glands are widely distributed over the body in both thick and thin skin. They are most numerous on the palms and soles, forehead, scalp, and axilla (arm pit), but are absent from the lips, glans penis, clitoris, labia minora, and nail bed. The secretory portion of eccrine sweat glands is located deep in the dermis or hypodermis.

- **Structure.** Simple coiled tubular glands consisting of an inner layer of cuboidal secretory cells surrounded by an outer layer of myoepithelial cells and a prominent basement membrane. Secretions are conveyed directly to the skin surface via a slender coiled duct composed of stratified cuboidal epithelium (Figure 11-4A–E).

- **Function.** Release a watery product by **merocrine secretion**. Although the electrolyte balance of the fluid produced by secretory cells is similar to blood plasma, almost all K^+, Na^+, and Cl^- is resorbed by the ducts so that sweat released onto the skin is hypotonic. Secretion is regulated by cholinergic sympathetic innervation. Evaporation of sweat cools blood in the superficial dermis, which is important in **thermoregulation**. In addition, cells in the ducts of eccrine glands **excrete** ions, urea, and lactic acid.

APOCRINE SWEAT GLANDS

Apocrine sweat glands are present only in thin, hairy skin of the axilla, the areola, and the anogenital region. Apocrine glands lie deeper in the dermis and hypodermis than do eccrine glands, and are usually present in association with hair follicles.

- **Structure.** Simple coiled tubular glands similar to eccrine glands but with larger lumens. The ducts open into hair follicles rather than directly onto the skin (Figure 11-4A and D).

- **Function.** Release a viscous, proteinaceous product by **merocrine secretion**. Secretions are initially odorless until metabolized by bacteria. Apocrine glands are small and insignificant in childhood and become functional at puberty. Secretion is stimulated by sympathetic adrenergic innervation in response to emotional stimuli and by sex hormones. The function of apocrine sweat gland secretions in humans is unknown. **Ceruminous glands** in the external auditory canal (which produce earwax) and the **glands of Moll** in the eyelids are modified apocrine glands, which function to clean and protect the auditory canal, and help prevent the evaporation of tears, respectively.

SEBACEOUS GLANDS

Sebaceous glands are present in the dermis and hypodermis throughout the body, except on the palms of the hands and the soles and dorsum of the feet; they are particularly abundant on the face and scalp.

- **Structure.** Simple branched acinar glands. Most sebaceous glands are associated with a hair follicle as a component of a **pilosebaceous unit** (Figure 11-4A and E). The duct opens into the hair follicle, except in skin without hair (i.e., on the glans penis, glans clitoris, labia minora, areola of the nipple, eyelids, and lips).

- **Function.** Release an oily, waxy mixture of fatty acids, triglycerides, cholesterol, and cell debris, known as **sebum**. As cells in sebaceous glands mature, they accumulate increasing amounts of sebum. The cells move toward the center of the gland and rupture, releasing their contents into hair follicles via **holocrine secretion**. Stem cells in the base of the gland and in hair follicles proliferate to replace the lost cells. Sebaceous glands begin to function at puberty and are stimulated by androgens in both men and women. Their secretions help lubricate skin and hair and protect against bacterial infection. **Meibomian glands** are specialized sebaceous glands on the eyelids, with secretions that slow the evaporation of tears.

PILOSEBACEOUS UNITS

Pilosebaceous units consist of a hair follicle, an arrector pili muscle, and several associated sebaceous glands.

- **Hair follicles.** Complex multilayered organs in the dermis and hypodermis that produce **hairs**, the filamentous structures composed of fused keratinocytes. Throughout a person's life, each hair follicle undergoes cyclical bouts of growth and hair production (**anagen** phase), degeneration (**catagen**), rest (**telogen**), and regrowth, such that each individual hair grows discontinuously. The duration of the hair-growth cycle varies on different parts of the body. For example, on the eyebrows, the entire cycle is completed in about 4 months, whereas on the scalp, the active anagen phase can be as long as 6 years.

  Hair follicles contain **stem cells** in discrete niches (the **bulge** and **isthmus**) that contribute to the cyclical regeneration of hair follicles, to the normal homeostatic maintenance of sebaceous glands, and to the repair of damaged epidermis after wounding. In humans, unlike other primates, hair contributes little to thermoregulation, but instead appears to serve a mechanosensory function.

- **Arrector pili muscles.** Bands of smooth muscle cells that originate in the connective tissue sheath of the hair follicle and run obliquely to attach to the papillary dermis. Contraction of the arrector pili muscle, which is stimulated by autonomic sympathetic innervation, causes the hair follicle and shaft to become more vertical (**piloerection**), producing "goose pimples" or "goose bumps," and assists in squeezing sebum into the hair follicle.

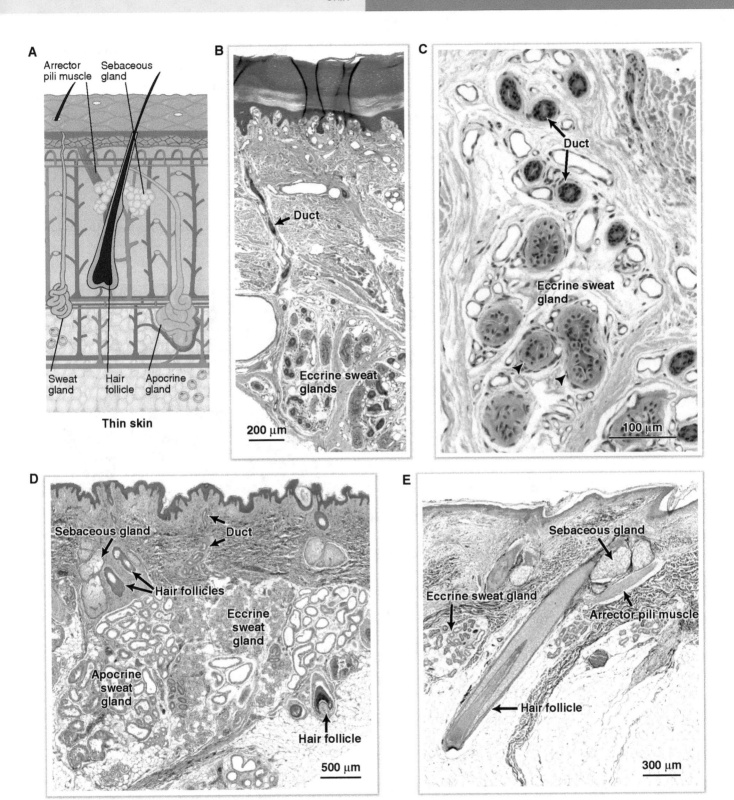

Figure 11-4: Epidermal appendages. **A.** Thin skin. **B.** Eccrine sweat glands in thick skin (fingertip). Eccrine sweat glands are simple coiled tubular glands in the dermis. Note that the ducts of eccrine glands stain more darkly than the secretory portions. **C.** Eccrine gland and duct shown at higher magnification than in part B. Note that the lightly stained cuboidal secretory cells are surrounded by myoepithelial cells, and that the darkly stained coiled ducts are composed of stratified cuboidal epithelium. Arrowheads, nuclei of myoepithelial cells. **D.** Apocrine and eccrine sweat glands, and sebaceous glands in thin skin. Apocrine glands are simple coiled tubular glands located deeper in the dermis and hypodermis than eccrine glands and have larger lumens. The ducts of eccrine sweat glands open directly onto the surface of the skin, whereas the ducts of apocrine glands open into hair follicles, above the ducts of sebaceous glands. **E.** Pilosebaceous unit in thin skin (scalp). A pilosebaceous unit is composed of a hair follicle, arrector pili muscle, and associated sebaceous glands. The ducts of the sebaceous glands open into the hair follicle. (Elastic stain)

STUDY QUESTIONS

Questions 1–3

Directions: Each of the numbered items or incomplete statements is followed by lettered options. Select the **one** lettered option that is **best** in each case.

Refer to the micrograph below to answer the following questions.

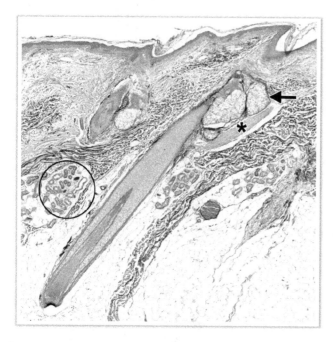

1. A 35-year-old man visits the clinic because he has a cyst on his scalp that has a foul-smelling drainage. The man reports that the drainage occurs spontaneously, and recently the cyst has enlarged and become painful. The cyst is associated with the structure in the micrograph indicated by the arrow. Which mechanism of secretion is normally used by this structure?

   A. Apocrine

   B. Autocrine

   C. Cytocrine

   D. Eccrine

   E. Holocrine

   F. Merocrine

2. What is the composition of the secretion produced by the encircled structures on the micrograph when the secretion is released on the skin surface?

   A. Hypertonic relative to plasma

   B. Hypotonic relative to plasma

   C. Isotonic with plasma

   D. Viscous and odorless

   E. Viscous and waxy

3. What is the predominant tissue in the structure shown in the micrograph indicated by an asterisk?

   A. Dense irregular connective tissue

   B. Dense regular connective tissue

   C. Nervous tissue

   D. Smooth muscle

   E. Striated muscle

Questions 4 and 5

Refer to the micrograph below to answer the following questions.

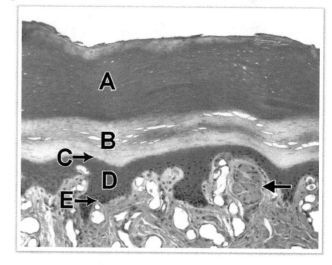

4. Which letter on the micrograph indicates the layer in which mitosis is most commonly observed?

5. What is the function of the structure and indicated by the arrow on the micrograph?

   A. Detection of cold and heat (thermoreception)

   B. Detection of deep pressure

   C. Detection of light touch

   D. Detection of pain (nociception)

   E. Detection of vibration

ANSWERS

1—E: The arrow indicates a sebaceous gland in the section of scalp shown in the micrograph. Sebum is released when cells in the gland rupture and release their contents into hair follicles, a mechanism termed holocrine secretion.

2—B: The circle on the micrograph encloses an eccrine sweat gland in the scalp. Eccrine glands release a watery product by merocrine secretion. Initially, the electrolyte balance of the fluid produced by the secretory cells is similar to blood plasma. However, almost all K^+, Na^+, and Cl^- is resorbed by the ducts, thus sweat released on skin is hypotonic.

3—D: The asterisk on the micrograph indicates an arrector pili muscle in the scalp. These muscles are bands of smooth muscle that originate in the connective tissue sheath of the hair follicle and attach to the papillary dermis. Contraction of an arrector pili muscle causes the hair follicle and shaft to become more vertical (piloerection), producing "goose bumps."

4—E: The epidermis is subjected to a great deal of wear and tear, and undergoes constant cellular turnover to replace dead or damaged cells. Layer E designates the stratum basale (germinativum), a single basal layer composed primarily of mitotically active keratinocytes. Scattered stem cells in the stratum basale and lower stratum spinosum (layer D) produce all of the keratinocytes in the epidermis.

5—C: The arrow on the micrograph indicates a Meissner's corpuscle in a dermal papilla in thick skin on the fingertip. Meissner's corpuscles respond to light touch, and are responsible for tactile discrimination.

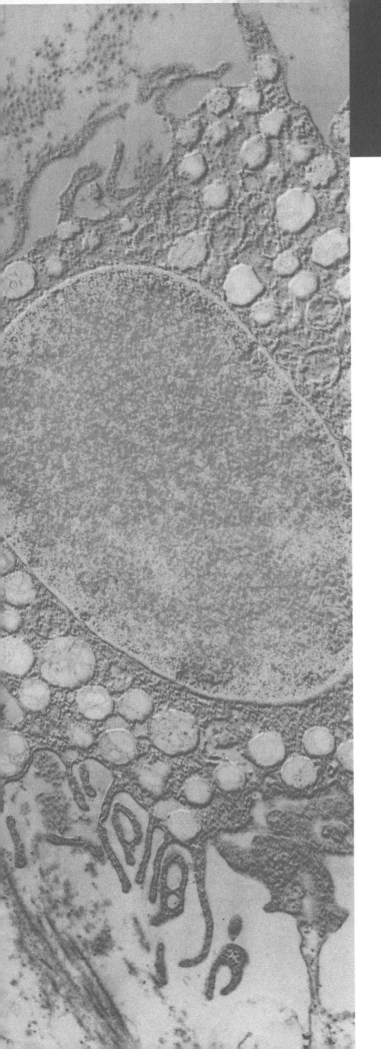

CHAPTER 12

DIGESTIVE SYSTEM

OVERVIEW

The digestive system consists of the **gastrointestinal (GI) tract** (or gut), which includes the oral cavity (mouth), esophagus, stomach, small intestine, and large intestine, and the **accessory organs**, which include the salivary glands, pancreas, liver, and gallbladder. The major function of the digestive system is to acquire nutrient molecules from ingested food and eliminate indigestible remnants. The digestive system also maintains a barrier between the contents of its lumen and body tissues and provides immune protection against infectious disease. The activities of the digestive system are regulated by the enteric nervous system and by both local and systemic hormones.

ORGANS OF THE DIGESTIVE SYSTEM

The digestive system consists of all organs of the GI tract and associated organs that participate in the process of digestion—the breakdown of ingested food into its nutrient molecules, the absorption of those molecules into the blood and lymphatic capillaries, and the elimination of waste products (Figure 12-1A).

GASTROINTESTINAL TRACT

The GI tract is a continuous hollow muscular tube structurally modified into discrete regions and organs that carry out specific functions, as follows:

Oral cavity (mouth). Contains the tongue, teeth, and minor salivary glands. The ducts of the major salivary glands open in the oral cavity, where the process of digestion is initiated.

Pharynx and esophagus. Transport food from the mouth to the stomach.

Stomach. Digests food and secretes hormones.

Small intestine (duodenum, jejunum, ileum). Completes the digestive process and absorbs nutrient molecules.

Large intestine (cecum, colon, rectum, anus). Absorbs water and electrolytes and compacts and eliminates feces.

ACCESSORY ORGANS

The following accessory organs are located outside of the GI tract proper and deliver their secretions, which aid digestion, to the gut via long ducts:

Salivary glands (parotid, submandibular, sublingual). Produce saliva.

Pancreas. Produces digestive enzymes, which act in the small intestine, and hormones, which are important for glucose and lipid metabolism.

Liver. Removes and secretes substances into the blood and produces bile.

Gallbladder. Concentrates and stores bile.

ORGANIZATION OF THE GASTROINTESTINAL TRACT

The wall of the GI tract distal to the pharynx consists of four concentric layers: the **mucosa, submucosa, muscularis externa**, and **serosa** or **adventitia**. The layers are similar throughout the length of the GI tract, but display regional modifications needed to perform specific functions (Figure 12-1B). The GI tract is innervated by enteric and sensory neurons.

LAYERS OF THE GASTROINTESTINAL TRACT

Beginning at the gut lumen and moving outward, the wall of the GI tract is composed of the following four layers:

Mucosa (mucous membrane). Consists of the epithelium that lines the lumen of the GI tract, the underlying lamina propria, and a thin layer of smooth muscle, called the muscularis mucosae.

- **Epithelium.** Simple columnar epithelium, except in the esophagus. The luminal surface may be highly folded to increase its surface area or invaginated to form tubular exocrine glands. The epithelium is composed of **absorptive cells, secretory cells, stem cells**, and **enteroendocrine cells**. The epithelium performs many of the essential functions of the GI tract.

- **Lamina propria.** A layer of loose connective tissue that underlies the basal lamina and supports the epithelium. The lamina propria contains blood and lymphatic vessels, and may also contain lymphatic aggregates, plasma cells, and glands. Plasma cells secrete antibodies, primarily IgA, into the intestinal lumen, which provides immune defense against viruses and bacteria.

- **Muscularis mucosae.** A thin, inner circular layer and an outer longitudinal layer of smooth muscle. Contractions of the muscularis mucosae enable local movements of the mucosa independent of other movements of the GI tract and increase contact of the mucosal layer with food.

Submucosa. A layer of dense, fibroelastic connective tissue with numerous blood vessels and lymphatics and **Meissner's submucosal nerve plexus**.

Muscularis externa. A thick layer of smooth muscle organized into an inner circular sublayer and an outer longitudinal sublayer. **Auerbach's myenteric nerve plexus** is sandwiched between the sublayers. Contractions of the muscularis externa generate peristaltic waves that mix and propel food in the GI tract.

Serosa or adventitia. A thin layer of loose connective tissue.

INNERVATION OF THE GASTROINTESTINAL TRACT

The motility and secretory activity of the GI tract are controlled by the **enteric nervous system**, a local, self-contained system consisting of millions of autonomic neurons organized into small ganglia in the submucosa (**Meissner's submucosal plexus**) and muscularis externa (**Auerbach's myenteric plexus**). The enteric nervous system functions largely independent of the central nervous system, but is modulated by the parasympathetic and sympathetic nervous systems. In general, the submucosal plexus governs the activity of glands and smooth muscle in the mucosa, whereas the myenteric plexus primarily controls the activity of the muscularis externa. Sensory neurons in the wall of the GI tract convey information about luminal contents, muscular contraction, and the secretory activity of the gut to nearby plexuses.

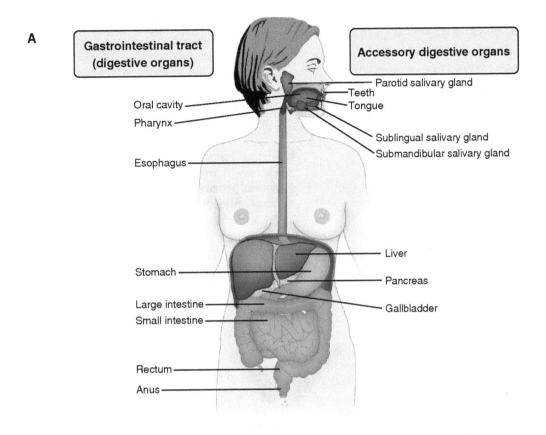

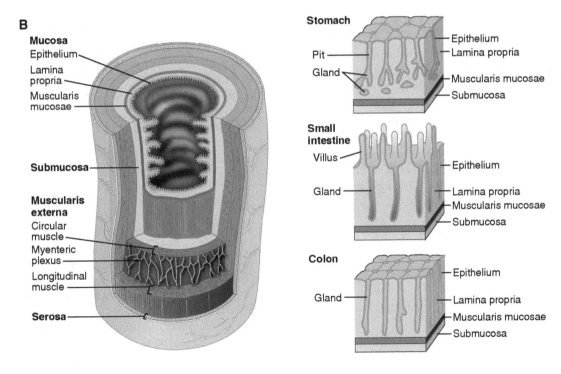

Figure 12-1: Location and histologic organization of the organs of the digestive system. **A.** Location of the organs of the gastrointestinal (GI) tract and accessory digestive organs. **B.** The figure on the left shows the overall organization of the wall of the organs of the gastrointestinal tract. The three panels on the right compare the arrangement of epithelium in the stomach (pits and glands), small intestine (villi and glands), and colon (glands only).

ORAL CAVITY

In contrast to more distal regions of the GI tract, the oral cavity (mouth) is lined with stratified squamous keratinized or partially keratinized (parakeratinized) epithelium, and has minor salivary glands in the mucosa and submucosa. Ducts of the major salivary glands also open in the oral cavity. Secretions of the salivary glands mix together in the oral cavity, initiating the chemical breakdown of starch. The mouth houses the tongue and teeth, accessory organs important for the mechanical breakdown of ingested food.

TONGUE

The tongue is a muscular organ in the oral cavity that is essential for speech and for mastication of food (Figure 12-2A). It is composed of interlacing bundles of striated skeletal muscle interspersed with adipose tissue, which provides great mobility for the manipulation of food, swallowing, and speech. The ventral surface of the tongue is covered by nonkeratinized stratified squamous epithelium, whereas the dorsal surface is covered by keratinized or parakeratinized epithelium (Figure 12-2B).

The anterior two-thirds of the dorsal surface of the tongue bears peg-like projections of the mucosa, called **lingual papillae**, which serve both mechanical and taste functions (Figure 12-2A–D). The following types of papillae have been identified:

- **Filiform papillae.** Narrow, conical structures found over the entire anterior two-thirds of the tongue. Filiform papillae lack taste buds but provide a roughness to the surface that produces friction for manipulating food in the mouth.
- **Fungiform papillae.** Mushroom-shaped structures scattered over the surface of the tongue and most numerous near the tip of the tongue. Fungiform papillae contain taste buds, primarily on their dorsal surface.
- **Circumvallate papillae.** Seven to twelve large, flattened domes, each surrounded by a narrow furrow and arranged in a V-shaped row anterior to a groove in the tongue, called the **sulcus terminalis**. Taste buds are located primarily in the lateral walls of the circumvallate papillae. **Von Ebner's glands**, minor salivary glands in the lamina propria, discharge serous secretions into the channels at the base of the papillae, which wash the taste buds, allowing them to receive and process new stimuli (Figure 12-2C).

The dorsal surface on the posterior third of the tongue (the **root**) lacks papillae, but is bumpy because **lingual tonsils** and smaller lymphoid nodules lie just deep to the mucosa (see Chapter 9).

TASTE BUDS Taste buds are intraepithelial sensory receptors that function in the perception of taste. They consist of an oval cluster of 50–80 spindle-shaped cells, including both **taste cells** and **support cells** (Figure 12-2C). Microvilli on the apical end of taste cells project into the oral cavity through an opening in the epithelium, called a **taste pore**, allowing molecules dissolved in the saliva to contact cell-surface taste receptors. Reception of taste stimuli activates afferent sensory fibers on the basal surface of taste cells. Taste cells turn over rapidly, with an average life span of 7–10 days, and are replaced by stem cells located basally in the taste buds.

TEETH

Teeth function to tear and grind food into smaller fragments, exposing more surface area to digestive enzymes. Adult humans have 32 **permanent teeth**, 20 of which are preceded by **primary** or **deciduous teeth** (also called baby teeth), which are shed between the ages of 6 and 12 years. Teeth are classified according to their shape and function as incisors, canines, premolars, and molars. Regardless of shape, each tooth consists of the following regions:

- **Crown.** The exposed portion above the gum. The crown is covered with **enamel**, the hardest material in the body. Enamel is composed almost entirely of **hydroxyapatite** and is similar to the calcified matrix of bone but without collagen.
- **Neck.** The constricted region that joins the crown to the root.
- **Root(s).** The portion that extends below the gums. Roots are covered with **cementum**, a calcified connective tissue, and are anchored in sockets in the jawbone (**alveoli**) by **periodontal ligaments**. An opening at the tip of each root, called an **apical foramen**, allows blood vessels and nerves to enter the tooth (Figure 12-2E).

The bulk of the tooth is formed by another calcified material, **dentin**, which surrounds a central **pulp cavity** containing soft connective tissue, blood vessels, and nerve fibers, collectively called **pulp**. Pulp supplies nutrients to the tooth tissues and provides sensation. The pulp cavity is lined with **odontoblasts** just deep to dentin, which secrete and maintain dentin throughout life. In contrast, the cells that produce enamel in developing teeth (**ameloblasts**) degenerate after the tooth emerges; thus, decayed or cracked areas of the tooth do not heal and must be restored with an artificial substance.

▽ **Dental carries** (cavities) are caused by mineral dissolution of the teeth by acids generated by bacteria in the oral cavity. Because of improved oral hygiene and the addition of fluoride to drinking water, caries have been reduced in developed countries. Fluoride integrates into the enamel and forms fluoroapatite, which protects the teeth from the harmful effects of bacterial acids.

Other than causing tooth decay and halitosis, the bacteria in the oral cavity are considered benign. However, in individuals with abnormalities of the heart valves, these bacteria may cause **infective endocarditis**. Minor breaks in the oral mucosa can allow bacteria access to the circulatory system and the opportunity to colonize damaged valves. The most frequent cause of infective endocarditis is a group of commensal bacteria called *Streptococcus viridans.* ▽

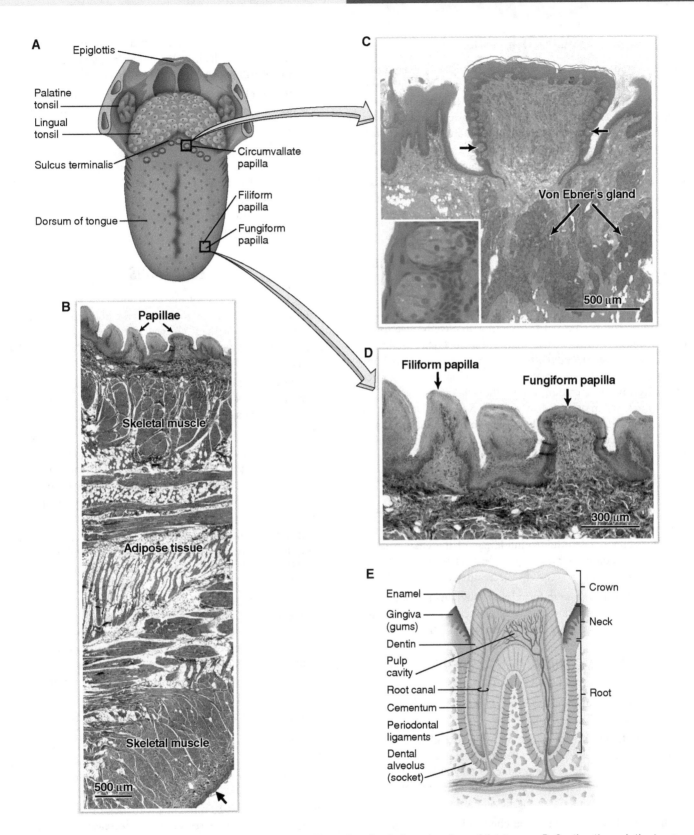

Figure 12-2: Histologic organization of accessory structures in the oral cavity. **A.** Dorsal surface of the tongue. **B.** Section through the tongue. Note the numerous papillae on the dorsal surface. The arrow indicates nonkeratinized stratified squamous epithelium on the ventral surface. **C.** Section through a circumvallate papilla on the dorsal surface of the tongue. Arrows point to taste buds, shown in the inset at higher magnification. **D.** Section through filiform and fungiform papillae on the dorsal surface of the tongue. Note that the lamina propria extends into the papillae. **E.** Longitudinal section of a molar tooth.

SALIVARY GLANDS

Salivary glands are accessory organs of the digestive system that produce **saliva**, a viscous solution that moistens and lubricates the oral cavity, initiates digestion, and provides antibacterial protection. There are three pairs of major salivary glands located outside the oral cavity, the **parotid, submandibular**, and **sublingual glands**, as well as numerous minor glands located in the oral mucosa and submucosa. Approximately 1.5 L of saliva is produced daily. Salivary glands are composed of varying proportions of **serous cells**, which secrete a watery mixture of digestive enzymes and ions, and **mucous cells**, which secrete mucus.

ORGANIZATION OF THE MAJOR SALIVARY GLANDS

All of the major salivary glands are **branched compound acinar** or **tubuloacinar glands** in which clusters of acini drain into a duct system that ultimately empties in the mouth (see Chapter 1). Acini are composed of serous cells, whereas tubules are composed of mucous cells. Each gland is covered with a connective tissue capsule, and is divided into lobes and lobules by **septa** originating from the capsule. **Myoepithelial cells** between the epithelium and basal lamina of secretory units and the initial part of the duct system help propel secretions toward the major ducts.

The duct system of the salivary glands consists of small ducts within lobules (**intralobular ducts**) that empty into larger ducts in the connective tissue between the lobules (**interlobular ducts**), and is organized as follows (Figure 12-3A):

- Secretory acini empty into **intercalated ducts** (intralobular) lined by cuboidal epithelial cells.

- Intercalated ducts merge to form **striated ducts** (intralobular), so named because the basal portion of the columnar duct cells appear striated due to infoldings of the basal plasma membrane and the accumulation of mitochondria. Epithelial cells in striated ducts resorb NaCl, creating hypotonic saliva.

- Intralobular ducts fuse to form larger **interlobular ducts** lined by columnar or stratified cuboidal epithelium, which fuse to form the main **excretory duct** that empties into the oral cavity.

DISTINGUISHING CHARACTERISTICS OF THE MAJOR SALIVARY GLANDS

The three major salivary glands can be identified by the following features:

- **Parotid glands.** Branched compound acinar glands located in each cheek below the ear. Parotid glands are composed entirely of serous acini that produce α-**amylase**, which initiates hydrolysis of carbohydrates, and **proline-rich proteins**, which have antibacterial properties (Figure 12-3B).

- **Submandibular glands.** Branched tubuloacinar glands located on either side of the neck below the mandible. Submandibular glands are composed primarily of serous acini, with approximately 10% containing mucous tubules capped with serous cells, a structure termed a **serous demilune**

(Figure 12-3C). In addition to α-amylase and proline-rich proteins, serous cells in submandibular glands produce the enzyme **lysozyme**, which hydrolyzes bacterial walls.

- **Sublingual glands.** Branched tubuloacinar glands located on the floor of the mouth. Sublingual glands are composed primarily of mucous cells, with serous cells present only in demilunes (Figure 12-3D).

▽ Individuals who have reduced saliva production as a result of disease or radiotherapy typically have dental caries and halitosis due to bacterial overgrowth, and also have difficulty swallowing and speaking due to inadequate lubrication. ▽

ESOPHAGUS

The esophagus is a straight muscular tube that transports food from the oral cavity to the stomach via the pharynx without significant metabolic changes. The wall of the esophagus possesses the same major layers as other regions of the GI tract and has the following distinguishing characteristics (Figure 12-3E):

- **Mucosa.** Lined by abrasion-resistant, **nonkeratinized stratified squamous epithelium**. Mucus-secreting **esophageal cardiac glands** are present in the lamina propria near the pharynx and the stomach.

- **Submucosa.** Contains mucus-secreting tubuloacinar glands called **esophageal glands**. The mucus lubricates the walls and aids in the passage of food. Aggregates of lymphoid cells are common in the submucosa near the gastroesophageal junction.

- **Muscularis externa.** Composed of both skeletal and smooth muscle. The proximal third is exclusively skeletal muscle; the middle third is both skeletal and smooth muscle; and the distal third is exclusively smooth muscle. Therefore, transport of food down the esophagus begins with voluntary peristalsis (skeletal muscle) and ends with involuntary peristalsis (smooth muscle). Tonic contractions of the muscularis externa at the upper and lower ends of the esophagus create **functional sphincters** that prevent reflux into the pharynx from the esophagus and into the esophagus from the stomach, respectively.

- **Adventitia/serosa.** The thin outer layer of the esophagus.

▽ The lower esophageal sphincter occasionally fails to close adequately and allows acidic gastric juices to flow back into the esophagus and attack the mucosa. The resulting inflammation and pain is known as **heartburn**, or **gastroesophageal reflux disease (GERD)**. Smoking, pregnancy, obesity, hiatal hernia, and smooth muscle relaxants can decrease the esophageal sphincter tone and predispose an individual to GERD. Recurrent reflux of gastric acids can cause esophagitis and induce dysplastic changes in the mucosa, which may result in adenocarcinoma of the lower esophagus. ▽

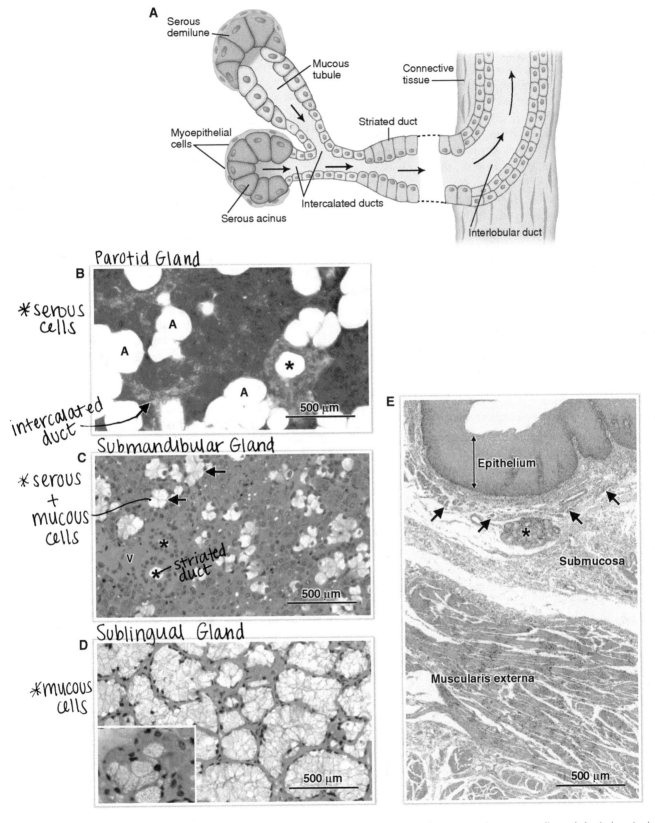

Figure 12-3: Structure of the salivary glands and the esophagus. **A.** Arrangement of serous and mucous cells and ducts in a typical salivary gland. Serous cells are found in acini and demilunes, whereas mucous cells are organized into tubules. Arrows indicate the direction of saliva flow. **B.** Section of a parotid gland. All secretory cells are serous (*dark purple*). The arrow indicates an intercalated duct; the asterisk indicates a small striated duct. A, adipocyte. **C.** Section of a submandibular gland composed of both serous (*pink*) and mucous cells (*arrows*). The asterisk indicates a striated duct. V, venule. **D.** Section of a sublingual gland composed primarily of mucous cells. The inset shows serous demilunes at higher magnification. **E.** Section through the wall of the esophagus. Note that the muscularis externa is exclusively smooth muscle, indicating that this section is taken from the distal portion of the esophagus. Arrows indicate the muscularis mucosae; the asterisk indicates an esophageal gland.

STOMACH

The stomach is a highly distensible, temporary storage organ where ingested food is mixed with gastric secretions to produce an acidic paste called **chyme**, which is released into the duodenum in small aliquots via the **pyloric sphincter**. Little or no absorption of food occurs in the stomach. The stomach is both an exocrine organ, secreting hydrochloric acid (HCl) and various digestive enzymes, and an endocrine organ, secreting hormones that regulate its function. It is divided into four anatomic regions, the **cardia**, **fundus**, **body**, and **pylorus**, three with distinctive histology (Figure 12-4A). The wall of the stomach is built from the same four layers as other regions of the GI tract. The mucosa and muscularis are uniquely modified for their specific roles, but other layers are unremarkable (Figure 12-4B).

GASTRIC PITS AND GLANDS

The stomach is lined with simple columnar epithelium composed entirely of mucous cells. This otherwise smooth lining is punctuated by millions of deep **gastric pits (foveolae)**. Multiple branched, tubular **glands** empty into the pits. Although both gastric pits and glands are formed by invagination of the epithelium into the lamina propria, these two regions of the mucosa can be distinguished as follows:

Gastric pits. Composed primarily of mucous cells, and are relatively uniform throughout all regions of the stomach.

Glands. Composed of various secretory and endocrine cells. The types of cells that comprise the glands, as well as the relative length of pits and glands, vary among stomach regions, reflecting important functional differences in the regions.

CELLS IN THE GASTRIC EPITHELIUM

The major functions of the gastric epithelium are secretion of HCl and enzymes for digestion and mucus to protect itself from the corrosive effects of the acid and enzymes. The mucosa consists of diverse cells necessary to perform these functions, as follows (Figure 12-4B and C):

Mucous surface cells. Cover the surface mucosa and line the pits (Figure 12-4B). Mucous surface cells secrete viscous **alkaline** mucus, which adheres to the epithelial surface and traps bicarbonate-rich fluid beneath, protecting the stomach lining from the harsh acidic contents in its lumen. Mucous surface cells are replaced approximately every 3 days.

Mucous neck cells. Line the entire gland in the cardia and pylorus. In the body and fundus, mucous neck cells are found primarily in the upper portion (**neck**) of the gastric glands, interspersed between parietal cells. These cells produce solu-

ble **acidic** mucus that mixes with and lubricates chyme as it is propelled along the GI tract.

Parietal cells. Large, round eosinophilic cells found primarily in the upper half of the gastric glands in the body and fundus. Parietal cells secrete HCl into the lumen. About 1–1.5 L of gastric juice is secreted daily, containing 0.4–0.5% HCl with a pH of 2. The export of protons occurs via an H^+, K^+–ATPase (proton pump), which functions by a mechanism similar to that of the sodium pump described in Chapter 1. In inactive cells, the pumps are stored in internal vesicles, which fuse with the surface when the cells are active, permitting H^+ export. Parietal cells also secrete the glycoprotein **gastric intrinsic factor**, which is essential for absorption of vitamin B_{12} in the ileum. The loss of parietal cells results in **pernicious anemia** caused by poor uptake of vitamin B_{12}. Secretory activity of the parietal cells is controlled by cholinergic parasympathetic stimulation and by **histamine** and **gastrin** produced by local enteroendocrine cells.

Chief (zymogenic, peptic) cells. Located in the base of the gastric glands and easily recognized by the presence of numerous eosinophilic secretory granules. Chief cells produce proteolytic enzymes, mainly **pepsinogen**, which is converted to its active form, **pepsin**, by gastric acid, as well as the enzyme **lipase** and the hormone **leptin**.

Enteroendocrine cells. Located in epithelium throughout the stomach and all other regions of the GI tract, and are considered part of the **diffuse neuroendocrine system (DNES)** (see Chapter 14). Enteroendocrine cells release their secretory products basally into the lamina propria, rather than apically into the lumen. Products of the enteroendocrine cells in the stomach include the following:

- **Gastrin.** Stimulates parietal cells to secrete HCl.
- **Glucagon.** Stimulates glycogenolysis in the liver.
- **Histamine.** Stimulates parietal cells to secrete HCl.
- **Somatostatin.** Inhibits the release of hormones produced locally; in the stomach this includes gastrin, glucagon and histamine.
- **Serotonin.** Increases gut motility.
- **Ghrelin.** Released when the stomach is empty and stimulates feelings of hunger.

Stem cells. Highly proliferative cells in the neck of the gland. Epithelial cells can only survive a few days in the stomach's harsh environment. The stem cells divide and replace the entire epithelium weekly. Newborn cells migrate up into the pits and differentiate into mucous surface cells, or migrate down into the glands and differentiate into mucous neck, parietal, chief, and enteroendocrine cells.

A

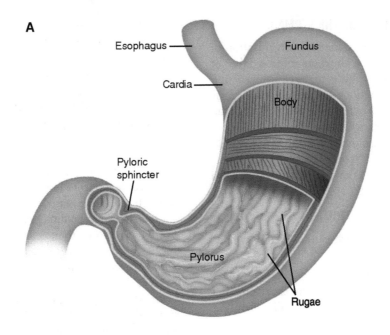

Esophagus

Fundus

Cardia

Body

Pyloric
sphincter

Pylorus

Rugae

B

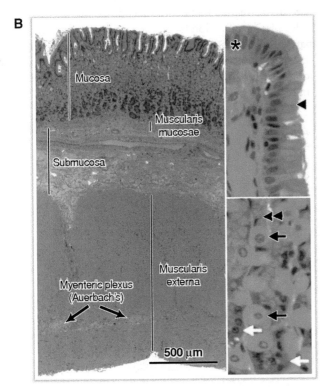

Mucosa

Muscularis
mucosae

Submucosa

Myenteric plexus
(Auerbach's)

Muscularis
externa

500 µm

C

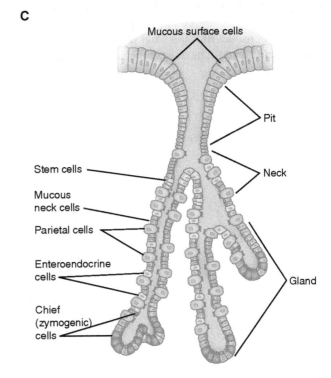

Mucous surface cells

Pit

Neck

Stem cells

Mucous
neck cells

Parietal cells

Enteroendocrine
cells

Gland

Chief
(zymogenic)
cells

Figure 12-4: Organization of the stomach. **A.** Cut-away drawing of the stomach showing the gross anatomic regions, the organization of smooth muscle layers in the muscularis externa, and the mucosal folds (rugae). **B.** Section through the body of the stomach showing the overall organization of layers in the wall. The inset on the upper right, at higher magnification, shows mucous surface cells covering the luminal surface (*asterisk*) and lining a pit (*arrowhead*). The inset on the lower right shows cells in a gastric gland at higher magnification. Mucous neck cells (*double arrowhead*), parietal cells (*black arrow*), and chief (zymogenic) cells (*white arrows*) can be readily distinguished. **C.** Small region of gastric epithelium showing the relationship between a gastric pit and the glands that empty into the pit and the types of cells that comprise the gastric epithelium. (*continued on page 181*)

MUSCULARIS EXTERNA OF THE STOMACH

The muscularis externa differs from the basic organization of the GI tract by the presence of a third, or middle, layer of oblique muscle fibers, which contributes to the churning action that mixes food with gastric secretions. In addition, the circular layer is thickened in the pylorus to form the **pyloric sphincter**. Contraction of the muscularis when the stomach is empty causes longitudinal folds in the mucosa and submucosa called **rugae**, which allow the stomach to expand to accommodate food and gastric juices and which flatten when the stomach is full (Figure 12-4D).

REGIONS OF THE STOMACH

The following regions of the stomach can be identified anatomically:

Cardia. A narrow, circular region about 1.5–3 cm wide at the gastroesophageal junction, characterized by short pits and mucus-secreting **cardiac glands**, similar to cardiac glands in the esophagus.

Fundus and body. Histologically similar regions characterized by long **gastric (fundic) glands** and relatively short pits. Gastric glands are composed primarily of **parietal cells**, which produce **HCl**, and **chief (zymogen) cells**, which produce **pepsinogen**. The body is the largest region of the stomach.

Pylorus. A funnel-shaped region that opens into the small intestine at the **pyloric sphincter**, which controls stomach emptying. The pylorus is characterized by long pits and mucus-secreting **pyloric glands**, similar to those in the cardia. The two regions can be distinguished, however; in the pylorus, the pits are long relative to the glands, whereas in the cardia, the pits and glands are approximately the same length (Figure 12-4E).

The stomach lining normally is protected from the acidic contents of its lumen by a thin layer of mucus. If this barrier breaks down, gastric acid can attack and erode the stomach wall, causing a **peptic (gastric) ulcer**, which is defined as an erosion of greater than 5 mm in diameter that penetrates through the muscularis mucosae. Peptic ulcers are extremely painful. If the ulcer perforates, the entire stomach wall can be involved, resulting in peritonitis, hemorrhage, and possibly death. Most recurring peptic ulcers are caused by *Helicobacter pylori*, a bacterium that subsists in the human digestive tract and can destroy the protective mucus layer. Antibiotics are now used in the treatment of peptic ulcer disease as well as drugs that reduce HCl production by blocking the histamine receptors on parietal cells or inhibiting the H+, K+–ATPase of these cells. The few types of ulcers not caused by *H. pylori* generally result from long-term use of nonsteroidal anti-inflammatory drugs (e.g., ibuprofen).

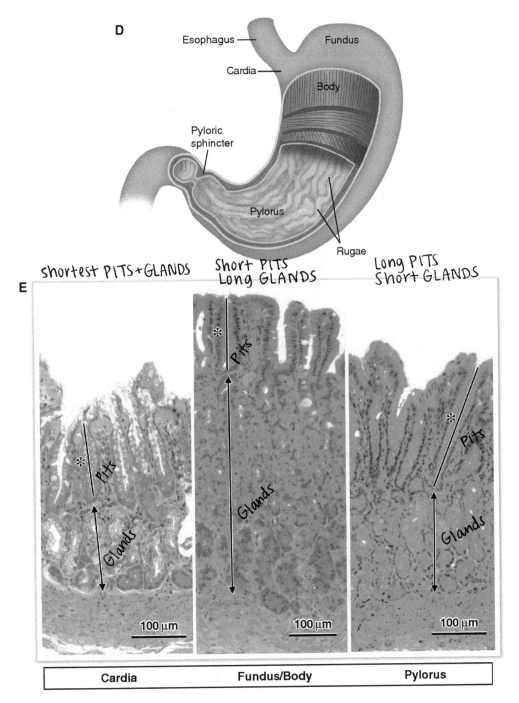

Figure 12-4: (*continued*) **D.** Cutaway drawing of the stomach showing the gross anatomic regions, the organization of the smooth muscle layers in the muscularis externa, and the muscosal folds (rugae). **E.** Sections through the cardia, fundus (or body), and the pylorus of the stomach. Note the differences in overall thickness, the relative lengths of the gastric pits (*asterisks*) and glands (*double-headed arrows*), and the relative numbers of mucous, parietal, and chief cells in the three regions.

SMALL INTESTINE

The small intestine is the region of the GI tract where digestive processes are completed and nutrients are absorbed. The small intestine also produces hormones that regulate diverse digestive activities. The terminal digestion of food requires enzymes produced by the pancreas and bile produced in the liver, as well as buffers and enzymes produced by epithelial cells and glands of the intestinal mucosa itself. The small intestine is the longest region of the GI tract (6–7 m long), extending from the pyloric sphincter to the large intestine, and is divided into three segments: the **duodenum**, **jejunum**, and **ileum**.

MUCOSA

The mucosa of the small intestine is typical of the GI tract, but possesses three modifications that increase the surface area over which absorption occurs, **plicae circulares**, **villi**, and **microvilli**, as illustrated in Figure 12-5A. As in the stomach, the mucosa of the small intestine possesses two histologically distinct regions—absorptive villi, which project into the lumen, and simple tubular (or branched tubular) **glands** called **crypts of Lieberkühn**, which project into the lamina propria and open between the villi. The lamina propria extends into the villi and surrounds the glands.

STRUCTURE AND FUNCTION OF VILLI The villus is the site of both terminal digestion and absorption of foodstuffs, and the mucosa of villi is specialized to perform these functions, as follows:

- **Epithelium.** Composed primarily of **absorptive columnar cells (enterocytes)** interspersed with some **goblet cells.**

- **Lamina propria.** Fills the core of villi and contains a rich capillary network, a central lymphatic vessel called a **lacteal**, and strands of smooth muscle.

- **Enterocytes.** Possess a dense array of microvilli on their luminal surface, termed the **brush border**. The brush border is covered with a glycocalyx containing various digestive enzymes (**brush-border enzymes**), which function to complete the digestion of carbohydrates and proteins. This arrangement allows enterocytes to hydrolyze macromolecules in the intestinal lumen, absorb the resulting amino acids and monosaccharides through sodium-linked cotransport, and release them into blood and lymph vessels in the lamina propria (see Chapter 1 for a discussion of solute transport).

- **Lacteals.** End blindly at the tip of the villus and function to absorb digested lipids. Pancreatic lipase and bile from the liver are required for lipid digestion; absorbed lipids are processed into **chylomicra** and are preferentially taken up by lacteals rather than by blood capillaries.

- **Smooth muscle.** Oriented vertically and attached to the basal lamina at the tip of the villus. Contraction of smooth muscle causes rhythmic movement of villi, which may aid in the transport of lymph from the lacteal and help mix the layer of fluid outside the epithelial cells.

Vessels in the villi eventually drain into the large blood and lymph vessels in the submucosa. The nutrient molecules carried by the vessels are transported to the liver, where they are used in the synthesis of essential proteins, carbohydrates, and lipids.

STRUCTURE AND FUNCTION OF GLANDS The epithelium of the intestinal glands is continuous with the epithelium of the villi, but is primarily involved in the secretion of substances into the intestinal lumen and renewal of the epithelium, rather than digestion and absorption of material from the lumen. Epithelial cells in the gland secrete a watery mixture containing mucus that serves as a carrier fluid for absorbing nutrients from chyme and lubricates the lumen. The epithelium of the glands is composed of the following cells (Figure 12-5A):

- **Goblet cells.** Secrete glycoproteins that lubricate the luminal surface and retain antimicrobial molecules such as defensins and secretory IgA near the epithelial surface. Goblet cells are least abundant in the duodenum and become more numerous toward the large intestine.

- **Paneth cells.** Located at the base of the intestinal glands and provide a protective function. Paneth cells secrete antimicrobial agents such as lysozyme, defensins, and phospholipase A_2. They are readily identified in histologic sections by their large eosinophilic secretory granules (Figure 12-5C).

- **Enteroendocrine cells.** Present in varying numbers throughout the small intestine and produce hormones that regulate diverse digestive activities, including the following:

  - **Cholecystokinin (CCK).** Stimulates the secretion of pancreatic enzymes; causes the gallbladder to contract and expel bile; and relaxes the hepatopancreatic sphincter to allow bile and pancreatic juice into the duodenum.

  - **Gastric inhibitory peptide (GIP).** Inhibits secretion from the gastric glands.

  - **Motilin.** Increases gut motility.

  - **Secretin.** Stimulates HCO_3^- secretion by the pancreatic ducts; increases bile output.

  - **Somatostatin.** Inhibits local release of CCK.

  - **Peptide YY.** Released in response to eating and reduces appetite.

- **Undifferentiated stem cells.** Present near the base of the glands and produce transit-amplifying progenitor cells that give rise to all cell types in the intestinal epithelium. The rapidly dividing progenitor cells line much of the walls of the glands. Proliferation stops at the top of the glands, and progenitors differentiate into enterocytes, goblet, or enteroendocrine cells and migrate toward the tip of the villus. Enterocytes and goblet cells of villi are replaced weekly; Paneth cells, which lie deep in the glands, are longer-lived and are replaced about every 30 days.

A

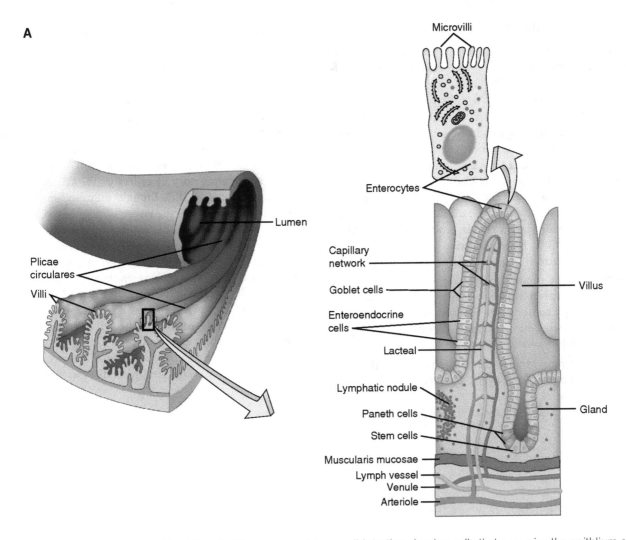

Figure 12-5: Organization of the small intestine. **A.** The mucosa of the small intestine showing cells that comprise the epithlium and the modifications that increase surface area: the plicae circulares, villi, and microvilli. Plicae circulares are permanent folds in the mucosa and submucosa, up to 1 cm high, arranged circularly around the lumen. Villi are finger-like extensions of the mucosa, about 0.5–1.5 mm long, and are the functional absorptive units of the small intestine. Microvilli are present on the luminal surface of all enterocytes, and are the primary sites of terminal digestion and absorption. (*continued on page 185*)

REGIONAL DIFFERENCES

The gross anatomic regions of the small intestine can be distinguished microscopically based on a few obvious structural differences in the mucosa and submucosa. The muscularis externa is similar throughout and is typical of the GI tract.

Duodenum. The submucosa contains **duodenal glands (Brunner's glands)**, compound tubular glands that secrete alkaline mucus that is rich in HCO_3^- (Figure 12-5B). The mucosal epithelial cells secrete almost all of the HCO_3^- that directly neutralizes the acidic chyme propelled from the stomach. When the mucus and HCO_3^- barrier fail, the intestinal wall can erode, resulting in a duodenal ulcer. Enteroendocrine cells in Brunner's glands produce **urogastrone**, a peptide that inhibits HCl production by the parietal cells. The bile and pancreatic ducts open into the duodenum.

Jejunum. Has the most highly developed plicae circulares, and is the main absorptive site in the GI tract. However, a typical histologic section of the jejunum is usually identified negatively—it lacks Brunner's glands (duodenum) or Peyer's patches (ileum) (Figure 12-5C).

Ileum. The lamina propria contains large aggregations of lymphoid follicles called **Peyer's patches** (Figure 12-5D).

Specialized epithelial cells, called **M cells**, are located over these patches (see Chapter 9). Peyer's patches and goblet cells are more numerous toward the last part of the small intestine.

▽ In healthy individuals, the mucosal immune system of the GI tract is always poised to respond to ingested pathogens, but ignores the trillions of normal commensal intestinal microbes. This homeostasis is disrupted in individuals with **inflammatory bowel disease (IBD)**, a group of chronic, relapsing inflammatory disorders of the GI tract. There are two main forms of IBD. **Crohn's disease**, an autoimmune disease that can affect the entire wall in any part of the GI tract, and **ulcerative colitis**, an inflammatory disease restricted to the mucosa of the colon and rectum. For reasons that are not fully understood (although genetic susceptibility and environmental factors such as smoking play a role), the normal epithelial barrier in the intestine breaks down in individuals with IBD, allowing commensal microbes in the lumen access to mucosal lymphoid tissue, which triggers an exaggerated local immune response. The resulting inflammation can cause mucosal damage, abdominal pain, diarrhea, and rectal bleeding, and in cases of long-term disease, there is increased risk of cancer. ▽

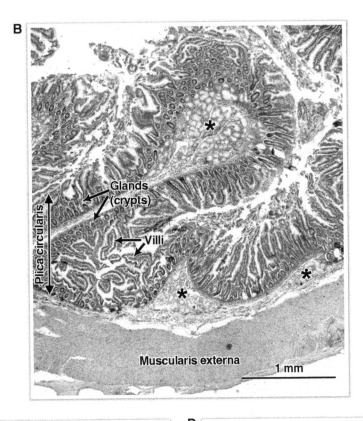

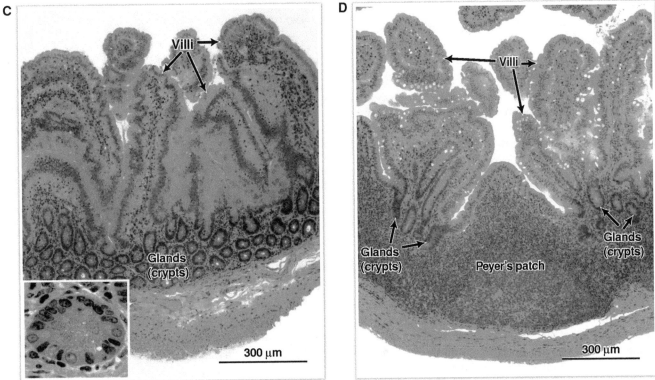

Figure 12-5: (*continued*) **B.** Section through a human duodenum. Note the size of the plica circularis relative to the villi. Asterisks indicate Brunner's glands in the submucosa, a distinguishing feature of the duodenum. **C.** Section through a primate jejunum. The inset shows Paneth cells in the base of a gland at higher magnification. **D.** Section through a primate ileum. Note the large lymphatic follicle labeled "Peyer's patch," a distinguishing feature of the ileum.

LARGE INTESTINE

The ileum delivers liquid containing the indigestible remains of foodstuffs to the large intestine. The large intestine absorbs water and electrolytes from the fluid, compacts and temporarily stores the waste, and eliminates it as feces. It also absorbs vitamins produced by intestinal bacteria and secretes mucus that lubricates the intestinal surface. The large intestine is subdivided into the cecum, colon, rectum, and anus; the appendix is a small appendage arising from the cecum. The large intestine is characterized by the absence of villi and plicae circulares and by the presence of straight tubular glands and bands of smooth muscle called the **taeniae coli** (Figure 12-6A and B).

MUCOSA OF THE COLON

The cecum and colon are indistinguishable histologically; therefore, both will be considered here as the colon. The lumen of the colon, similar to that of the small intestine, is lined with simple columnar epithelium but is not folded and contains no villi. The flat luminal surface of the colon is penetrated by deep tubular glands containing the following cell types (Figure 12-6A):

Absorptive columnar cells (enterocytes). The predominant cells in the surface epithelium between glands, although a few are found in the glands. Unlike enterocytes in the small intestine, absorptive cells in the large intestine lack brush-border enzymes and play no role in the digestive breakdown of food. Instead, they simply absorb water, salts, and vitamins produced by intestinal bacteria, including vitamins B_1 (thiamin), B_2 (riboflavin), B_{12}, and K.

Goblet cells. The predominant cells of the gland. Goblet cells secrete mucus that functions to lubricate the intestinal surface, are more numerous than in the small intestine, and increase in number along the length of the colon.

Enteroendocrine cells. Present in the glands in small numbers.

Stem cells. Progenitor cells located at the base of the glands that produce all the cells in the epithelium in a process similar to that in the small intestine. The epithelium of the large intestine is replaced weekly.

▽ Because of its rapid turnover, the epithelium in the intestinal (and gastric) mucosa is particularly sensitive to processes that interfere with cell division. For example, radiotherapy or chemotherapy targeted at rapidly dividing malignant cells also causes loss of intestinal epithelium, resulting in nausea, vomiting, and diarrhea, and increased risk of systemic infection. The rapid turnover of intestinal epithelium puts it at increased risk of developing malignancies. **Colorectal cancer** is one of the most common cancers, with approximately 1.2 million new cases worldwide every year. ▼

The lamina propria is rich in lymphoid cells and lymphoid nodules, but is otherwise unremarkable. The muscularis mucosae and the submucosa are typical of the GI tract and have no significant distinguishing characteristics.

THE MUSCULARIS EXTERNA OF THE COLON

The muscularis externa of the colon is unusual in that the outer longitudinal layer of smooth muscle is not of uniform thickness; instead, it is gathered into three thick, equidistant, longitudinal bands known as **taeniae coli**. The taeniae coli maintain constant tonus, which gives the colon its characteristic puckers, called **haustra**. The serosa along the sides of the taeniae forms small sacs filled with adipose tissue, called **omental (epiploic) appendages** (Figure 12-6A and B).

RECTUM AND ANUS

The rectum temporarily stores feces, which the anus transports to the exterior. The rectum resembles the colon, except that the glands of the rectum are shallower and it lacks taeniae coli. The epithelium of the anus changes from simple columnar near the rectum to nonkeratinized stratified squamous, and then to keratinized near the external opening. The external orifice is controlled by an internal sphincter composed of smooth muscle and an external sphincter composed of skeletal muscle, which are under autonomic and voluntary control, respectively.

▽ **Hirschsprung's disease**, also known as **congenital aganglionic megacolon**, is caused by failure of the neural crest cells to migrate into the wall of the gut during embryonic development. As a result, the distal end of the colon is devoid of myenteric neurons (Auerbach's plexus). Intestinal motility is impaired, resulting in chronic constipation and dilatation of the colon. ▼

APPENDIX

The appendix is a small, blind-ending extension of the proximal cecum. It has the same basic structure as the large intestine but with a small, irregular lumen and shorter, less dense glands. The main distinguishing feature of the appendix is the abundance of lymphoid follicles in the mucosa and submucosa, which are especially prominent in children and decrease with age (Figure 12-6C). The appendix has no known function.

▽ Because the appendix is a narrow closed sac, infectious agents can become trapped in its lumen and cause inflammation accompanied by swelling and pain, a condition called **appendicitis**. Appendicitis is most common in children and teenagers, but can occur at any age. Acute appendicitis is a medical emergency, usually requiring removal of the appendix. If left untreated, the appendix can rupture and cause peritonitis and even death. ▼

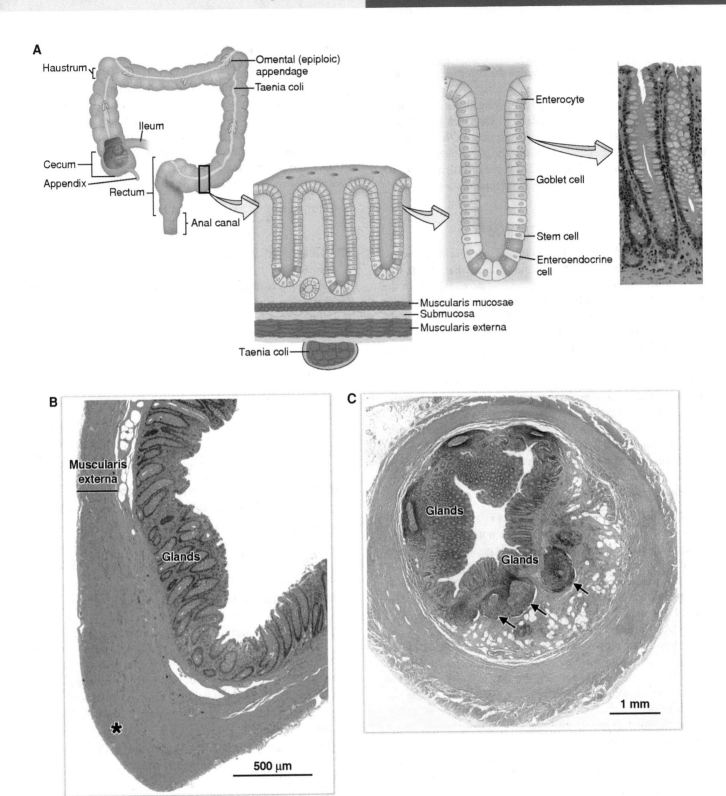

Figure 12-6: Organization of the large intestine. **A.** Distinguishing characteristics of the large intestine, including the gross anatomic regions and external features (*left*), glands and the taeniae coli in the wall of the colon (*middle*), and the distribution of cell types in a single intestinal gland (*right*). The micrograph on the right shows two glands in the colon. Note that mucous cells are the predominant cells in the glands. **B.** Section through the wall of a primate colon. The asterisk marks one band of the taeniae coli. **C.** Section through the appendix. Arrows point to large lymphatic aggregates.

PANCREAS

The pancreas is an accessory organ of the digestive system and possesses both exocrine and endocrine functions. The exocrine pancreas produces an alkaline fluid containing digestive enzymes, which are delivered to the duodenum. **Islets of Langerhans (pancreatic islets)** in the endocrine pancreas produce hormones important for the metabolism of glucose and lipids. The pancreas is covered with a thin connective tissue capsule, from which fine septa extend and organize the gland into lobules containing both exocrine and endocrine cells (Figure 12-7A).

EXOCRINE PANCREAS

STRUCTURE OF THE EXOCRINE PANCREAS The exocrine pancreas consists of a collection of compound tubuloacinar glands that secrete into a system of ducts, which ultimately empty into the duodenum at the **hepatopancreatic ampulla**. Each acinus is composed of 45–50 pyramidal-shaped epithelial cells, radially oriented around a central lumen and surrounded by a basal lamina. **Acinar cells (zymogen cells)** are typical protein-secreting cells with extensive rough endoplasmic reticulum (RER), a round nucleus located basally, a prominent Golgi apparatus, and numerous secretory (zymogen) granules located apically. The number of zymogen granules varies with food intake and is highest in individuals who have fasted. Each acinus is drained by an **intercalated duct** lined by pale, low cuboidal cells; the initial cells of the intercalated duct, **centroacinar cells**, penetrate into the lumen of the acinus and are a distinguishing feature of the exocrine pancreas (Figure 12-7A and C).

FUNCTION OF THE EXOCRINE PANCREAS The exocrine pancreas daily produces 1.5–2.5 L of pancreatic juice, a watery alkaline liquid containing digestive enzymes. Acinar cells produce the enzymes in the pancreatic juice, including **proteases** (e.g., trypsinogen, chymotrypsinogen, procarboxypeptidase), **lipases, α-amylase, and nucleases** (DNase and RNase). Centroacinar and epithelial cells of the small ducts produce the watery fluid, rich in HCO_3^-, into which the enzymes are released.

Proteases are stored as enzymatically inert **zymogens** in secretory granules in acinar cells, and remain inactive when released into the alkaline pancreatic fluids. The proenzymes become active only upon reaching the duodenum, where the brush-border enzyme **enteropeptidase** (formerly called **enterokinase**) cleaves trypsinogen to its active form trypsin, which then catalyzes cleavage of the other proenzymes in a cascade. The HCO_3^- neutralizes the acid in the duodenum, enabling the pancreatic enzymes to function at their optimal pH. These mechanisms prevent the pancreas from self-digestion and allow for an efficient use of the enzymes within the duodenum.

REGULATION OF THE EXOCRINE PANCREAS The secretion of pancreatic enzymes from acinar cells is stimulated by **CCK**, whereas the production of watery fluid containing HCO_3^- from centroacinar cells and intercalated ducts is stimulated by **secretin**. Both hormones are secreted by enteroendocrine cells in the duodenum and jejunum in response to gastric chyme. In addition, vagal cholinergic parasympathetic stimulation promotes the release of pancreatic enzymes.

▽ Pancreatic enzymes occasionally become active within the cytoplasm of the acinar cells, causing autodigestion of pancreatic tissue, a painful and sometimes fatal disorder called **acute pancreatitis**. This condition is often associated with excessive use of alcohol and with gallstones (e.g., blockage of the main pancreatic duct). Pancreatitis also can be caused by a viral infection or can be an adverse effect of drugs. ▼

ENDOCRINE PANCREAS

STRUCTURE OF THE ENDOCRINE PANCREAS The endocrine pancreas is composed of 1–2 million **islets of Langerhans (pancreatic islets)**, which comprise only 1–2% of the gland and are scattered among acini of the exocrine pancreas. Each islet is a spherical aggregate of several hundred polygonal cells enveloped by a thin layer of reticular connective tissue. Cells are arranged in irregular cords separated by fenestrated capillaries, and appear as typical protein-secreting endocrine cells (see Chapter 14) (Figure 12-7B). At least five types of cells can be distinguished by immunostaining with hormone-specific antibodies.

FUNCTION OF THE ENDOCRINE PANCREAS Hormones produced by cells in the islets of Langerhans play essential roles in glucose and lipid metabolism and regulation of blood glucose levels, as follows:

- **α Cells** (≈20% of cells). Secrete **glucagon**, which stimulates glycogen breakdown and the release of glucose into blood.
- **β Cells** (≈70%). Secrete **insulin**, which stimulates glucose uptake into tissues and also regulates lipid metabolism.
- **δ Cells** (5–10%). Secrete **somatostatin**, which inhibits hormone release by the α and β cells.
- **Other hormone-producing cells** are present in small numbers.

▽ **Diabetes mellitus** is a complex, chronic syndrome in which disruption of insulin signaling produces abnormally high blood glucose levels. If the disease remains untreated, dysregulated glucose and lipid levels can cause circulatory disorders, renal failure, blindness, gangrene, stroke, and myocardial infarcts. The two major forms of diabetes are as follows:

- **Type 1 diabetes mellitus** results from the loss or inactivity of β cells and the consequent lack of insulin production, most frequently due to autoimmune destruction of β cells early in life. The disease typically appears before the age of 20, and can be treated with exogenous insulin. Thus, it is called **insulin-dependent**, or juvenile-onset diabetes.
- **Type 2 diabetes mellitus** results from the failure of tissues to respond to insulin, which is being produced normally. Thus, type 2 diabetes is known as **non–insulin-dependent**. It is more common than type 1 diabetes and is typically associated with people who are overweight and older than age 40. ▼

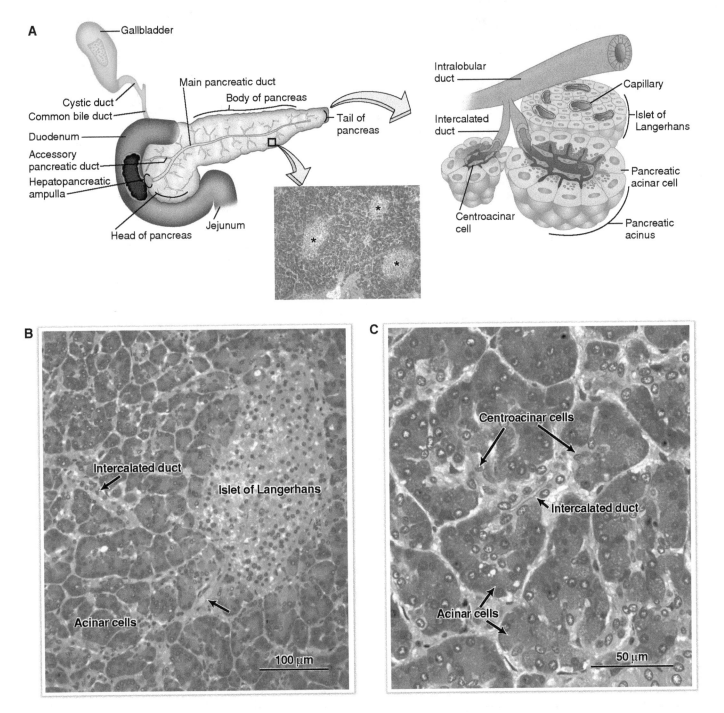

Figure 12-7: Organization of the pancreas. **A.** The anatomic arrangement of the gallbladder and pancreas and their associated ducts, which deliver secretions to the duodenum (*left*), and the organization of the exocrine and endocrine cells, and ducts in the pancreas (*right*). The inset shows a representative section through the pancreas at low magnification. Asterisks indicate the islets of Langerhans. The pancreas is subdivided into three indistinct regions: the head, the body and the tail. Pancreatic islets are scattered among exocrine acini in all three regions, and are most abundant in the tail. **B.** Representative section of the pancreas at higher magnification than in part A. Note the difference between the organization of acinar cells (exocrine) and cells in the islet of Langerhans (endocrine). The arrow indicates a small blood vessel. **C.** Section through the exocrine pancreas showing acinar cells, centroacinar cells, and intercalated ducts at higher magnification than in part B. Note the abundant zymogen granules located apically in acinar cells.

LIVER AND GALLBLADDER

The liver and gallbladder are accessory organs associated with the GI tract. The liver produces bile, a fat emulsifier important for lipid digestion, which is stored in the **gallbladder.** In addition, the liver provides many other vital metabolic and regulatory functions. It is encased in a thin, connective tissue capsule (**Glisson's capsule**) covered with the visceral peritoneum, and is composed of anastomosing plates of parenchymal cells (**hepatocytes**) interspersed between capillaries (**hepatic sinusoids**) (Figure 12-8A and B). The portal vein delivers nutrient-rich blood from the intestines to the sinusoids. Hepatocytes filter this blood, removing some components (e.g., toxins and drugs) and synthesizing proteins, carbohydrates, and lipids, which either are released back into the bloodstream or stored. The liver has an impressive ability to regenerate lost tissue.

LIVER

STRUCTURE OF THE LIVER The following cells and structures can be observed in histologic sections of the liver:

- **Portal triads.** Regions containing branches of **hepatic portal venules**, **hepatic arterioles**, and **bile ducts**, as well as nerves and lymphatics, together surrounded by dense connective tissue (Figure 12-8A–C). Bile ducts in triads are lined with cuboidal epithelium; they enlarge and merge to form the right and left **hepatic ducts**, which exit the liver.

- **Terminal hepatic venules (central veins).** Single venules distributed throughout the liver that are not surrounded by connective tissue. Terminal hepatic venules merge to form larger hepatic veins, which ultimately drain into the inferior vena cava (Figure 8A, B, and D).

- **Hepatocytes.** Large cells with abundant RER and smooth endoplasmic reticulum (SER), Golgi apparatus, and mitochondria that form the parenchyma of the liver; some hepatocytes are binucleate. Hepatocytes are arranged in anastomosing **plates**, with a thickness of one or two cells.

- **Hepatic sinusoids.** Large, highly permeable sinusoidal capillaries that lie between the plates of hepatocytes (Figure 12-8A and E). Mononuclear macrophages, called **Kupffer cells**, are located along the sinusoid walls and function to remove bacteria that invade from the intestines.

- **Space of Disse.** A narrow space separating the hepatocytes and sinusoids. Hepatocytes project numerous short microvilli into the space of Disse, increasing the area for the exchange between the hepatocytes and the bloodstream (Figure 12-8A). **Hepatic stellate cells** reside in the space of Disse and store fat and fat-soluble vitamin A.

- **Bile canaliculi.** A network of anastomosing channels within the plates of hepatocytes. The lateral membranes of neighboring hepatocytes form occluding junctions that seal off small spaces between the hepatocytes, creating the **canaliculi** (Figure 12-8E). Canaliculi drain into bile ducts in the portal triads.

FUNCTION OF THE LIVER Hepatocytes perform all of the vital functions of the liver, producing bile and regulating the composition of blood plasma. Nutrient-rich blood from the portal venules and oxygen-rich blood from the hepatic arterioles in the portal triads flows into the sinusoids and contacts hepatocytes in the space of Disse. The functions of hepatocytes are as follows:

- **Synthesis.** Hepatocytes produce most **plasma components**, including albumin, transferrin, prothrombin, fibrinogen, and all of the globulins except γ-globulin; **glycogen**; **lipids and lipoproteins**, including cholesterol, low-density lipoprotein, very low density lipoprotein, and fatty acids; and **hormones**, including angiotensinogen, thrombopoietin, and insulin-like growth factor.

- **Storage.** Hepatocytes store glucose in the form of glycogen and lipids as fat droplets; they break down and release these molecules into the bloodstream as needed.

- **Modification and elimination.** Hepatocytes import and chemically modify molecules from the plasma to generate products that are often less toxic and more soluble for excretion. Modification reactions include oxidation by cytochrome P_{450} in the SER; reduction in the mitochondria; conjugation with glucuronic acid or sulfate in the SER; and methylation and hydrolysis. The modified molecules are released back into the bloodstream and eventually exit the body via urine or are released into bile and excreted. For example, hepatocytes convert ammonia (NH_3) produced by the removal of amine groups from amino acids, which is toxic, to less toxic urea, which is excreted in urine. In contrast, bilirubin, which is derived from hemoglobin during turnover of senescent red blood cells by macrophages, is glucuronidated and secreted in bile.

- **Lipid digestion.** Hepatocytes secrete bile salts, which are stored in the gallbladder until needed in the duodenum. Bile emulsifies fat and facilitates lipase action. Hepatocytes reabsorb bile salts returned to the liver via the portal vein (**enterohepatic recirculation**) and synthesize new salts as necessary.

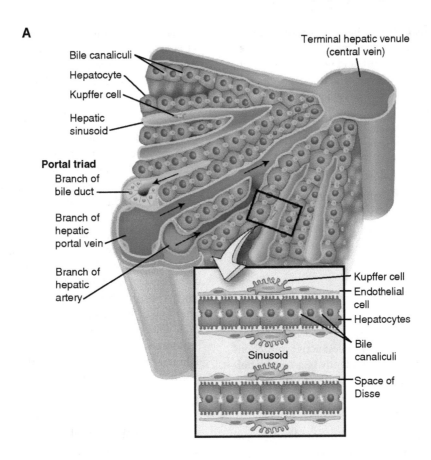

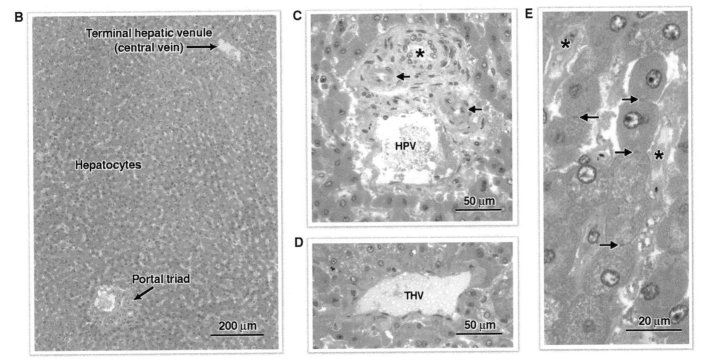

Figure 12-8: Organization of the liver and gallbladder. **A.** Organization of hepatocytes, blood vessels, and bile ducts in the liver. Arrows indicate the direction of fluid flow in the blood vessels and bile ducts. Note that blood and bile flow in opposite directions. The inset shows a sinusoid at higher magnification. **B.** Section of human liver. **C.** Section of a portal triad shown at higher magnification than in part B. Note that the constituents of the portal triad, branches of the hepatic portal vein/venule (HPV) and hepatic arterioles (*arrows*) and a bile duct (*asterisk*) are surrounded by connective tissue. **D.** Section of a dterminal hepatic vein/venule (THV, often called a central vein) shown at the same magnification as part C. Note that the terminal hepatic veins are not surrounded by connective tissue. **E.** Section of hepatocytes and sinusoids (*asterisks*) shown at higher magnification than parts C and D. Arrows indicate the bile canaliculi, cut in cross-section. (*continued on page 193*)

BLOOD FLOW AND THE HEPATIC ZONES As blood flows from a portal triad to a terminal hepatic venule, the amount of O_2 decreases, CO_2 increases, and hepatocytes absorb and secrete plasma components, thereby establishing chemical gradients along this route. Accordingly, the properties of hepatocytes change from the triad to the terminal hepatic venule. The microanatomy of the human liver appears irregular, but the predictable change in the properties of hepatocytes along the sinusoids allows three functional zones to be recognized. The properties of hepatocytes vary among the zones, as follows (Figure 12-8F):

Zone 1. Closest to the portal triad. Hepatocytes in zone 1 receive the freshest blood, richest in O_2 and nutrients, and are most active in glycogen and protein synthesis, endocytosis, glucuronidation, and bile production. Cells in this zone have the most lysosomes, the largest mitochondria, and the highest regenerative capacity.

Zone 2. Transitions between zones 1 and 3 with intermediate cellular characteristics.

Zone 3. Closest to the terminal hepatic venule. These cells receive blood low in O_2 and have thin, less active mitochondria. Fat is deposited here most abundantly (e.g., in alcoholics or in obese individuals). Zone 3 cells are rich in SER, cytochrome P_{450} enzymes, and other drug-modifying activities, which makes them susceptible to damage.

▽ Excess iron absorbed by patients with **hereditary hemochromatosis** enters hepatocytes by endocytosis and, therefore, accumulates first in zone 1. An overdose of acetaminophen produces hepatic necrosis because zone 3 cells convert it to a toxic agent. Inflammation in the liver can activate stellate cells, which divide and support liver regeneration. However, these cells may become overactive and secrete excessive amounts of collagen. The resulting **fibrosis** impedes blood flow and impairs all hepatocyte functions. Extensive fibrosis, the loss of hepatocytes, and disruption of blood circulation defines **cirrhosis**, the 12th most common cause of death in the United States. ▽

GALLBLADDER

The gallbladder is located under the liver, and stores and concentrates bile to be released to the duodenum as needed. The wall of the gallbladder is composed of simple columnar epithelium bearing apical microvilli, a lamina propria, a thin muscularis with bundles of smooth muscle arranged in different orientations, and a serosa or adventitia. The mucosa is highly folded when the gallbladder is empty (Figure 12-8G). Bile exits the liver in the common hepatic duct, which joins the **cystic duct** from the gallbladder to form the common bile duct. The common bile duct in turn joins with the pancreatic duct to empty into the duodenum at the hepatopancreatic ampulla. The flow of bile is controlled by hepatopancreatic sphincters. CCK released by the enteroendocrine cells in the duodenum in response to a fatty meal causes the smooth muscle cells of the sphincter to relax and the smooth muscle of the gallbladder to contract and inject bile into the duodenum.

▽ As an individual ages, it is common for bile salts or cholesterol (which is normally excreted in bile) to precipitate and form **gallstones**. Gallstones are twice as common in women as in men and become problematic only when the stones are small enough to exit the gallbladder but large enough to block a duct. The result of duct blockage is tremendous pain. The most common treatment for gallstones is surgical removal of the gallbladder. After the gallbladder is removed, bile is delivered to the duodenum in an unregulated fashion. ▽

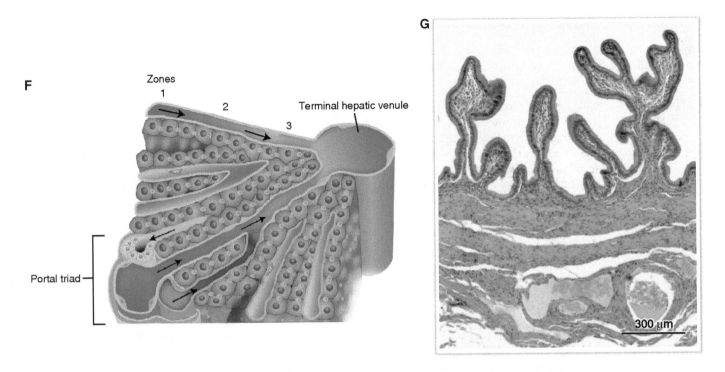

F

Zones
1
2
3
Terminal hepatic venule

Portal triad

G

300 μm

Figure 12-8: (*continued*) **F.** Schematic of the liver showing the blood flow and location of hepatic zones. **G.** Section through the wall of the gallbladder. Note the highly folded epithelium, which allows the gallbladder to expand to accommodate stored bile.

STUDY QUESTIONS

Directions: Each of the numbered items or incomplete statements is followed by lettered options. Select the **one** lettered option that is **best** in each case.

1. The mutation responsible for cystic fibrosis produces a defect in Cl^- transport across epithelial cells, which also reduces the transmembrane movements of cations such as Na^+. This results in the production of saliva containing a higher than normal concentration of NaCl. Which of the following type of cells in salivary glands is most likely affected in this case?

   A. Intercalated duct cells

   B. Interlobular duct cells

   C. Mucous secretory cells

   D. Serous secretory cells

   E. Striated duct cells

2. A 48-year-old man is seen by his physician because he is having difficulty swallowing. An endoscopic examination of the esophagus shows abnormal growths beginning immediately below the pharynx and continuing down more than two-thirds of the length of the organ. Biopsy specimens are obtained at two sites, one from the top and one from the bottom of the esophagus, and are sent to the histopathology laboratory. The pathologist is able to verify the locations of the biopsy specimens by observation of which tissue layer?

   A. Adventitia

   B. Mucosa

   C. Muscularis externa

   D. Submucosa

3. A 23-year-old man has been diagnosed with vitamin B_{12} deficiency and begins taking oral supplements containing vitamin B_{12}. The patient's symptoms, however, do not improve, and serum analysis indicates that he is not absorbing the supplement. The physician suspects that the patient may be suffering from an autoimmune condition and orders laboratory studies. The studies show the presence of antibodies in the serum that are reactive against gastric cells. Which is the most likely target of these autoantibodies?

   A. Chief cells

   B. Enteroendocrine cells

   C. Lymphatic lacteals

   D. Parietal cells

   E. Surface mucous cells

4. Which intestinal cells produce antibacterial compounds?

   A. Enterocytes

   B. Enteroendocrine cells

   C. Goblet cells

   D. Microfold cells

   E. Paneth cells

Questions 5 and 6

Refer to the diagram below illustrating the microanatomy of the liver when answering the following questions.

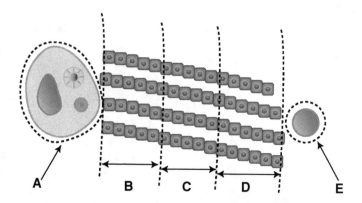

5. A child is brought to the Emergency Department after accidentally swallowing a large number of acetaminophen tablets, which are converted into a toxic arylating agent by cytochrome P_{450}. After physical examination and laboratory studies are completed, the child is diagnosed with acute liver failure. What is the most likely location of the initial liver damage in this child?

6. What is the location of cells that display the greatest endocytic activity?

7. A 38-year-old man is seen in the Emergency Department because he has experienced recent onset of severe and constant abdominal pain. The patient explains that he has been drinking heavily for years, often beginning at noon, and consequently has been unable to retain a steady job. The physician suspects that the patient is experiencing acute pancreatitis and orders laboratory studies. The presence of which substance in the blood sample would support the suspected diagnosis?

   A. Albumin

   B. Bile salts

   C. Cholecystokinin

   D. Glucose

   E. Insulin

   F. Lipase

8. The section in the micrograph shown below was collected
from which organ and the region of that organ?

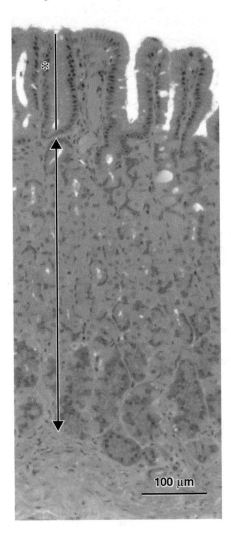

100 μm

A. Large intestine, colon

B. Small intestine, duodenum

C. Small intestine, ileum

D. Small intestine, jejunum

E. Stomach, body or fundus

F. Stomach, cardia

G. Stomach, pylorus

ANSWERS

1—E: The striated duct cells actively resorb NaCl, which creates hypotonic saliva and helps recover water to the circulatory system. The saliva of individuals with cystic fibrosis is "saltier" than normal. Several screening tests based on measuring the NaCl concentration of epithelial secretions have been developed to detect cystic fibrosis in newborns. Testing sweat is a common method, as are genetic tests. Parents often identify infants with the disease because the infant "tastes salty."

2—C: The muscularis externa of the esophagus contains only skeletal muscle in the upper third of the tube and only smooth muscle in the lower third, so the biopsy specimen containing skeletal muscle must be the one taken from the upper portion and the specimen containing only smooth muscle in the muscularis externa must be taken from the inferior portion.

3—D: Parietal cells produce gastric intrinsic factor, which is required for the absorption of vitamin B_{12}. Autoimmune attacks against parietal cells, or intrinsic factor itself, cause pernicious anemia because of the body's inability to absorb vitamin B_{12}. The patient's condition is known as autoimmune atrophic gastritis.

4—E: Paneth cells, found near the base of glands in the intestine, produce and secrete a variety of antibacterial compounds, including defensins, lactoferrin, and lysozyme.

5—D: The highest activity of cytochrome P_{450} is found in the cells of zone 3, which are near the terminal hepatic venule in the area labeled "D" in the diagram.

6—B: Cells of zone 1 are the most active in endocytosis; it is in this zone that metal oxides, whose uptake requires endocytosis, are deposited in the greatest amount.

7—F: Lipase is one of the enzymes produced in large quantities by the exocrine pancreas. Pancreatic lipase is released into the duodenum via pancreatic ducts and normally is not found in the blood. Bile salts are found in the blood as a result of liver or gallbladder disease. Albumin, cholecystokinin, glucose, and insulin normally are found in blood. Lipase and other pancreatic enzymes can leak into the blood when pancreatic tissue is damaged, such as during an attack of acute pancreatitis.

8—E: The micrograph is a section through the body or fundus of the stomach. Note the long glands (double-headed arrow) and relatively short pits (indicated by an asterisk) and the abundance of parietal cells and chief cells in the glands.

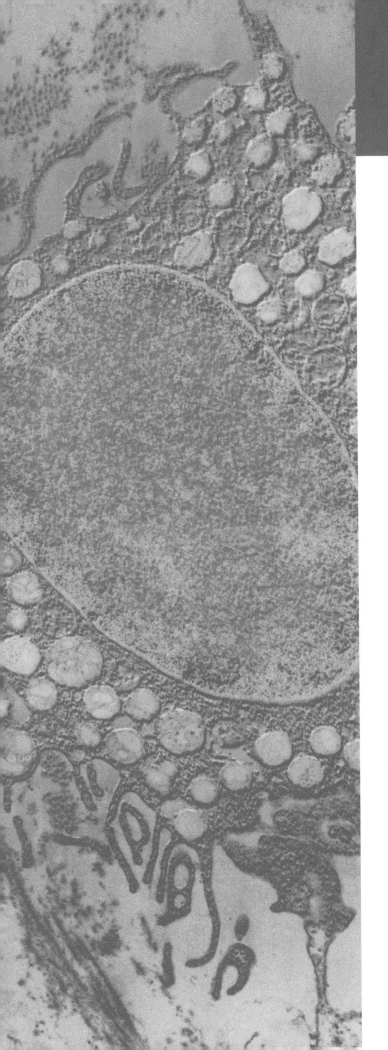

CHAPTER 13

URINARY SYSTEM

OVERVIEW

The urinary system is composed of the **kidneys**, **ureters**, **urinary bladder**, and **urethra**. The kidney is a homeostatic filtering mechanism responsible for regulating body water, ions, and pH, as well as for excreting many soluble wastes from the blood. These functions center on the process of urine being produced by the kidneys followed by its elimination from the body via the ureters, bladder, and urethra.

- The functional units of the kidney are **nephrons** and **collecting ducts**. The nephron starts at a capillary bed, the **glomerulus**, which is located inside a **renal corpuscle**. Urine begins as a **filtrate** of blood plasma, which is driven into the renal corpuscle by blood pressure.

- The filtrate flows down a series of tubules in the nephron and the collecting duct system, which recovers solute and water in a process known as **concentrating** the urine.

- The continuous adjustment of the volume and ionic composition of blood is linked to the processes responsible for urine concentration.

- Two countercurrent systems are found in the kidney. The countercurrent multiplier of the **loop of Henle** is important for urine concentration, and the countercurrent exchanger of the **vasa recta** supplies blood to the kidney medulla.

- Regulation of flow through the tubules is accomplished in part by components of the nephron known as the **juxtaglomerular apparatus**.

The kidneys serve as endocrine organs. They produce **renin**, a protease that affects systemic blood pressure, and **erythropoietin**, a cytokine that stimulates red blood cell production. The kidneys also provide the final hydroxylation reaction required to produce active **vitamins D$_2$ and D$_3$**, hormones involved in Ca^{2+} regulation and bone physiology.

ANATOMY OF THE KIDNEY

The body contains one pair of kidneys, located retroperitoneally in the abdominal cavity on either side of the spine. A coronal section of the kidney shows the gross anatomic features, the **cortex, medulla,** and **calyces,** which provide the framework and nomenclature used in the description of the microanatomy and function of the kidney (Figure 13-1A and B).

CORTEX

The cortex of the kidney contains **renal corpuscles** and three kinds of tubules, the **proximal, distal, and collecting tubules,** each of which has a highly convoluted structure. The cortex adjacent to the medullary areas is called the **juxtamedullary cortex.**

MEDULLA

The kidney medulla is divided into anywhere from 6 to 18 conical structures, called **renal pyramids**. The medulla contains **collecting ducts** and **thick** and **thin limbs** of the **loop of Henle.** Extensions of the medulla, called **medullary rays,** are organized around the collecting tubules and project into the cortex. The tubules in the medullary rays are arranged in parallel lines and stand out against the background of the convoluted tubules.

CALYCES

The tips of the renal pyramids project into spaces called minor calyces, into which urine flows. The minor calyces merge together to form the major calyces, which in turn merge in the **renal pelvis,** at the beginning of the **ureter.**

NEPHRON AND COLLECTING DUCTS

Each kidney contains about 1 million nephrons, which arise in the embryo from the urogenital ridge. The nephron begins at the renal corpuscle and ends at the distal convoluted tubule, where it joins a collecting tubule that merges with a collecting duct. The latter two structures are derived from mesonephric duct tissue.

RENAL CORPUSCLE

The filter unit of the kidney consists of an elaborate blood filtration unit, the glomerulus, enclosed in a capsule (Figure 13-1C and D). About 180–200 L of **filtrate** is formed daily from blood passing through the capillaries.

- **Capsule.** A simple layer of squamous epithelial **parietal cells** forms a hollow sphere, known as **Bowman's capsule.** It is pierced at one end, the **vascular pole,** by an **afferent** and **efferent** arteriole. Roughly opposite this, the capsule opens at the **urinary pole,** which is the beginning of the **proximal convoluted tubule (PCT).**

- **Glomerulus.** The filtration apparatus is composed of **fenestrated capillaries,** pericytes called **glomerular mesangial cells,** and **podocyte cells,** all of which are difficult to distinguish in standard histologic sections. The mesangial cells are considered to be pericytes, which are capillary-associated cells (see description in Chapter 8). They are scattered between capillary branches and are involved in turning over the basal lamina, controlling capillary diameter, and secreting various vasoactive compounds and cytokines. Podocytes extend elaborate processes that completely surround the capillaries. The fluid that exits the capillaries and escapes beyond the podocyte processes and into the **urinary space** inside the capsule is called the **initial filtrate.**

- **Filtration apparatus.** There are three components of the glomerular filter: the **fenestrated endothelium,** the thick **basal lamina,** and the filtration slits between the podocyte processes, called **pedicles.** These components prevent the passage of proteins with the mass of albumin (68 kDa) or larger; the filtrate resembles protein-poor blood plasma.

- **Glomerular filtration rate (GFR).** Because the efferent arteriole has a smaller diameter than the afferent arteriole, blood pressure in the glomerular capillaries is about three times higher than the pressure in other capillary beds, and is responsible for driving fluid into the urinary space. This outward force is opposed both by the higher osmotic pressure of blood in the glomerular capillaries, which contains a significantly higher concentration of protein, and by fluid pressure in Bowman's space. The net filtration pressure is about 10 mm Hg, resulting in a total GFR of approximately 125 mL/min from both kidneys.

▽ The filtration barrier is labile. The momentary increase in blood pressure that occurs as a person stands up causes a brief increase in the amount of albumin normally entering the urine, called orthostatic **proteinuria.** An extended period of proteinuria usually indicates a pathologic condition in the kidneys, although it can result from excess protein in the serum, such as a plasmacytoma, which secretes immunoglobulins. Proteinuria caused by complications in the kidney is usually related to a defect in either the glomerular filtration apparatus or reabsorption of protein from the proximal tubules. In **diabetes mellitus,** the rate of entry of protein into the urine may decline at first, but it typically increases dramatically as the filtration barrier becomes damaged. Mutations affecting **nephrin,** a transmembrane protein present in the podocyte slit membrane, are associated with severe proteinuria that occurs soon after birth and often results in kidney failure. Patients with these complications require dialysis or kidney transplantation. These mutations are rare, but have been observed with increased frequency in people of Finnish ancestry. ▽

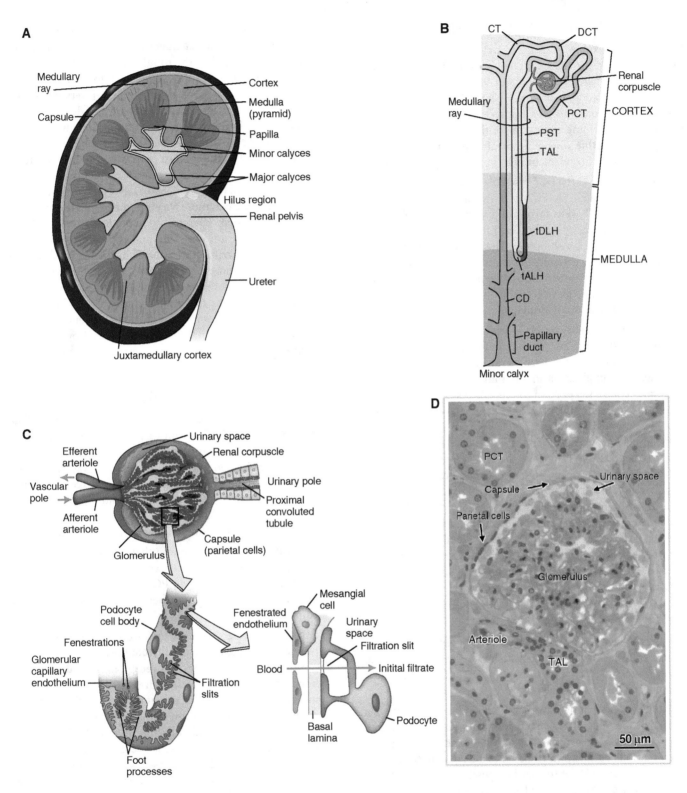

Figure 13-1: Anatomy of the kidney and the renal corpuscle. **A.** Section of the kidney with the medullary pyramids surrounded by cortex. Urine production begins in the cortex and ends when fluid exits a medullary papilla and enters a minor calyx. **B.** Kidney tubules. The nephron begins at the renal corpuscle in the cortex, where the initial filtration of blood occurs. The nephron continues as a series of tubules, the proximal convoluted tubule (PCT), proximal straight tubule (PST), thin descending and ascending limbs of the loop of Henle (tDLH, tALH), thick ascending limb (TAL), and distal convoluted tubule (DCT). The collecting tubule (CT) and collecting duct (CD) have a separate embryonic origin. **C.** Renal corpuscle. The initial filtrate of blood occurs inside Bowman's capsule. The glomerulus filter unit has three components: (1) the fenestrated endothelium of the capillary bed; (2) a thick basal lamina; and (3) the filtration slits between the foot processes of the podocyte cells. Glomerular mesangial cells are pericytes associated with endothelial cells. **D.** Renal corpuscle. The urinary space is visible under the simple parietal cell layer of Bowman's capsule; the three cell types of the glomerulus cannot be individually distinguished. A portion of the TAL is visible near one of the arterioles associated with this corpuscle. The PCTs are the most common feature of the cortex of the kidney.

PROXIMAL CONVOLUTED TUBULE

Kidney tubules connect renal corpuscles to minor calyces and **reabsorb** most of the 180 L of filtrate produced daily. As filtrate flows through these tubules, about 99.5% of the volume is returned to the blood, including 1.4 kg of NaCl and 0.25 kg of glucose daily. Along the way, transport processes are used to **regulate** blood volume, ionic composition, and pH. The filtrate finally becomes **urine** when it flows into a minor calyx.

The PCT is the first element in this recovery system.

■ **Structure.** Cells of the PCT resemble enterocytes of the small intestine, although the PCT cells are shorter (Figure 13-2A and B). The PCT cells form a simple columnar epithelium with a low apical brush border; junctional complexes link cells beneath the brush border. This section of the nephron follows a twisting convoluted path so that tubule profiles are sectioned in many orientations. When a PCT enters a **medullary ray**, it courses parallel to the other elements, where it is called a **proximal straight tubule**.

■ **Function.** PCT cells use Na^+–dependent cotransporters to move **solutes** (e.g., sugars and amino acids) from the PCT lumen and out of the basolateral membranes toward the **peritubular capillaries** in the thin connective tissue surrounding the basal portions of the tubules (see Chapter 1 for a complete description of this mechanism). Water follows the solute flow, and peritubular capillaries recover about 65% of the initial filtrate volume and Na^+, as well as almost all of the glucose and amino acids, into the capillaries. **Proteins** in the filtrate are taken up by endocytosis and degraded in lysosomes. **Organic acids** and **bases** that diffuse out of the capillaries are transported across the cells and into the filtrate, a process that removes some metabolites and toxins as well as drugs (e.g., penicillin) from the blood. **Blood pH** is regulated by the secretion of NH_4^+ and H^+ into the filtrate and recovery of HCO_3^-. The PCT cells contain the **hydroxylase** activity that produces active vitamins D_2 and D_3.

LOOP OF HENLE

The loop of Henle matches fluid flow with the distribution of ion pumps to produce the high NaCl concentration in the medulla that is used in the final concentration of urine. The loop of Henle is composed of thick and thin limbs (Figure 13-2C and D).

■ **Structure.** A straight proximal tubule eventually becomes a **thin descending limb**, which continues its descent before making a 180° turn at the **loop of Henle**. The tubule is now a **thin ascending limb**, which becomes a **thick ascending limb**. Approximately 15% of the renal corpuscles occur in the **juxtamedullary cortex**, and these have long descending limbs that penetrate deep into a medullary pyramid. **Cortical** proximal tubules arising from corpuscles far from a medullary area form loops that only enter the top of the medulla or may complete the loop in a medullary ray outside the medulla proper. The thick ascending limb eventually exits the medullary ray and returns to the renal corpuscle of its origin near the afferent and efferent arterioles at the vascular pole. At that point, cells in a portion of the tubule wall facing the corpuscle are taller and more numerous, forming a structure called the **macula densa**.

■ **Function.** The tubules of the thick ascending limb in the medulla contain many **sodium pumps** that actively transport Na^+ from the fluid in the tubule into the interstitial space between the tubules in the medulla; Cl^- follows, resulting in the export of NaCl in the medullary interstitium. The antiparallel, or countercurrent, flow in the descending and ascending limbs means that fresh filtrate along with some of the exported Na^+ that leaks back into the tubules is continuously supplied to the sodium pumps in the thick ascending limb. These pumps actively remove a fraction of the Na^+ from the fluid in this region, and as the Na^+ concentration in the tubule increases, more Na^+ is pumped into the interstitial space. This system operates as a **countercurrent multiplier** that uses active ion pumping and geometry to produce a steady-state concentration of NaCl in the medulla, which is as high as four times that in blood. Because the cells in the thick ascending limb are impermeable to water, fluid exiting the medulla in these tubes is hypotonic. Urea is also present at a high concentration in the medulla. As much as 20% of the NaCl in the original filtrate is recovered by the loop of Henle; only 10% of the water is returned due to the impermeability of the thick ascending limb.

DISTAL CONVOLUTED TUBULE

The thick ascending limb continues in the cortex as a **distal convoluted tubule (DCT)**. From this point on, the recovery of water and ions is continuously regulated by hormones to meet the body's requirements. The DCT originally was described as an extension of the thick ascending limb that coursed until it joined a collecting duct. Ultrastructural and biochemical studies now show that the distal convoluted tubule contains three distinct segments (not distinguished in this discussion).

■ **Structure.** DCT cells are shorter than the cells of the PCT and lack a brush border. As a consequence, the DCTs have relatively larger lumen than do PCTs (Figure 13-2A). The cells are held tightly together at their apical ends by junctional complexes.

■ **Function.** The recovery of Ca^{2+} from the filtrate in these tubules is stimulated by **parathyroid** hormone. The terminal portion of the DCT contains receptors for **aldosterone**, a hormone produced by the adrenal glands. Aldosterone increases the expression of Na^+ and K^+ channels in the apical membranes and the sodium pump in the basolateral membranes of the DCT and collecting duct cells. These membrane changes produce increased recovery of Na^+ from the filtrate, accompanied by increased water recovery. Although aldosterone stimulation only accounts, at most, for a few percent of total Na^+ recovery, lack of this hormone is fatal.

▽ **Diuretics** are drugs that inhibit reabsorption from the kidney tubules, resulting in increased delivery of Na^+ and water to the urine. They are commonly used to reduce blood pressure, which occurs as a consequence of reducing the total volume of water in the body. Diuretics are a diverse set of compounds acting on different tubule segments. For instance, **loop diuretics** (e.g., furosemide) inhibit a Na^+ cotransporter in the thick ascending limb, whereas the **thiazides** inhibit a different cotransporter expressed in the DCTs. ▽

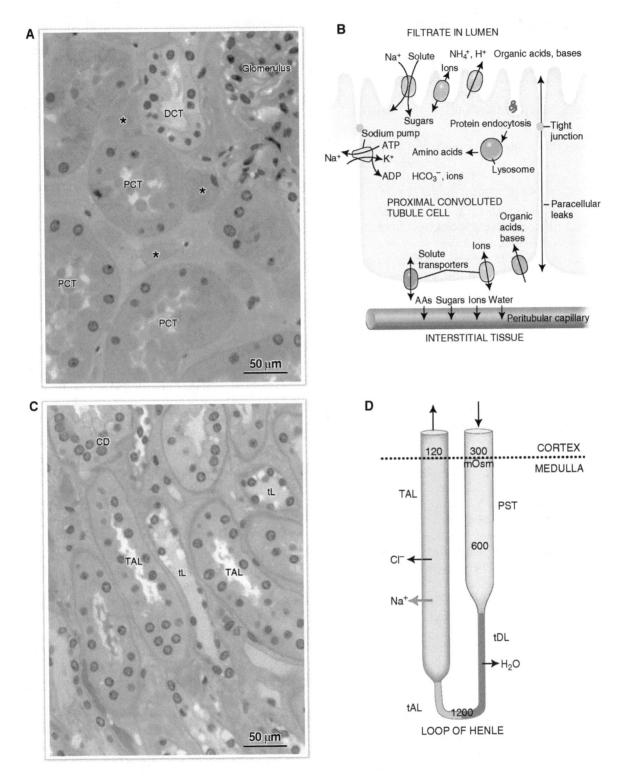

Figure 13-2: Proximal and distal tubules and the loop of Henle. **A.** Proximal and distal tubules in the cortex. The brush border of the proximal convoluted tubule (PCT) cells may accumulate as blebs in the lumen of the tubules during fixation. The distal convoluted tubule (DCT) cells lack a brush border. Solutes and fluid recovered by the tubules flow into capillaries and venules (*asterisks*) in the thin interstitial connective tissue. **B.** Proximal tubule transport processes. The proximal convoluted tubules utilize the sodium pump to generate a Na$^+$ electrochemical potential that drives the reabsorption of solutes and water from the filtrate. Proteins that escape into the filtrate are specifically removed by endocytosis and degraded in lysosomes. Solutes and water that exit the basolateral membranes enter capillaries in the interstitial tissue. AAs, amino acids; ADP, adenosine diphosphate; ATP, adenosine triphosphate. **C.** The loop of Henle in the medulla. The thin limbs (tL) of the loop of Henle are lined with a simple low epithelium. The thick ascending limbs (TAL) are lined with a cuboidal epithelium. The collecting ducts (CD) are also present in the medulla and are lined with tall, narrow cells. **D.** Countercurrent multiplier system. Sodium pumps in the TAL actively remove Na$^+$ from the filtrate (*red arrow*); Cl$^-$ follows passively. The sodium pumps continuously move a fraction of the Na$^+$ that is inside the tubules out and generate a high Na$^+$ concentration in the interstitial tissue outside the tubules.

COLLECTING TUBULES AND COLLECTING DUCTS

Typically, several DCTs merge with a single collecting tubule in the cortex. The collecting tubule enters a medullary ray and continues into the medulla, where it is then called a medullary collecting duct. The **final concentration** of the filtrate occurs in these collecting ducts and is under hormonal control.

▪ **Structure.** Cells of the collecting tubules are initially cuboidal and become taller as the tubules approach the collecting ducts. Cells in the medullary collecting ducts are typically columnar and pale; the boundaries between adjacent cells are often easily visible (Figure 13-3A). There are two cell types in these tubules, **principal** and **intercalated**, which express different ion transport activities, but they cannot be distinguished in standard sections using the light microscope. Collecting ducts merge to form papillary ducts, which open into a minor calyx near the tips of the renal pyramid. **Papillary ducts** are lined by **urothelium** and, at this point, the filtrate now has become urine.

▪ **Function.** Similar to the terminal portions of the DCTs, the collecting tubule and the collecting duct cells contain aldosterone receptors. **Aldosterone** is a hormone that promotes the recovery of Na$^+$ and, therefore, water from the filtrate in these tubules. Collecting ducts also contain receptors for **antidiuretic hormone** (**ADH**), also known as **vasopressin**, which is produced in the hypothalamus and released from nerve endings in the neurohypophysis (see Chapter 14). An increase in the osmolarity of blood plasma in the hypothalamus causes the release of ADH, which binds to collecting duct cells and causes the fusion of endocytic vesicles containing **aquaporin** proteins with the plasma membrane (Figure 13-3B). Aquaporins facilitate the movement of water from the filtrate to cross the cells and enter the interstitial fluid of the medulla. This process is driven by the osmotic pressure that is produced by the high NaCl concentration in the interstitium, and results in the return of water to the blood and the excretion of concentrated urine. Low levels of ADH result in the excretion of dilute urine.

JUXTAGLOMERULAR APPARATUS

The **juxtaglomerular apparatus** (**JGA**) is a portion of the thick ascending limb that courses along the vascular pole of the renal corpuscle and aids in controlling the GFR. The rate of filtrate flowing through the kidney tubules is regulated by several mechanisms to maintain adequate solute and water recovery. For instance, smooth muscle in the afferent arterioles supplying the glomerular capillaries automatically reacts to counter changes in systemic blood pressure and maintains a relatively constant rate of initial filtrate formation. A more sophisticated feedback system is provided by the JGA, which monitors fluid in the thick ascending limb and adjusts the rate of formation of initial filtrate. The operation of this system influences blood pressure throughout the body.

COMPONENTS OF THE JUXTAGLOMERULAR APPARATUS

The JGA consists of the three types of cells that are found at the vascular pole of each renal corpuscle, and described as follows (Figure 13-3C and D):

▪ **Macula densa cells.** An accumulation of cells of the thick ascending limb form a "dense spot," or macula densa. These cells monitor the NaCl concentration in the filtrate and generate responses when the value decreases out of an optimal range.

▪ **Juxtaglomerular cells.** Modified smooth muscle cells, called juxtaglomerular cells, are found in the tunica media of the afferent arteriole. These cells contain the protease **renin** that is stored in vesicles.

▪ **Extraglomerular mesangial cells.** Extraglomerular mesangial cells are located outside the renal corpuscle. These cells are believed to be involved in some of the signaling activities of the JGA.

RESPONSES TO HIGH OR LOW NaCl CONCENTRATIONS

The cells of the macula densa monitor the NaCl concentration in the filtrate and adjust the rate of formation of the initial filtrate. If the concentration is high, the fluid is flowing too rapidly through the nephron and the GFR is reduced; if the concentration is low, the GFR is increased (Figure 13-3E).

▪ **Response to high NaCl.** The macula densa cells signal the afferent arteriole to constrict, which slows the rate of initial filtrate formation and allows the tubules downstream more time to process the filtrate. Signaling appears to involve the release of adenosine triphosphate or adenosine, either by the macula densa cells or by the extraglomerular mesangial cells.

▪ **Response to low NaCl.** The macula densa cells signal the juxtaglomerular cells to release **renin** into the blood. This protease converts **angiotensinogen**, produced in the liver, to **angiotensin I**, which is then further processed to **angiotensin II** by the **angiotensin-converting enzyme** (**ACE**). The angiotensin molecules cause constriction of the arteries and arterioles. The efferent arterioles are more sensitive to angiotensin II than the afferent arterioles, and their greater constriction increases the blood pressure in the glomeruli, increasing the GFR. Generally, angiotensin II increases systemic blood pressure by its action on arterioles throughout the body. It also stimulates the release of aldosterone from the adrenal glands, which acts on the DCTs and the collecting ducts to increase Na$^+$ reabsorption, which increases water recovery and, therefore, also increases systemic blood pressure. Renin secretion is also increased by sympathetic nerve stimulation of the juxtaglomerular cells.

▽ The powerful effects that the renin-angiotensin-aldosterone system exerts on systemic blood pressure have caused its components to be useful targets for drugs that treat and control **hypertension** (high blood pressure). Drugs aimed at inhibiting ACE, also known as **ACE inhibitors**, are important agents in widespread clinical use. ▽

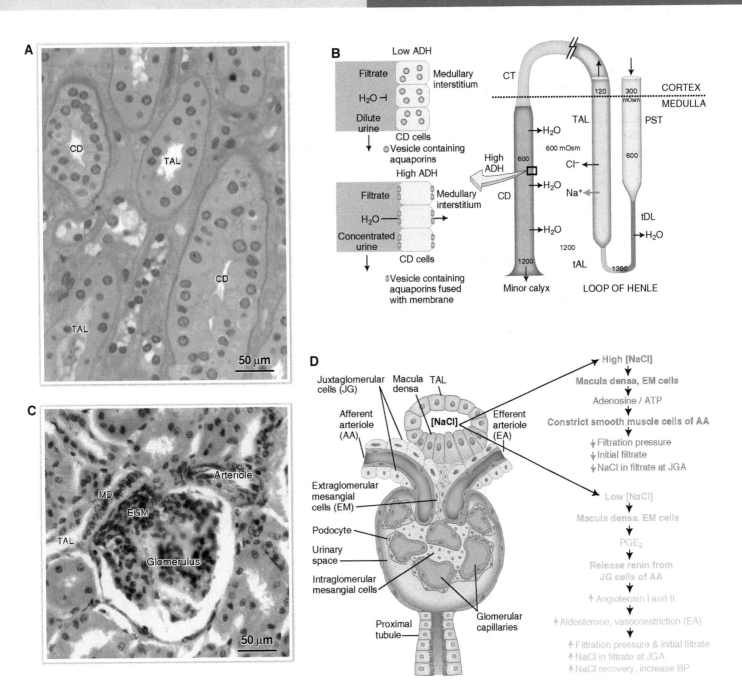

Figure 13-3: Collecting ducts and the juxtaglomerular apparatus (JGA). **A.** Collecting ducts (CD) in the medulla. The cells of the collecting ducts are tall and pale, and their boundaries are more distinct than those of the thick ascending limbs (TAL). **B.** Antidiuretic hormone (ADH) controls water flow across collecting ducts. Fluid entering the medulla in collecting ducts is hypotonic. If the body needs water, ADH is released from the pituitary gland. High ADH levels cause the cytoplasmic vesicles containing the aquaporin molecules to fuse with the cell membranes of the collecting ducts and water from the filtrate flows into the medullary tissue, resulting in a concentrated urine. If the ADH levels are low, dilute urine carries off excess water. **C.** The JGA. The macula densa (MD) and extraglomerular mesangial cells (EGM) are often visible in thicker sections, as this one taken from rodent tissue. An arteriole associated with the glomerulus is also visible. **D.** Functions of the JGA. The macula densa cells monitor the concentration of NaCl in the filtrate. If the NaCl concentration is too high, filtrate production is slowed by constricting the afferent arteriole (AA). If the NaCl concentration is too low, filtrate production is increased by constricting the efferent arteriole (EA) via a reaction that begins with the release of renin from the juxtaglomerular cells (JG). ATP, adenosine triphosphate; BP, blood pressure; PGE$_2$, prostaglandin-E$_2$. Prostaglandins are signaling molecules that are derived from membrane lipids. They act locally and produce a wide variety of effects in different tissues.

RENAL BLOOD SUPPLY

The kidneys, in their capacity as a homeostatic filter, receive about 20% of the cardiac output, far in excess of their metabolic needs. Most of the arterial blood entering the kidneys is first routed to a renal corpuscle to create the initial filtrate. The efferent arteriole then supplies **peritubular capillaries** in the cortex or medulla to take up reabsorbed fluid and solutes. Notable features of this blood supply include the high pressure operation of the glomerular capillaries located between two arterioles, described previously, and the geometry of the vessels that supply the medulla without removing the high Na^+ concentration produced by the loop of Henle, which will be discussed below.

STRUCTURE OF THE RENAL BLOOD SUPPLY

The arterial blood distribution in the kidneys is illustrated in Figure 13-4A. The venous return mirrors the arteries. The **segmental** branches of the renal artery are **end arteries**; if one artery is blocked and blood flow ceases, the portion of the kidney where the blood supply is blocked dies. The arrangement and function of the vasculature that arises from an efferent arteriole depends on the location of the renal corpuscle.

■ The **peritubular capillaries** supplied by the afferent arterioles exiting from the **cortical renal corpuscles** are located in the thin interstitial tissue surrounding the proximal and distal tubules, similar to the arrangement that is found in other organs. **Fibroblast** cells in the interstitium of the cortex are the body's principal source of **erythropoietin**.

■ The **efferent arterioles** from the **juxtamedullary renal corpuscles** enter the medulla and supply the **vasa recta**, which are long straight looping capillaries (Figure 13-4B and C). The looping geometry dictates a **countercurrent** flow of blood in these vessels. Blood in the descending portions loses water and picks up NaCl as it travels toward the medullary tip. After the loop, the ascending blood equilibrates with the medullary interstitium as it moves towards the cortex, losing NaCl and gaining water. As a result, the vasa recta supplies the medulla with blood, but only removes a portion of the NaCl actively transported into the tissue by the loop of Henle. This system is called a **countercurrent exchanger** because it employs diffusion rather than an active process, ion transport, to accomplish solute exchange.

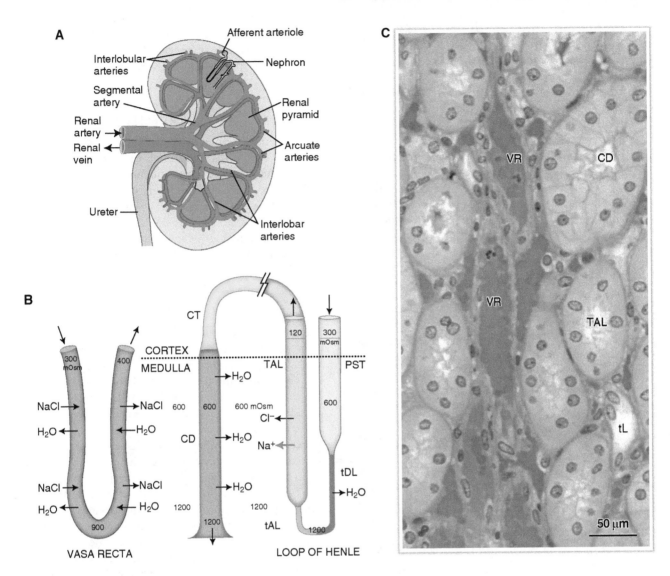

Figure 13-4: Blood supply in the kidney and the extrarenal collecting system. **A.** Renal arteries and veins. All arteries entering the parenchyma of the kidney eventually supply an afferent arteriole entering a renal corpuscle. The venous system mirrors the arteries. **B.** Vasa recta and kidney tubules. Efferent arterioles from the juxtamedullary glomeruli enter the medullary tissue and form long, looping capillaries of the vasa recta. This countercurrent system exchanges water and solutes, allowing the vasa recta to supply the medulla with blood without removing a large amount of NaCl used for the concentration of urine from the collecting ducts (CD). PST, proximal straight tubule; tAL, thin ascending limb; TAL, thick ascending limb. **C.** Vasa recta in the medulla. Large blood-filled elements of the vasa recta (VR) parallel the collecting ducts (CD) and the thick ascending limbs (TAL). Two neutrophils are visible in the vessels. (*continued on page 207*)

EXTRARENAL COLLECTING SYSTEM

It is commonly assumed that no further processing of the urine occurs after it enters the minor calyces in the kidney, although recent evidence suggests that small molecules may move across the epithelium lining the bladder. Certainly, the principal job of the **ureters**, **urinary bladder**, and **urethra** is to store urine until it is convenient to release it. Storage time may be considerable, and urine can contain toxic concentrations of ammonia and urea, varying between 50 and 1200 mOsm, with a pH ranging from 4.5 to 10. The chemical resistance of these structures is provided by the epithelial lining, the **urothelium**. The epithelium rests on a lamina propria to form a **mucosa**, which is surrounded by the **muscularis** and **adventitial** layers.

UROTHELIUM

The epithelium covering most of the extrarenal surfaces is specialized to prevent leakage of urine back to the body while undergoing considerable expansion. This tissue originally was called **transitional epithelium**, a term that now is being replaced by urothelium or uroepithelium.

Urothelium is a stratified epithelium. The luminal surface is covered with a single layer of **umbrella cells**, which are large, often binucleated cells (Figure 13-4D). Umbrella cells produce plasma membranes with unusual lipid and protein components that prevent diffusion of urine into the cells. In a relaxed state, these cells contain numerous disk-shaped vesicles that can be inserted in the membrane when stretched by increased urine volume in the lumen. Several layers of intermediate cells and a single layer of basal cells are located beneath the umbrella cells. The urothelium normally turns over extremely slowly, at a rate of about 3 to 6 months, but it can regenerate quickly if damaged.

URETER

The ureters include the initial structures lined by urothelium: the minor and major calyces, the renal pelvis, and the ureters proper, which connect each kidney to the urinary bladder. The stratified urothelium is a thickness of two to three layers of cells in the minor calyces and becomes a thickness of four to five layers of cells in the ureters (Figure 13-4E). The lamina propria is a dense connective tissue; there is no submucosa in the ureter. The muscularis of the ureter consists of bundles of smooth muscle cells surrounded by connective tissue; this muscle is arranged in two layers: an inner longitudinal layer and an outer circular layer. As the ureter approaches the bladder, a third layer of longitudinal muscle appears. The smooth muscle undergoes periodic peristaltic contractions to move urine to the bladder.

URINARY BLADDER

There are at least six layers of urothelial cells in a relaxed bladder (Figure 13-4F). A thin submucosa lies between the lamina propria and the muscularis. The muscularis, called the **detrusor** muscle, contains three layers with varying orientations; the middle layer is the thickest. The ureters enter the bladder at an oblique angle, and as the bladder fills with urine and expands, the openings of the ureters are pinched closed. The upper portion of the bladder's adventitia is covered with a mesothelium.

URETHRA

The urethra conducts urine from the bladder outside the body when the detrusor muscle contracts. The mucosa of the urethra is highly folded and lined with urothelium at the beginning and stratified squamous at the end, where the lumen becomes continuous with the skin. In males, the initial **prostatic** portion of the urethra is lined with urothelium. As the urethra passes through the prostate gland, the epithelium changes to pseudostratified columnar and may contain areas of stratified columnar epithelium. The portion of the urethra that extends through the penis is about 15 cm in length and is initially lined with pseudostratified columnar epithelium (see Chapter 16). In females, the urethra is shorter, about 4 to 5 cm long, and, as in males, becomes lined with stratified epithelium at the end. The relatively short length of the female urethra is assumed to contribute to the higher incidence of urinary tract infections in women compared to men; for instance, pathogens would not have to travel as far to get to the bladder in women.

▽ A congenital defect in the development of the ureter is frequently associated with urinary tract infections in children. If a ureter is too short, its opening into the bladder is not pinched closed as the bladder fills and urine refluxes into the kidneys. This is known as **primary vesicoureteral reflux**. A blockage anywhere in the urinary system also can result in reflux, termed secondary vesicoureteral reflux. Persistent reflux can damage the kidneys. Vesicoureteral reflux is present in about 30% of patients who are diagnosed with the initial episode of a urinary tract infection caused by bacteria. This is among the most common acute bacterial infections diagnosed in children. ▽

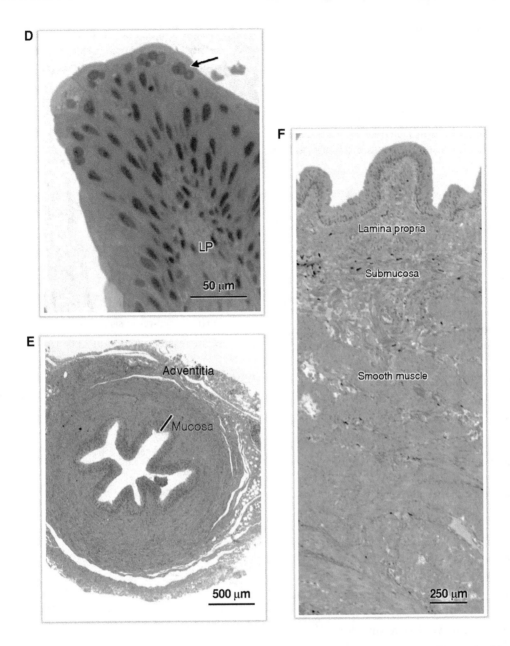

Figure 13-4: (*continued*) **D.** Urothelium. The stratified urothelium of the ureter is several cell layers thick. The umbrella cells at the surface are often binucleated or multinucleated (*arrow*). This epithelium sits on a lamina propria (LP), the dense connective tissue in the ureter and bladder. **E.** Ureter. The folded lumen of the ureter allows it to expand when filled with fluid. **F.** Urinary bladder. The bladder contains a thick wall of smooth muscle, arranged in three layers of varying orientation. The urothelium at the top of the section is supported by a dense lamina propria and an additional submucosal layer.

STUDY QUESTIONS

Directions: Each of the numbered items or incomplete statements is followed by lettered options. Select the **one** lettered option that is **best** in each case.

1. If the [NaCl] of the fluid in a thick ascending limb entering the juxtaglomerular apparatus is sensed by macula densa cells as being too high, a response is initiated that results in what change to a nearby structure?

   A. The afferent arteriole constricts

   B. The efferent arteriole constricts

   C. The extraglomerular mesangial cells migrate into the renal corpuscle

   D. The juxtaglomerular cells release renin

   E. The podocytes contract

2. Potentially therapeutic drugs may prove ineffective in clinical trials because they are modified by hepatocytes, released back into the blood, and then specifically eliminated by which cells?

   A. Collecting duct cells

   B. Distal convoluted tubule cells

   C. Kupffer cells

   D. Other hepatocytes

   E. Proximal convoluted tubule cells

3. A researcher studying kidney function in live rats is using microsurgical techniques. She injects a small amount of a fluorescent, water-soluble dye into the blood of an afferent arteriole entering a cortical renal corpuscle in the kidney. She then observes the dye's movement into the renal corpuscle and beyond. What is the most likely route that this dye will follow?

   A. All of the dye will travel down the proximal convoluted tubule

   B. All of the dye will leave via the efferent arteriole and then enter a stellate vein

   C. Some of the dye will travel down the proximal convoluted tubule and the remainder will exit via the efferent arteriole and then enter the stellate vein

   D. Some of the dye will travel down the proximal convoluted tubule and the remainder will exit via the efferent arteriole and then enter peritubular capillaries

   E. Some of the dye will travel down the proximal convoluted tubule and the remainder will exit via the efferent arteriole and then enter the vasa recta

4. What would be indicated by the presence of numerous disk-shaped vesicles in the cytoplasm of umbrella cells in the urothelium of the urinary bladder?

   A. High levels of antidiuretic hormone in the blood

   B. High levels of aldosterone in the blood

   C. The bladder is empty

   D. The bladder is full

5. A 48-year-old patient is seen in the clinic for recent onset of hypertension. Examination indicates stenosis of the right renal artery. What is the most likely explanation for the patient's hypertension?

   A. Decreased renin secretion from the right kidney

   B. Decreased secretion of antidiuretic hormone (ADH) from the pituitary

   C. Decreased urine production by the patient

   D. Increased renin secretion from the right kidney

   E. Increased secretion of ADH from the pituitary

   F. Increased urine production by the patient

6. The blood filtration unit of the kidney is formed by collaboration between endothelial cells and which other cells?

   A. Juxtaglomerular cells

   B. Mesangial cells

   C. Parietal cells

   D. Podocyte cells

   E. Proximal convoluted tubule cells

ANSWERS

1—A: A high concentration of NaCl in the fluid of the thick ascending limb indicates that flow through the nephron is too rapid. A signal is sent (the release of adenosine triphosphate) that causes the afferent arteriole to constrict, which will reduce the pressure gradient across the glomerular capillaries, decrease the formation of initial filtrate, and slow the rate of fluid movement through the nephron.

2—E: Hepatocytes can modify compounds present in the blood, such as therapeutic drugs, and then release them back into the circulation. Several kinds of modification reactions occur, for instance the addition of glucuronic acid (see Chapter 12). Proximal convoluted tubule cells contain transport systems that catalyze the movement of organic acids and bases from the blood into the filtrate. These acids and bases can include glucuronate-modified drugs, as well as toxins and metabolites that may or may not have been processed by the liver.

3—D: Blood pressure will force some of the dye out of the glomerular capillaries and into the initial filtrate, which then will travel down the proximal convoluted tubule. The remaining dye will exit the glomerulus in the efferent arteriole, which then will supply peritubular capillaries in the cortex. It is only the small number of renal corpuscles in the juxtamedullary cortex that supply the vasa recta.

4—C: The umbrella cells at the surface of urothelium contain disk-shaped vesicles that serve as reserve membrane that can be added to the surface as increasing fluid pressure expands the extrarenal structures. Thus, the presence of numerous vesicles in the cytoplasm of the bladder's umbrella cells would indicate that the bladder is empty and relaxed.

5—D: Stenosis (narrowing) of the right renal artery will chronically reduce blood pressure in the right kidney and, as a consequence, stimulate it to release high levels of renin. The overproduction of renin will produce hypertension in this patient, because the right kidney attempts to increase its blood flow. Renin production from the left kidney, which is experiencing high blood pressure, will decrease. Urine production should not be a problem because people can donate a kidney for transplantation and live a normal life with a single kidney. Increased antidiuretic hormone secretion would raise blood pressure, but that would not be expected to happen in this patient.

6—D: The filter unit of the kidney is composed of fenestrated endothelial cells, the basal lamina, and the filtration slits produced by podocyte cells. Mesangial cells are involved with the turnover of the basal lamina, but they do not directly contribute to the filtration unit.

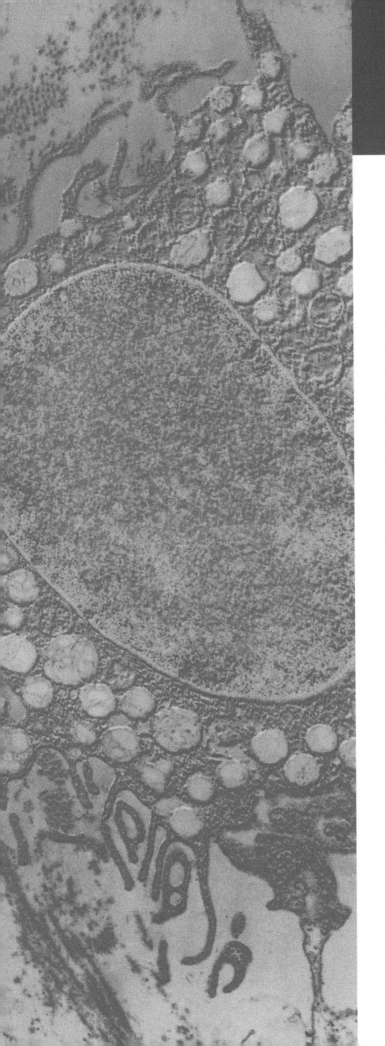

CHAPTER 14

ENDOCRINE SYSTEM

OVERVIEW

In previous chapters of this book, we have discussed the structure and function of the cells, tissues, and organs that comprise the human body. Two organ systems have evolved that integrate and coordinate the activities of these components—the nervous system and the endocrine system. The key activity of both systems is signaling. Signaling by the nervous system is rapid and specifically targeted, whereas signaling by the endocrine system is usually slower and more wide ranging. The activities of the two systems are closely linked and complementary.

The endocrine system consists of all the cells, tissues, and organs that secrete signaling molecules called **hormones**, a diverse array of peptides, lipids, and amino acid derivatives. Hormones are released into the circulation, bind to receptors on or in target cells, and function to regulate their activity. Hormone secretion itself is highly regulated by various mechanisms, including hormonal signals from the hypothalamus and the

211

pituitary gland, and by negative feedback (either direct or indirect) from target cells. The physiologic processes regulated by hormones include blood pressure and composition, metabolism and energy balance, and growth, development, and reproduction.

ENDOCRINE SIGNALING

Endocrine cells synthesize and secrete signaling molecules called **hormones**, which function to regulate the activity of other cells and organs to achieve a state of metabolic or physiologic balance called **homeostasis**. Most hormones are released into blood and lymphatic channels and circulate throughout the body to affect distant target cells, a process that is known as classic **endocrine signaling**. In other instances, hormones only affect neighboring cells (**paracrine signaling**) or act on the secretory cells themselves (**autocrine signaling**). Hormones exert their effects by binding to and activating **hormone receptors** on or in specific populations of **target cells**. Thus, although hormones may circulate widely, their effects are restricted to cells that express the appropriate receptors.

DISTRIBUTION OF ENDOCRINE CELLS

Endocrine cells are found in the following sites (Figure 14-1A):

- **Endocrine glands.** Endocrine cells comprise the entire parenchyma of specialized organs, such as the **pituitary (hypophysis), adrenal, thyroid, parathyroid**, and **pineal glands**, whose sole function is hormone production.

- **Discrete clusters of cells.** Groups of hormone-producing cells are found in organs that have other functions, such as the **pancreas, ovary, placenta**, and **testis** (see Chapters 12, 15, and 16).

- **Isolated individual cells.** Single hormone-producing cells of the **diffuse neuroendocrine system (DNES)** are interspersed among cells in the epithelial lining of other organs, primarily those in the digestive system and respiratory tract.

FEATURES OF ENDOCRINE GLANDS

The following features characterize endocrine organs:

- **Absence of ducts.** Unlike exocrine glands, which release secretory products into the ducts for delivery to an epithelial surface, endocrine organs lack ducts and release hormones into the bloodstream or lymphatic channels.

- **Abundant blood supply.** To facilitate the release of hormones into the bloodstream, endocrine glands are richly vascularized and capillaries are usually fenestrated.

- **Abundant parenchyma and scarce stroma.** Most endocrine glands are composed primarily of parenchymal cells, which usually are polyhedral epithelial cells arranged in anastomosing cords with at least one surface adjacent to a capillary or small lymphatic vessel. Other than the fine reticular fibers that support the vasculature and hormone-producing cells, there is minimal connective tissue.

CHARACTERISTICS OF HORMONES AND RECEPTORS

Most hormones can be grouped into three categories, based on their chemical structure.

1. **Peptide hormones.** This is the largest group of hormones and includes those produced in the hypothalamic nuclei, the anterior pituitary, and the islets of Langerhans. Peptide hormones have the following properties:

   - Synthesized in the rough endoplasmic reticulum (RER) of endocrine cells, typically as larger precursor molecules called **preprohormones**, which undergo posttranslational processing in the RER and Golgi apparatus.

   - Stored in secretory granules and released by exocytosis.

   - Water soluble and unable to cross the lipid bilayer of the cell membrane. They bind to receptors on the target cell surface, primarily **G-protein-coupled receptors** or **receptor-protein kinases**.

   - Binding of a hormone to a receptor on a target cell results in activation of an **intracellular signaling cascade**, which regulates the expression of genes required to mediate the response of the target cell (Figure 14-1B).

2. **Steroid hormones.** This group includes hormones produced in the gonads, adrenal cortex, and placenta. Steroid hormones have the following properties:

   - Synthesized, on demand, in the smooth endoplasmic reticulum (SER) and mitochondria from cholesterol, and are not stored.

   - Lipid soluble. Exported from endocrine cells and enter target cells by diffusing across the plasma membrane. Because of their low solubility in water, steroid hormones require a carrier protein in blood.

   - Bind to receptors in the cytoplasm. The hormone–receptor complex functions as a **transcription factor** to regulate gene transcription (Figure 14-1C).

3. **Amino-acid derivatives** include the catecholamines and thyroid hormones.

   - Synthesized from tyrosine.

   - Catecholamines are water soluble and bind to cell-surface receptors.

   - Thyroid hormones are lipid soluble; they require a carrier protein, diffuse through membranes, and bind to nuclear receptors.

REGULATION OF HORMONE SECRETION

In most cases, hormone secretion is regulated by a process termed **negative feedback** or **feedback inhibition**. When a hormone activates its target cell, an inhibitory signal is generated, either directly or indirectly, and halts hormone production. In a few cases, if the hormone level is insufficient to elicit an adequate response in the target cell, a **positive feedback** signal is released, triggering an increase in hormone secretion. These positive and negative feedback mechanisms regulate endocrine secretion to maintain homeostasis. Many hormones are secreted with a distinct circadian or pulsatile pattern, which is important for normal endocrine function.

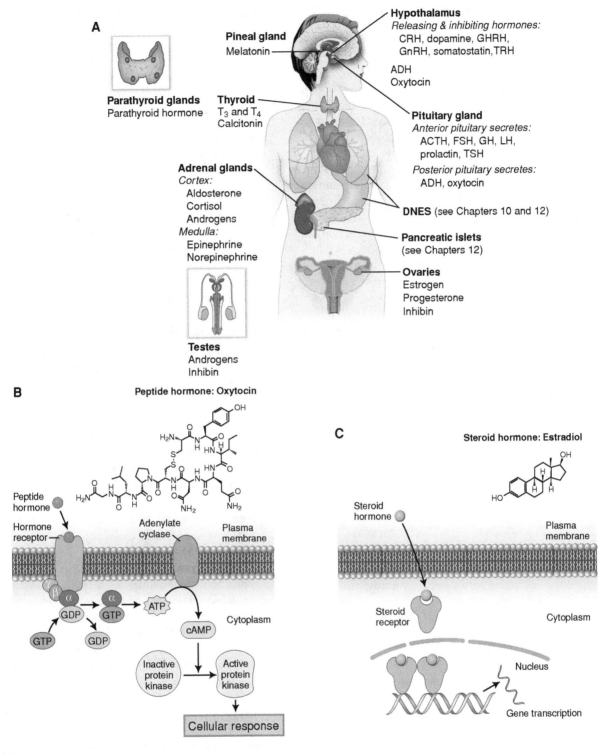

Figure 14-1: Location of endocrine organs and mechanisms of endocrine signaling in the human body. **A.** Location of endocrine organs and their major hormones. ACTH, adrenocorticotropic hormone; ADH, antidiuretic hormone; CRH, corticotropin-releasing hormone; DNES, diffuse neuroendocrine system; FSH, follicle-stimulating hormone; GH, growth hormone; GHRH, growth hormone-releasing hormone; GnRH, gonadotropin-releasing hormone; LH, luteinizing hormone; TRH, thyrotropin-releasing hormone; TSH, thyroid-stimulating hormone; T$_3$, triiodothyronine; T$_4$, thyroxine. **B.** Example of peptide hormone signaling through a G-protein-coupled receptor. When hormone binds to the receptor, the α-subunit dissociates from the G-protein and activates adenylate cyclase, which triggers the intracellular signaling cascade required for cellular response. The inset shows the structure of a representative peptide hormone (oxytocin). ATP, adenosine triphosphate; cAMP, cyclic adenosine monophosphate; GDP, guanosine diphosphate; GTP, guanosine triphosphate. **C.** Steroid hormone signaling. Hormone diffuses through the plasma membrane and binds to a cytoplasmic receptor. The hormone-receptor complex moves to the nucleus and functions as a transcription factor to regulate gene expression. The inset shows the structure of a representative steroid hormone (estradiol).

PITUITARY GLAND

The pituitary gland (**hypophysis**) is often called the "master gland." It orchestrates the activities of many other endocrine glands by producing hormones that are responsible, either directly or indirectly, for regulating growth, reproduction, ion and water balance, and metabolism. In fact, the pituitary is part of a "master pair," as it works in tandem with the **hypothalamus**, to which it is intimately connected. The activities of these two structures are tightly integrated and known as the **hypothalamic-pituitary axis (HP axis)** or the **hypothalamohypophyseal system**. The connections between the hypothalamus and the pituitary gland enable neural input from the internal and external environment to regulate hormone secretion from the pituitary, which in turn affects the function of other cells and organs in the body. A considerable variety of hormones are produced and released into the blood supply of the hypothalamohypophyseal system. All of the hormones are peptides and are stored in granules.

STRUCTURE OF THE PITUITARY GLAND

The pituitary is a small, bean-shaped gland about the size of a pea, and is located at the base of the brain in a cavity called the **sella turcica** (Figure 14-2A). The pituitary has two distinct components or **lobes** (Figure 14-2B):

- **Anterior pituitary (adenohypophysis).** The glandular anterior pituitary comprises 75–80% of the pituitary. The anterior pituitary is further subdivided into the **pars distalis (pars anterior), pars intermedia**, and **pars tuberalis**.
- **Posterior pituitary (neurohypophysis).** The neural **posterior pituitary** comprises 20–25% of the pituitary. The posterior pituitary is subdivided into the large **pars nervosa** and the smaller **infundibulum (infundibular, neural, or pituitary stalk)**, which is attached to the hypothalamus at the **median eminence**.

The anterior and posterior lobes of the pituitary differ in embryonic origin, structure, and function, but are joined into a single gland that is covered with a dense connective tissue capsule derived from the dura mater.

RELATIONSHIP OF THE PITUITARY GLAND AND HYPOTHALAMUS

The pituitary gland is physically and functionally connected to the hypothalamus by neural and vascular pathways in the infundibulum (Figure 14-2A). It is through these connections that the actions of the nervous and endocrine systems are integrated and coordinated. The hypothalamus receives and integrates neural input from the internal and external environment and from other regions of the brain. It stimulates the appropriate endocrine response, either by secreting **releasing or inhibiting hormones** (also called factors) that regulate hormone secretion by the anterior pituitary, or by secreting hormones themselves (e.g., oxytocin or antidiuretic hormone, known as ADH). Secretion of almost all pituitary hormones is controlled by either hormonal or neural signals from the hypothalamus. In many cases, feedback inhibition from the target reduces production of the hypothalamic-releasing or hypothalamic-inhibiting hormones, in addition to directly diminishing production of the pituitary hormone.

EMBRYOGENESIS OF THE PITUITARY GLAND

The anterior and posterior lobes of the pituitary gland originate from different embryonic structures (Figure 14-2C).

- **Anterior pituitary.** Develops as an outgrowth of surface ectoderm lining the roof of the mouth, called **Rathke's pouch**. The outgrowth subsequently detaches, and the anterior wall thickens to become the **pars distalis**. The posterior wall differentiates to the **pars intermedia**, and the dorsolateral portions extend around the infundibulum as the **pars tuberalis**.
- **Posterior pituitary.** Develops as a downgrowth from the floor of the diencephalon and remains attached to the brain via the infundibulum. The posterior pituitary, therefore, can be considered an extension of the hypothalamus.

BLOOD SUPPLY OF THE PITUITARY GLAND

The pituitary gland is richly vascularized by the superior and inferior hypophyseal arteries, which give rise to the hypophyseal portal system and to the capillaries in the pars nervosa, respectively (Figure 14-2B).

- **Hypophyseal portal system.** Consists of a **primary capillary plexus** in the median eminence and infundibulum and a **secondary capillary plexus** in the pars distalis, connected by long **hypophyseal portal veins** in the pars tuberalis. The portal system forms an essential link between the hypothalamus and the endocrine system, and is the route by which hypothalamic-releasing and -inhibiting hormones reach their target cells in the pars distalis to control pituitary function. Capillaries in primary and secondary plexuses are fenestrated, which facilitates the release of both hypothalamic and pituitary hormones.
- **Pars nervosa vasculature.** The pars nervosa is supplied by an additional capillary plexus, which is connected to the plexus in the pars distalis by short portal veins. Blood from both the anterior and posterior lobes of the pituitary drain into the **cavernous sinus** and then into the systemic circulation (Figure 14-2B).

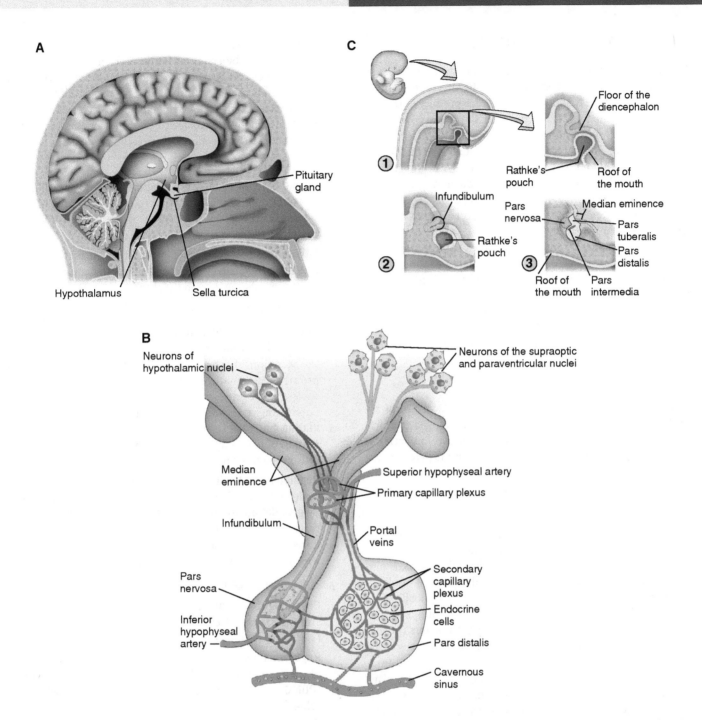

A. Pituitary gland

Hypothalamus   Sella turcica

C.
Floor of the diencephalon
Rathke's pouch   Roof of the mouth
Infundibulum   Pars nervosa   Median eminence
Rathke's pouch   Pars tuberalis   Pars distalis
Roof of the mouth   Pars intermedia

B.
Neurons of hypothalamic nuclei
Neurons of the supraoptic and paraventricular nuclei
Median eminence
Superior hypophyseal artery
Primary capillary plexus
Infundibulum
Portal veins
Pars nervosa
Secondary capillary plexus
Endocrine cells
Inferior hypophyseal artery
Pars distalis
Cavernous sinus

Figure 14-2: Location, organization, and development of the pituitary gland. **A.** Location of the pituitary gland. The pituitary is attached to the hypothalamus at the base of the brain. **B.** Sites of hormone production, storage, and release, and the blood supply of the pituitary gland. Releasing and inhibiting hormones (*purple*) are produced in the hypothalamic nuclei and are stored and released in the median eminence; hormones reach the pars distalis via the hypophyseal portal system. Anterior pituitary hormones (*green*) are produced and secreted by cells in the pars distalis. Posterior pituitary hormones (*orange*) are produced in the hypothalamic supraoptic and paraventricular nuclei and stored and released in the pars nervosa. **C.** Embryonic origin of the pituitary gland. The anterior pituitary gland develops as an outgrowth of the roof the mouth, which eventually detaches from the oral epithelium. The posterior pituitary gland develops as a downgrowth from the floor of the diencephalon and remains attached to the brain via the infundibulum.

ANTERIOR PITUITARY GLAND

The anterior pituitary gland (adenohypophysis) consists of three regions—the **pars distalis (pars anterior)**, the **pars intermedia**, and the **pars tuberalis**—all of which produce hormones (Figure 14-3A and B). Cells in the anterior pituitary synthesize a broad spectrum of hormones; most are **tropins** that regulate the function of other endocrine cells, including secretory cells in the gonads, thyroid, and adrenal cortex (Figure 14-3A). Anterior pituitary hormones influence reproduction, growth and metabolism, and ion balance. Their secretion is regulated by complex interactions involving the hypothalamus, anterior pituitary, and target organs.

Cells in the anterior pituitary produce peptide hormones and, therefore, have well developed RER and an abundance of secretory granules. The cells were originally classified as **acidophils** or **basophils**, based on the staining properties of their granules (Figure 14-3A and C). Immunostaining with hormone-specific antibodies now has shown that cells in the pars distalis produce at least six different hormones and that most cells produce only a single hormone (Figure 14-3A).

PARS DISTALIS

The largest and most important region of the anterior pituitary is the pars distalis. Cells in this region are arranged in thick, branching cords that border on wide fenestrated capillaries. A network of fine reticular fibers supports both the parenchymal cells and the capillaries (Figure 14-3C).

The following hormones are produced in the pars distalis:

SOMATOTROPIN

Physiologic function. Somatotropin, also known as **growth hormone (GH)**, promotes body growth by increasing protein synthesis and cellular proliferation throughout the body. GH also has acute metabolic effects that oppose the effects of insulin—increasing blood glucose levels and stimulating fat mobilization. Some effects of somatotropin are indirect. Somatotropin stimulates the liver to produce **insulin-like growth factors** (primarily **IGF-1** or **somatomedin C**), which then mediate many of the effects attributed to somatotropin. For example, IGF-1 stimulates mitosis of chondrocytes in the epiphyseal growth plate, thereby promoting long bone growth. Somatotropin levels (and hence IGF-1 levels) decrease at puberty, which contributes to closure of the epiphyseal plate (see Chapter 4).

Regulation. Somatotropin secretion is regulated by the balance between the stimulatory factor, **growth hormone-releasing hormone (GHRH)**, and the inhibitory factor, **somatostatin**, both of which are produced by neurosecretory cells in the hypothalamus. GH is released in pulsatile bursts, primarily at night. It also is released in response to acute stress, which serves to increase blood glucose levels.

PROLACTIN

Physiologic function. Prolactin promotes the development of mammary glands during pregnancy and stimulates milk production after parturition. Prolactin-producing cells and their secretory granules increase in size and numbers during pregnancy and lactation.

Regulation. Prolactin is the only anterior pituitary hormone that is not regulated by a releasing hormone from the hypothalamus. Instead, prolactin secretion is tonically inhibited by **dopamine** (also known as **prolactin inhibitory factor**, or **PIF**) from the hypothalamus, and is released in response to decreased dopamine levels. In addition, **oxytocin** released from the posterior pituitary during breast-feeding stimulates prolactin production.

THYROID-STIMULATING HORMONE

Physiologic function. Thyroid-stimulating hormone (TSH), or **thyrotropin**, regulates thyroid function by stimulating the synthesis, storage, and liberation of the thyroid hormones thyroxine (T_4) and triiodothyronine (T_3).

Regulation. Secretion of TSH is stimulated by **thyrotropin-releasing hormone (TRH)** from the hypothalamus and inhibited by circulating thyroid hormones.

FOLLICLE-STIMULATING HORMONE AND LUTEINIZING HORMONE

Physiologic function. In females, follicle-stimulating hormone (**FSH**) stimulates the growth of ovarian follicles and the secretion of estrogen (see Chapter 15). In males, FSH promotes spermatogenesis and stimulates Sertoli cells to secrete androgen-binding protein (ABP) (see Chapter 16). Luteinizing hormone (**LH**) (known as **interstitial cell-stimulating hormone** in males) induces ovulation and the formation of corpora lutea in the ovary, and stimulates Leydig cells in the testis to synthesize and release testosterone.

Regulation. Individual cells in the pars distalis can synthesize both FSH and LH. Secretion of both hormones is stimulated by **gonadotropin-releasing hormone (GnRH)** from the hypothalamus, and is inhibited by estrogens and androgens produced in the ovaries and testes, respectively. In addition, inhibin, which is also produced in the gonads, inhibits FSH secretion.

ADRENOCORTICOTROPIC HORMONE

Physiologic function. Adrenocorticotropic hormone (**ACTH**) stimulates the adrenal cortex to produce glucocorticoids (e.g., cortisol and corticosterone).

Regulation. ACTH is derived from posttranslational cleavage of a large precursor molecule called pro-opiomelanocortin (**POMC**). Secretion of ACTH is stimulated by **corticotrophin-releasing hormone (CRH)** from the hypothalamus and inhibited by circulating glucocorticoids.

PARS INTERMEDIA

The pars intermedia is rudimentary and ill-defined in humans and consists of a few vestigial colloid-filled cysts lined by basophilic cuboidal epithelium (Figure 14-3B). Cells in the pars intermedia are immunoreactive for corticotropic hormones and most likely produce minor derivatives of POMC, such as β-lipotropic pituitary hormone (β-LPH), α-melanocyte-stimulating hormone (α-MSH), or β-endorphin.

PARS TUBERALIS

The pars tuberalis consists of a thin collar of cuboidal basophilic epithelial cells on the anterior surface of the infundibulum. Cells in the pars tuberalis secrete FSH and LH.

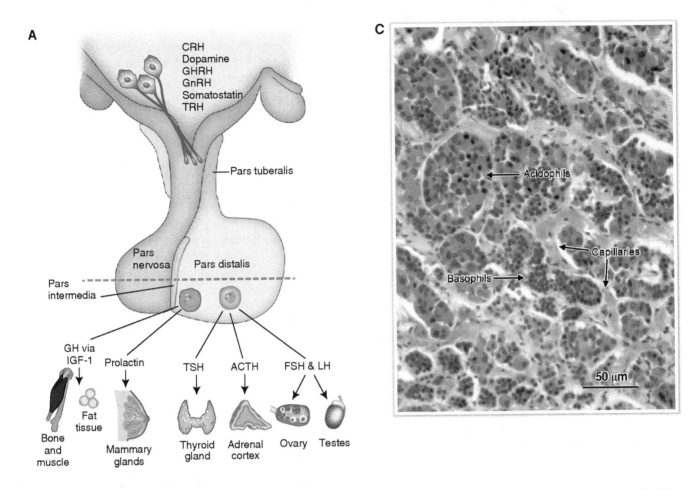

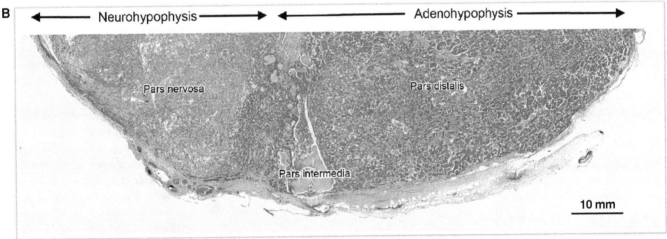

Figure 14-3: Structure and function of the anterior pituitary gland. **A.** Hormones produced in the pars distalis by basophils (*blue*) and acidophils (*orange*), the target organs affected by each hormone, and the hypothalamic-releasing and hypothalamic-inhibiting hormones that regulate the function of the pars distalis. The dashed line indicates the level of the histologic section shown in part B. ACTH, adrenocorticotropic hormone; CRH, corticotropin-releasing hormone; FSH, follicle-stimulating hormone; GH, growth hormone; GHRH, growth hormone-releasing hormone; GnRH, gonadotropin-releasing hormone; IGF-1, insulin-like growth factor-1; LH, luteinizing hormone; TRH, thyrotropin-releasing hormone. **B.** Sagittal section of the pituitary gland at the level of the dashed line in part A. Note the striking difference in appearance of the anterior and posterior pituitary glands and the presence of colloid in the pars intermedia. **C.** Section of the pars distalis at high magnification. Note the difference in staining properties of acidophils and basophils and the abundance of capillaries.

POSTERIOR PITUITARY GLAND

The posterior pituitary gland (**neurohypophysis**) develops as a downgrowth from the diencephalon. It consists of the large pars nervosa and the smaller infundibulum, and is connected to the hypothalamus by the median eminence. Although there are no hormone-producing cells in the posterior pituitary, hormones synthesized in the hypothalamic nuclei are stored and released in both the median eminence and the pars nervosa (Figure 14-4A).

MEDIAN EMINENCE

Hypothalamic-releasing and -inhibiting hormones that regulate the function of the anterior pituitary are stored in axons in the median eminence and are discharged into the primary capillary plexus in the median eminence. These hormones include CRH, dopamine, GHRH, GnRH, somatostatin, and TRH (Figure 14-1A).

INFUNDIBULUM AND PARS NERVOSA

The pars nervosa secretes two hormones, **antidiuretic hormone (ADH)** and **oxytocin**, which are synthesized primarily by neurosecretory cells in the **supraoptic and paraventricular nuclei of the hypothalamus**, respectively. Axons of these neurons extend in the **hypothalamohypophyseal tract** in the infundibulum and terminate within the network of fenestrated capillaries of the pars nervosa. These unmyelinated axons and their terminals comprise the parenchyma of the pars nervosa, occupying approximately 75% of the lobe. The remaining 25% consists of supporting glial cells similar to astrocytes, called **pituicytes** (Figure 14-4B).

ADH and oxytocin are small cyclic peptides, each containing just nine amino acids. Both are synthesized as a larger prohormone molecule that also contains **neurophysin**, a carrier protein that assists in transporting the hormone to the axon terminals. The prohormone is cleaved to its active form, is stored in secretory granules in the axon terminals, and is released by action potentials triggered by stimuli impinging on neurons in the supraoptic and paraventricular nuclei.

ANTIDIURETIC HORMONE

- **Physiologic function.** ADH is the principal hormone controlling water balance in the body. It acts to concentrate urine by increasing the permeability of the collecting tubules of the kidneys to water, thereby causing water to be resorbed rather than excreted (see Chapter 13). ADH is also called **vasopressin** because it causes constriction of arterioles.

- **Regulation.** Secretion of ADH is controlled via signals from the hypothalamus, triggered primarily in response to changes in the osmolarity of body fluids and blood volume.

OXYTOCIN

- **Physiologic function.** Oxytocin causes contraction of the uterine smooth muscle at childbirth and during orgasm, and contraction of myoepithelial cells in the mammary glands during breast-feeding, which assists in milk let-down and ejection. In addition, oxytocin promotes maternal behavior and stimulates prolactin secretion, and may play a role in pair bonding and building trust in both men and women.

- **Regulation.** Distension of the cervix and uterus during labor and stimulation of the nipples during suckling send neural signals to the hypothalamus, which trigger oxytocin production and release.

▽ A synthetic version of oxytocin (e.g., Pitocin) is used clinically to help induce labor and to control postpartum bleeding. ▽

DISORDERS OF PITUITARY FUNCTION

DISORDERS OF THE ANTERIOR PITUITARY GLAND

Most disorders of the pituitary gland result from slow-growing tumors (**adenomas**) that arise from pituitary endocrine cells. Pituitary adenomas usually are composed of a single cell type and produce a single hormone, although some adenomas appear to produce no hormone (Figure 14-4C). Pituitary adenomas are relatively common; high resolution imaging has revealed that approximately 20% of otherwise normal healthy adults have small pituitary tumors but experience no associated symptoms. Three general categories of disorders caused by pituitary tumors are as follows:

- **Hypersecretion.** The clinical symptoms that result from excess hormone production by an adenoma depend on the cell type from which the tumor is derived. For example, the most frequent type of hyperfunctioning adenomas secrete excessive amounts of prolactin, resulting in amenorrhea, galactorrhea, loss of libido, and infertility. In contrast, hypersecretion of ACTH is the most common cause of the adrenal disorder known as Cushing's disease (see Disorders of Adrenal Gland Function, below).

- **Hyposecretion.** Decreased secretion of hormones from the pars distalis can result from disorders that interfere with the delivery of hypothalamic-releasing hormones or from damage to the pars distalis itself.

- **Tumor mass effects.** As pituitary adenomas enlarge, they can cause various clinical symptoms that are not directly related to endocrine dysfunction. For example, because the pituitary lies near the optic nerve and optic chiasm, adenomas in this region may compress optic fibers and cause difficulties with vision (Figure 14-4D). Similar to any intracranial mass, a pituitary adenoma can cause increased intracranial pressure, resulting in headaches and nausea and vomiting.

SYNDROMES OF THE POSTERIOR PITUITARY GLAND

Clinically important syndromes of the posterior pituitary gland involve excess or inadequate production of ADH.

- **Hypersecretion.** Pathologic overproduction of ADH, termed **syndrome of inappropriate ADH (SIADH)**, results in the retention of solute-free water and hyponatremia (low Na$^+$); in extreme cases, it can be fatal.

- **Hyposecretion.** Individuals with a lesion of the hypothalamus that reduces the production of ADH lose the capacity to concentrate urine, a condition known as **central diabetes insipidus**. The affected individual excretes excessive amounts of dilute urine and can become dehydrated.

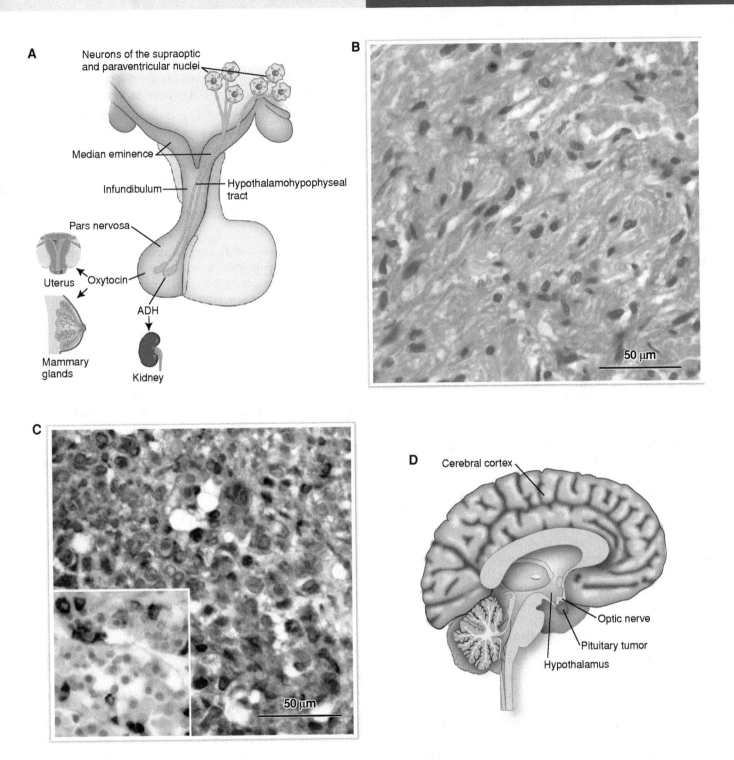

Figure 14-4: Structure and function of the posterior pituitary gland and pituitary adenomas. **A.** Posterior pituitary hormones and target organs affected by each hormone. Hormones are produced in the hypothalamic paraventricular and supraoptic nuclei and are transported via the hypothalamohypophyseal tract to the pars nervosa, where they are stored and released. ADH, antidiuretic hormone. **B.** Section of the pars nervosa. The pars nervosa consists of the axons and nerve terminals of hormone-producing cells, as well as glia cells, called pituicytes. Cell bodies of the neurosecretory cells reside in the hypothalamus. **C.** Section of a pituitary adenoma immunostained for prolactin. The inset shows a section of normal pituitary at the same magnification, also stained for prolactin. Note that almost all cells in the adenoma produce prolactin (*brown stain*). Sections courtesy of Amy Lowichik, MD, University of Utah School of Medicine, Salt Lake City, Utah. **D.** Pituitary adenoma impinging on the optic nerve. Benign pituitary tumors in this location can result in difficulties with vision.

THYROID GLAND

The thyroid gland is a single bilobed gland, located anterior to the trachea and inferior to the larynx. It originates as an outgrowth of the embryonic pharynx, and is the first endocrine organ to develop. The thyroid begins to function at about 10 weeks' gestation; its secretions are essential for normal development. The thyroid is covered with a thin, connective tissue capsule, from which septa penetrate into the parenchyma, dividing it into irregular lobules. The parenchyma consists of millions of spherical glandular units called **follicles**, which secrete the thyroid hormones **thyroxine (tetraiodothyronine, T_4)** and **triiodothyronine (T_3)**, whose main function is to regulate basal metabolic rate. In addition, the thyroid contains a small population of **parafollicular cells** derived from the neural crest that secrete **calcitonin**, a minor hormone involved in calcium homeostasis. The thyroid is the only endocrine organ that stores a large quantity of hormone (2–3 month supply). Another unusual feature of the thyroid is that it stores hormone extracellularly, as a macromolecular precursor called **thyroglobulin**.

THYROID FOLLICLES

The thyroid follicle is the structural and functional unit of the thyroid gland. Each follicle is a small sphere-shaped storage compartment (50–900 μm in diameter), with walls composed of simple epithelial cells ranging in height from flattened to cuboidal (Figure 14-5A and B). The height of the epithelium reflects the functional state of the follicular cells; cuboidal cells are metabolically active, whereas flattened cells are quiescent. Follicular cells have microvilli on their apical surface and are joined at their apical edges by tight junctions. Each follicle is surrounded by basal lamina and is separated from other follicles by fine reticular connective tissue containing a rich network of fenestrated capillaries. The lumen of each follicle is filled with amorphous colloid containing the glycoprotein **thyroglobulin**, which plays an important role in the synthesis and storage of thyroid hormones.

THYROID HORMONES

Synthesis and secretion. The thyroid hormones T_3 and T_4 are iodinated derivatives of the amino acid tyrosine, and are produced from the glycoprotein thyroglobulin (Figure 14-5C). Briefly, follicular cells synthesize thyroglobulin and secrete it into the follicular lumen, where tyrosine residues on thyroglobulin are iodinated and the iodinated thyroglobulin is stored as a colloid. When thyroid hormones are needed, follicular cells phagocytose the colloid, cleave it into T_3 and T_4, and release the active hormones into the bloodstream. Roughly 90% of the secreted thyroid hormone is T_4; the remaining 10% is T_3, which is more potent and acts more rapidly than T_4. T_4 can be deiodinated to T_3 (or to an inactive form of T_3) in peripheral tissues, which will allow local regulation of hormone activity.

Physiologic function. Thyroid hormones have widespread, diverse, long-lasting effects, and influence the activity of nearly every organ system in the body. The hormones stimulate the transcription of an array of genes whose combined effect is a generalized increase in cellular metabolism. Thyroid hormones are essential for normal growth and development during childhood.

▽ Insufficient hormone function during fetal life or early childhood results in mental retardation, short stature, and coarse facial features, a condition formerly called **cretinism**, but now more properly called **congenital** or **neonatal hypothyroidism**. ▽

Regulation. Secretion of thyroid hormones is regulated by a classic negative feedback loop in which TRH produced in the hypothalamus stimulates the release of TSH from the anterior pituitary, which then triggers the secretion of T_3 and T_4 from the thyroid. Circulating T_3 and T_4 feedback and inhibit the release of both TRH and TSH. The amount of inhibition is proportional to the concentration of circulating thyroid hormones (Figure 14-5D).

▽ **Thyroid disorders** are common throughout the world. Both hypothyroidism and hyperthyroidism can result in chronic enlargement of the gland, termed a **goiter**. The most frequent cause of goiter worldwide is hypothyroidism that results from insufficient dietary iodine. Lack of iodine impedes the synthesis and release of thyroid hormones, causing TSH levels to increase and stimulate additional colloid production and compensatory growth of the gland. In developed countries, where iodine is commonly added to table salt, iodine deficiency is rare. The most common form of hypothyroidism in areas where iodine levels are sufficient is **Hashimoto's thyroiditis**, an autoimmune disorder that destroys thyroid follicular cells and is most prevalent in women between 45 and 65 years of age. The most common form of hyperthyroidism, also an autoimmune disorder, is **Graves' disease**, in which autoantibodies to the TSH receptor bind to the receptor and chronically stimulate follicular cells to synthesize and release hormones.

Clinical symptoms of thyroid disorders result from an imbalance of anabolism and catabolism. **Hypothyroidism** is characterized by weight gain, lack of energy, and fatigue, sluggishness, and constipation. **Hyperthyroidism** is characterized by weight loss, heat intolerance (due to increased basal metabolic rate), tremor, rapid pulse, and bulging eyes (**exophthalmos**). ▽

PARAFOLLICULAR CELLS

Structure. Parafollicular cells (also known as **C cells**) differ from follicular cells in their origin, location, appearance, and function. These cells are derived from the neural crest and migrate into the developing thyroid to reside between the follicular cells and basal lamina or in the interstitium (Figure 14-5A). Parafollicular cells are larger than thyroid cells. They stain less intensely and store their secretory product, the polypeptide hormone **calcitonin**, in granules. Because of their origin and location, parafollicular cells can be considered as part of the DNES.

Physiologic function. In animals, calcitonin functions to reduce the levels of Ca^{2+} in the blood, principally by decreasing the activity of osteoclasts, which reduces bone resorption and the release of Ca^{2+} (see Chapter 4). In addition, calcitonin promotes the excretion of Ca^{2+} by the kidneys. Curiously, calcitonin appears to have only weak effects on Ca^{2+} homeostasis in humans. There are no serious clinical consequences from calcitonin excess or deficiency.

Regulation. There are no hormones that stimulate or inhibit calcitonin secretion. Instead, secretion is triggered by an increase in blood Ca^{2+} levels and inhibited by decreased Ca^{2+}.

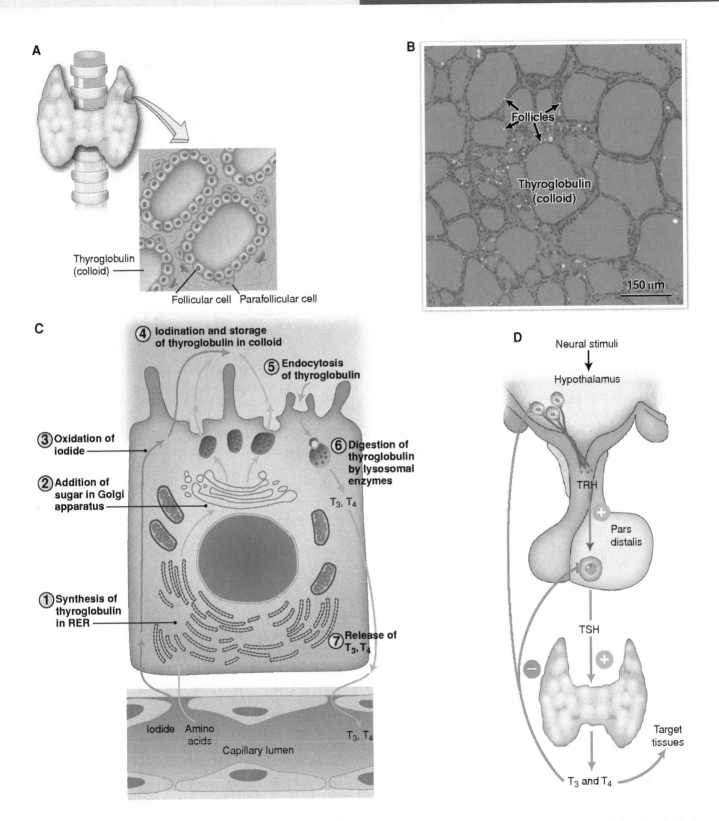

Figure 14-5: Structure and function of the thyroid gland. **A.** Location of the thyroid gland and organization of thyroid follicles. Parafollicular cells lie outside the follicular epithelium and are difficult to distinguish without immunostaining: **B.** Section of the thyroid gland. Note the variation of the height of the cells in the different follicles, which reflects their differing functional states. **C.** Steps in the synthesis (*red and green arrows*) and release (*blue and orange arrows*) of thyroid hormones. **D.** Regulation of thyroid hormone synthesis and release. Thyroid-releasing hormone (TRH) from the hypothalamus stimulates secretion of thyroid-stimulating hormone (TSH) from the pars distalis, which triggers the production and secretion of thyroid hormones. The hormones feedback to inhibit secretion of both TRH and TSH. T_3, triiodothyronine; T_4, thyroxine.

PARATHYROID GLANDS

The parathyroid glands are small endocrine organs located on the back of the thyroid gland, within its connective tissue sheath (Figure 14-6A). **Parathyroid hormone (PTH)**, together with vitamin D, regulates Ca^{2+} and phosphate homeostasis, and is essential for life. Most individuals have four parathyroid glands, but the number can vary from two to six. Each gland is ensheathed by a thin, connective tissue capsule, which extends fine septa that incompletely divide the gland into lobules. Parathyroid glands contain two types of parenchymal cells: **chief cells** and **oxyphil cells**. Adipocytes begin to appear in parathyroid glands about the time of puberty and increase with age; eventually as much as 50% of each gland is comprised of adipocytes (Figure 14-6B).

CHIEF CELLS

- **Structure.** Chief cells, the principal functional cells in parathyroid glands, are small cells characterized by round nuclei, abundant RER, and secretory granules. They are arranged in cords or sheets, surrounded by a rich capillary network and loose reticular fibers.

- **Physiologic function.** Chief cells secrete the peptide hormone **PTH**, which functions to increase blood Ca^{2+} levels and promote phosphate excretion. As described in Chapter 4, PTH binds to receptors on osteoblasts, which ultimately results in an increase in the number and activity of osteoclasts; the activated osteoclasts degrade bone matrix and release Ca^{2+} and phosphate into the blood. PTH also stimulates Ca^{2+} reabsorption and phosphate excretion by the kidney tubules, and promotes the absorption of dietary Ca^{2+} from the small intestine by stimulating the production of vitamin D, which is required for Ca^{2+} uptake.

- **Regulation.** Similar to calcitonin, there are no hormones that stimulate or inhibit PTH secretion. PTH secretion is triggered by low blood Ca^{2+} levels, and is reduced as Ca^{2+} levels return to normal.

▽ Both increased and decreased secretion of PTH can have serious clinical manifestations. Excessive secretion of PTH can arise from disease of the parathyroid gland, **primary hyperparathyroidism**, most commonly caused by a PTH-secreting adenoma, or from a disorder unrelated to the parathyroids, **secondary hyperparathyroidism**; for example, a kidney disease that hinders Ca^{2+} resorption will cause blood Ca^{2+} levels to decrease, resulting in the continual secretion of PTH to restore the Ca^{2+} levels in blood. Chronic hyperparathyroidism results in hypercalcemia (elevated blood Ca^{2+} levels), decalcification of bone, and kidney stones. Paradoxically, intermittent administration of PTH can stimulate an increase in bone mass, and appears to be an effective therapy in the treatment of osteoporosis.

Hypoparathyroidism most commonly is the result of disease processes that destroy the glands or from injury or removal of the glands during surgery. The resulting hypocalcemia can cause increased bone density, mental confusion, memory loss, tetany, and convulsions, and can be life-threatening. Calcium infusions, oral calcium supplements, and vitamin D are used to restore normal calcium levels in patients with hypoparathyroidism. ▽

OXYPHIL CELLS

Oxyphil cells are less numerous than chief cells and occur as a single cell or in small clusters. They are slightly larger and more eosinophilic than chief cells, and are tightly packed with mitochondria and glycogen granules. Oxyphil cells first appear about the time of puberty and increase in numbers with aging. They have no known secretory product. Oxyphil cells may represent chief cells that have become quiescent.

PINEAL GLAND

The pineal gland (**pineal body** or **epiphysis cerebri**) is a single, tiny endocrine organ shaped like a pinecone and located between the superior colliculi in the brain (Figure 14-6C). It is composed of parenchymal cells, called **pinealocytes**. Pinealocytes produce **melatonin**, a hormone that functions to transduce information about environmental daylight and darkness into biologic rhythms.

- **Structure.** The pineal gland is covered with pia mater, and is attached to the roof of the diencephalon; surprisingly, it is not innervated by central neurons. Pinealocytes are arranged in cords or rosettes, separated into incomplete lobules by septa extending from the pia covering. The pineal gland is relatively nondescript when seen in standard histologic sections, and is most easily identified by the presence of hydroxyapatite concretions of unknown function, called **brain sand** (Figure 14-6D).

- **Physiologic function and regulation. Melatonin** secretion from pinealocytes is promoted by darkness and inhibited by daylight. Information about ambient light conditions is conveyed to the pineal gland via a circuitous route that involves specialized light-sensitive retinal ganglion cells, the hypothalamic suprachiasmatic nucleus, and the superior cervical ganglion (Figure 14-6E). Postganglionic superior cervical ganglion axons terminate on pinealocytes and regulate melatonin secretion by releasing norepinephrine. The resulting circadian changes in melatonin levels cause rhythmic changes in the activity of other endocrine organs, including the hypothalamus, pituitary, and gonads.

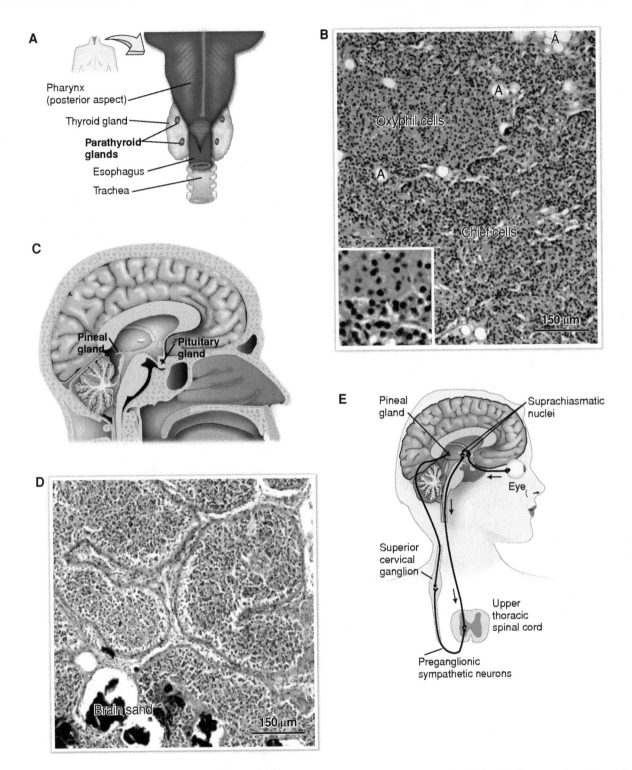

Figure 14-6: Location and histology of the parathyroid and pineal glands. **A.** Location of the parathyroid glands. Four small parathyroid glands lie at the posterior of the thyroid gland, embedded in its connective tissue sheath. **B.** Section of a parathyroid gland from an older person; note the abundance of oxyphil cells and adipocytes (A). The inset shows oxyphil and chief cells at higher magnification. **C.** Location of the pineal gland. **D.** Section of the pineal gland, which is most easily recognized by the presence of brain sand. **E.** Neural pathway by which ambient light modulates melatonin secretion from the pineal gland.

ADRENAL GLANDS

The adrenal (**suprarenal**) glands are paired organs located at the top of each kidney (Figure 14-7A). In adults, each gland usually is about 4–6 cm long and weighs about 4 g but can vary in size and weight, depending on the individual's age and physiologic condition. The adrenal glands play an important role in the maintenance of body water and Na$^+$ balance, in the control of blood pressure, and in regulating the body's adaptive response to stress.

STRUCTURE OF THE ADRENAL GLANDS

Each adrenal gland is composed of two distinct regions that differ in origin, structure, and function (Figure 14-7A and B).

Adrenal cortex. The larger, outer adrenal cortex is derived from mesoderm. It secretes an array of steroid hormones and is regulated largely by ACTH. Hormones produced by the adrenal cortex are essential for life.

Adrenal medulla. The inner adrenal medulla is derived from the neural crest. It secretes catecholamines and is regulated by the sympathetic nervous system. Hormones secreted by the adrenal medulla are important but are not essential.

Each adrenal gland is covered with a dense connective tissue capsule. Fine septa accompanied by blood vessels and nerves extend from the capsule into the gland. Parenchymal cells in both the cortex and medulla are arranged in cords along fenestrated capillaries.

BLOOD SUPPLY OF THE ADRENAL GLANDS

The blood supply to the adrenal glands is arranged as follows:

Cortex. Short arteries form a network of capillaries between cells and drain into capillaries in the adrenal medulla.

Medulla. Long arteries pass through the cortex to form an extensive capillary network in the adrenal medulla.

Because of this arrangement, cells in the deeper cortical layers and in the medulla are exposed to steroids produced in more superficial cortical layers, which can influence their function (Figure 14-7B).

ADRENAL CORTEX

The adrenal cortex produces a diversity of steroid hormones. Its parenchymal cells have an ultrastructural appearance that is characteristic of steroid-secreting cells—abundant SER and mitochondria (where steroids are synthesized) and lipid droplets (where cholesterol is stored). Steroids are synthesized and released on demand and are not stored in granules. Steroid hormones are low molecular weight and lipid soluble and, as a result, they are released by diffusing through the plasma membrane without requiring exocytosis.

The secretory products of the adrenal cortex are classified into three general categories, **mineralocorticoids, glucocorticoids,** and **androgens,** named for the primary effects that the hormones mediate.

The adrenal cortex is subdivided into three concentric zones or layers that differ in their histologic appearance and secreted hormones (Figure 14-7B and C). From superficial to deep, the three zones of the cortex and their major secretory products are as follows:

1. **Zona glomerulosa**
   - **Structure.** The zona glomerulosa is composed of ovoid cords of small, eosinophilic cells and comprises approximately 15% of the total volume of the gland.
   - **Physiologic function.** The primary secretion of the zona glomerulosa is **aldosterone**, a mineralocorticoid that regulates electrolyte and water balance and blood pressure. The main action of aldosterone is to stimulate epithelial cells in the distal renal tubules to reabsorb Na$^+$ and excrete K$^+$. Aldosterone produces similar effects in other epithelia, including the gastric mucosa, colon, and salivary and sweat glands. When Na$^+$ is resorbed, water is drawn into the circulatory system, which increases blood volume and blood pressure. In the absence of aldosterone, the depletion of Na$^+$ and the retention of K$^+$ can be fatal.
   - **Regulation.** Aldosterone secretion is controlled primarily by renin-angiotensin signaling, as discussed in Chapter 13.

2. **Zona fasciculata**
   - **Structure.** The zona fasciculata consists of long, straight cords of large cuboidal cells, one to two cells thick. The zona fasciculata comprises approximately 75–80% of the total volume of the adrenal gland.
   - **Physiologic function.** The zona fasciculata produces the glucocorticoids **cortisol** (the main glucocorticoid in humans) and **corticosterone,** as well as some comparatively weak androgens such as **dehydroepiandrosterone (DHEA).** Glucocorticoids increase blood glucose levels by stimulating gluconeogenesis in the liver and also mobilize amino acids and fatty acids from muscle and adipose tissue. In addition, glucocorticoids modulate cytokine production and can have a general anti-inflammatory effect.
   - **Regulation.** Secretion by the zona fasciculata is regulated by a classic negative feedback loop involving the CRH, ACTH, and glucocorticoids. Cortisol is secreted in a circadian pattern. In the morning, high levels of glucocorticoids help the body become alert and active; low levels decrease activity prior to sleep. Cortisol also is secreted in response to stress. These daily and stress-related changes are mediated by neural inputs to the CRH-producing cells in the hypothalamus.

3. **Zona reticularis**
   - **Structure.** The zona reticularis is a narrow layer of small, darkly stained eosinophilic cells arranged in irregular anastomosing cords, and comprises approximately 7% of the volume of the adrenal gland.
   - **Physiologic functions.** The zona reticularis secretes androgens, principally DHEA, and small amounts of glucocorticoids.
   - **Regulation.** Similar to the zona fasciculata, secretion by the zona reticularis is regulated by CRH and ACTH.

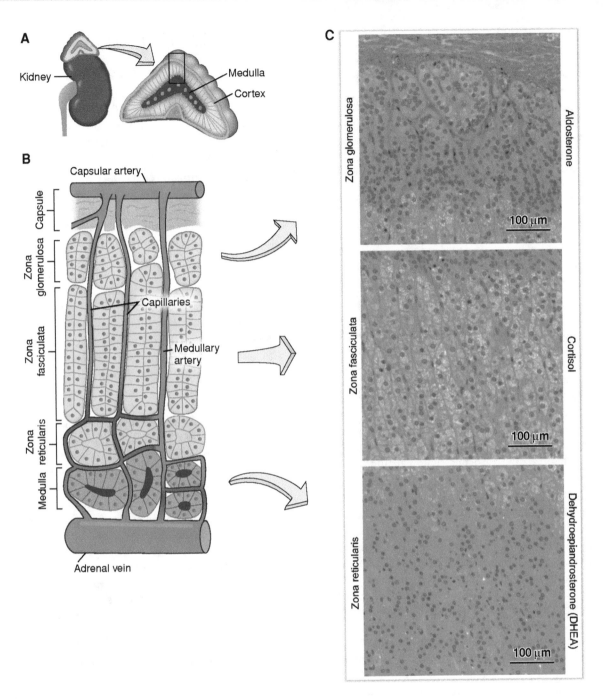

Figure 14-7: Structure and function of the adrenal glands. **A.** Location and overall organization of the adrenal gland. The area within the box is shown at higher magnification in part B. **B.** Organization of secretory cells in the adrenal cortex and medulla and the vasculature of the gland. Note that the cortex is supplied by a network of capillaries that drain into capillaries in the medulla, whereas other capillaries in the medulla are supplied directly by long arteries that pass through the cortex. Capillaries in the medulla, along with those from the cortex, form the medullary veins, which join to form the adrenal (suprarenal) vein. **C.** Section of each zone of the adrenal cortex. Note the differences in organization and staining properties of the cells in the three zones and the predominant hormone produced in each zone. Cells in the zona fasciculata contain abundant lipid droplets, which are extracted during tissue preparation. This gives the cells in this layer a distinctive, vacuolated appearance and they are sometimes referred to as spongiocytes. The capsule is toward the top in all panels. (*continued on page 227*)

ADRENAL MEDULLA

Structure. The adrenal medulla is composed of large poly-hedral cells arranged irregularly in clusters or in short cords that border on fenestrated capillaries (Figure 14-7D and E). The parenchymal cells originate from the neural crest cells that migrate into the developing gland at about 6 to 7 weeks' gestation. The cells are innervated by sympathetic pregan-glionic axons in the splanchnic nerve and secrete catecho-lamines and, therefore, can be considered as modified postganglionic sympathetic neurons (Figure 14-7F). Cells in the adrenal medulla are often referred to as **chromaffin cells** because their secretory granules stain intensely brown or black when exposed to potassium dichromate (the chro-maffin reaction). Clusters of neuroendocrine cells, similar to those in the adrenal medulla, are also found widely dis-persed as **paraganglia** closely associated with autonomic ganglia.

Physiologic function. Cells in the adrenal medulla secrete **catecholamines**; 75–80% of the cells secrete **epinephrine** (also known as **adrenaline**), and the remaining cells secrete **norepinephrine (noradrenaline)**. Catecholamines are secreted as part of the "fight-or-flight" response and have wide-spread effects that boost the supply of O_2 and glucose to the brain and muscles at the expense of nonemergency func-tions such as digestion. Thus, the catecholamines increase cardiac output, heart rate, and the release of glucose from the liver, while causing vasoconstriction in the skin and gut.

Regulation. During normal activity, the adrenal medulla con-tinuously secretes small quantities of catecholamines; however, in response to extreme fear, stress, or trauma, it can release quantities more than 100 times the resting level. These emo-tional reactions trigger the sympathetic preganglionic neurons that innervate the adrenal medullary cells to release acetylcho-line and stimulate catecholamine secretion. In addition, corti-sol stimulates the synthesis of epinephrine. Because cortical capillaries travel through the medulla en route to the adrenal vein, cells in the adrenal medulla are exposed to the high con-centrations of cortisol produced in the adrenal cortex in response to stress, which helps to coordinate the stress response of the two endocrine regions (i.e., adrenal cortex and adrenal medulla).

DISORDERS OF ADRENAL GLAND FUNCTION

DISORDERS OF THE ADRENAL CORTEX

Most common disorders of adrenal cortical function result from an excess or a deficiency of all the adrenal cortical steroid hormones, but the clinical manifestations are largely caused by disturbances in glucocorticoid secretion from the zona fascicu-lata, as illustrated in the following examples:

- **Cushing's syndrome** is a disorder caused by **excess cortisol in the blood**. The most common form, **Cushing's disease**, is due to pituitary adenomas that produce excess ACTH; how-ever, increased circulating cortisol also can result from gluco-corticoid drugs or from tumors that produce cortisol. Individuals with Cushing's syndrome experience weight gain (especially in the face, neck, and trunk), hyperglycemia, hypertension, muscle wasting, and mental disturbances. Cushing's disease is not limited to humans, however, and also is relatively common in domestic horses and dogs.

- **Addison's disease (chronic adrenal insufficiency)** is a rare dis-order in which the adrenal cortex does not secrete sufficient glucocorticoids and mineralocorticoids. It is usually caused by damage to the adrenal cortex, most commonly from autoim-mune attack, but it may also be the result of hemorrhage, tumor, or infection, such as tuberculosis or the human immunodefi-ciency virus (HIV). The symptoms of Addison's disease are low blood pressure, fatigue, chronic diarrhea, weight loss, and dark-ening of the skin. Symptoms may develop slowly and can be difficult to recognize, but the disease is fatal if patients are not treated with glucocorticoids. Exogenous glucocorticoids sup-press ACTH secretion and production of native glucocorticoids; therefore, patients prescribed large doses of glucocorticoids therapeutically may develop acute hypoadrenal crisis if the ther-apy is stopped abruptly. Moreover, because cortisol is released in response to virtually all types of stress (e.g., trauma, illness, tem-perature changes, mental stress), even minor illnesses can become life-threatening in the absence of cortisol.

DISORDERS OF THE ADRENAL MEDULLA

Pheochromocytoma is a rare neuroendocrine tumor of the adrenal medulla. The tumor secretes excessive amounts of catecholamines, causing elevated heart rate, elevated blood pressure, anxiety, severe headache, excessive sweating, and weight loss. Most pheochromocytomas are benign and are treated by surgical removal.

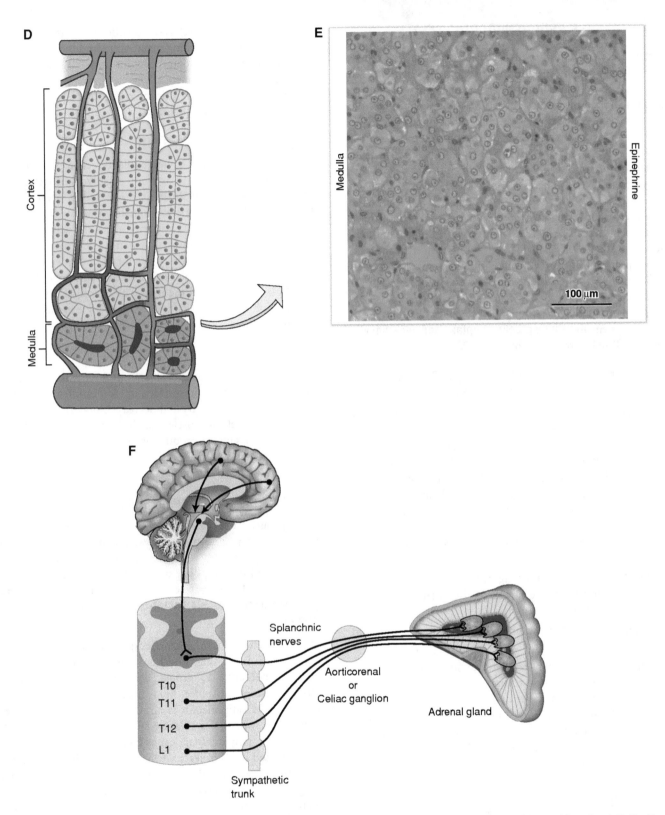

Figure 14-7: (*continued*) **D.** Organization of secretory cells in the adrenal cortex and medulla and the vasculature of the gland. **E.** Section of the adrenal medulla, which secretes predominantly epinephrine. **F.** Innervation of secretory cells in the adrenal medulla. Cells are innervated by sympathetic preganglionic neurons in the splanchnic nerves and, therefore, can be considered as modified postganglionic sympathetic neurons.

DIFFUSE NEUROENDOCRINE SYSTEM

In addition to hormone-producing cells in the endocrine organs and in the gonads and pancreas, endocrine cells are found scattered singly or in small clusters in many organs, especially in the epithelium of the respiratory system and digestive tract, and also in the pancreatic acini and ducts, the urogenital tract, hypophysis, skin (Merkel cells), and thyroid gland (parafollicular cells). At least 35 types of cells have been identified, which differ in their embryonic origin and secretory products. Most are derived from the neural crest or from the endodermal cells of the embryonic gut, and most secrete a single hormone, although in some cases individual cells produce more than one hormone.

Together, these dispersed cells comprise the **diffuse neuroendocrine system (DNES)**, so named because most of the cells secrete polypeptides or biogenic amines identical to neurotransmitters in the central nervous system (e.g., serotonin or 5-hydroxytryptamine). DNES cells also are referred to as **gastroenteropancreatic (GEP) cells**. **Enteroendocrine cells** in the digestive system and **amine precursor uptake and decarboxylation (APUD) cells** are subsets of the DNES.

STRUCTURE AND FUNCTION OF THE DIFFUSE NEUROENDOCRINE SYSTEM

DNES cells sit on a basal lamina and release their secretory products basally into the lamina propria (Figure 14-8A and B). In some cases, such as somatostatin produced in the stomach and intestine, the secreted factor acts locally on nearby target cells (paracrine signaling); in other cases, such as glucagon, which also is secreted by the stomach and intestine, the secreted factor enters the circulation to affect distant target cells (endocrine signaling).

Morphologically, there are two types of DNES cells, described as follows:

- **Open DNES cells** extend long, thin apical processes with microvilli that reach the lumen of the organ (e.g., in the digestive and respiratory tract).

- **Closed DNES cells** are covered by other epithelial cells and do not contact the lumen.

Both types of cells may be regulated by stimuli reaching their basal surface (e.g., via a nerve ending, a circulating hormone, or local paracrine or autocrine factors), but only open cells receive additional stimuli at their apical surface. Open cells can monitor the changing contents of the lumen (e.g., the pH, chemical composition, or air quality) and respond with the release of the appropriate hormones.

ADIPOSE TISSUE

Although the main functions of adipose tissue are to regulate the level of lipid in the blood and to store fat, adipose tissue is also an important endocrine organ that produces many hormones and cytokines, referred to collectively as **adipokines**, which function in regulating metabolism and the immune system. Two of these adipokines, **leptin** and **adiponectin**, have been well studied and function together in maintaining energy balance. Macrophages in adipose tissue of obese individuals produce **inflammatory cytokines**.

LEPTIN

Leptin is secreted by adipocytes, binds to receptors in the brain and peripheral tissues, and acts in a negative feedback loop to **maintain energy homeostasis**. It is secreted in proportion to the amount of body fat, thereby signaling the amount of energy stored in adipose tissue. Leptin modulates the activity of neurons in hypothalamic nuclei, which secrete neuropeptides involved in regulating feeding behavior and body weight. Leptin also increases the basal metabolic rate and stimulates fatty acid oxidation in muscle and the liver. The overall effects of leptin are to inhibit food intake, increase energy expenditure, and deplete fat stores.

ADIPONECTIN

Adiponectin is an abundant **plasma protein** with actions that are complementary to those of leptin. Adiponectin functions to increase insulin sensitivity and decrease blood glucose levels. It acts on both skeletal muscle and the liver to increase glucose uptake, reduce glucose production, and increase fatty acid oxidation. Circulating concentrations of adiponectin are inversely correlated with body fat.

MACROPHAGE FACTORS

Macrophages are resident cells in the stroma of adipose tissues. In individuals with normal body weight, macrophages function to block inflammatory responses and promote tissue repair. In obese individuals, adipose tissue becomes infiltrated with activated macrophages that secrete proinflammatory cytokines, such as **tumor necrosis factor-α (TNFα)** and **interleukin-6 (IL-6)** (Figure 14-8C).

▽ Secretion and function of adipokines are dysregulated in obese individuals and may contribute to the development of **metabolic syndrome**, a condition characterized by obesity, insulin resistance, lipid disorders, and hypertension. For example, adiponectin is abnormally decreased in obese individuals, whereas circulating levels of leptin are unusually high. Obese individuals appear to be resistant or insensitive to the increased levels of leptin, which do not reduce food intake or increase energy expenditure, as seen in individuals of normal weight. This resembles the situation for individuals with type 2 diabetes mellitus (i.e., non–insulin-dependent diabetes) who are resistant to the effects of insulin. Moreover, the increased levels of TNFα and IL-6 produced by macrophages result in a low-grade chronic inflammatory state, which has been linked to the development of insulin resistance and diabetes. ▼

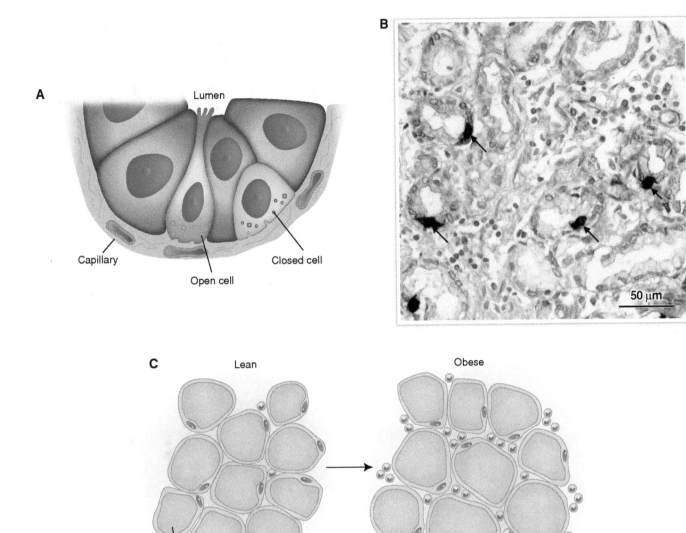

Figure 14-8: Cells of the diffuse neuroendocrine system (DNES) and adipose tissue. **A.** DNES cells. Open cells (*blue*) extend from the basal lamina to lumen and, therefore, can be regulated by signals from both their apical and basal surfaces, whereas closed cells (*green*) can only receive signals at their basal surface. Both types of cells release hormone basally. **B.** Section of glands in the stomach stained with antibodies to gastrin (*arrows*). Gastrin-secreting cells are part of the DNES. Section courtesy Frederick Clayton, MD, University of Utah School of Medicine, Salt Lake City, Utah. **C.** Differences between adipose tissue in lean and obese individuals. Note that adipocytes in obese individuals enlarge, increase secretion of leptin, decrease secretion of adiponectin, and become infiltrated with activated macrophages, which secrete inflammatory cytokines. IL-6, interleukin-6; TNFα, tumor necrosis factor-α.

STUDY QUESTIONS

Directions: Each of the numbered items or incomplete statements is followed by lettered options. Select the **one** lettered option that is **best** in each case.

Questions 1 and 2

Refer to the illustration below to answer the following questions.

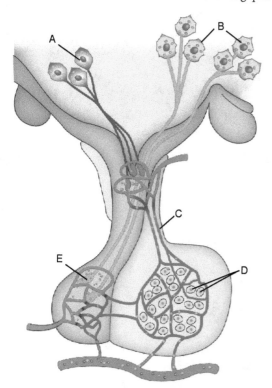

1. Which letter on the illustration best designates the cells that secrete hormones that *directly* regulate the function of the gonads?

2. What would be an effect of a lesion that destroys the cells labeled B?

A. Disruption in the rhythmic activity of other endocrine organs

B. Inability to concentrate urine

C. Increased secretion of prolactin

D. Lethargy and weight gain

E. Reduced secretion of adrenal cortical steroids

3. A woman is brought to the emergency department complaining of severe pain resulting from kidney stones. After evaluating the patient, the physician discovers that the woman has broken several bones recently, suggesting that her bones are fragile. Laboratory studies show elevated levels of Ca^{2+} in the blood. An MRI and subsequent biopsy specimen show an adenoma in which endocrine gland or region of gland?

A. Adrenal cortex

B. Adrenal medulla

C. Anterior pituitary gland

D. Follicular cells in the thyroid gland

E. Parafollicular cells in the thyroid gland

F. Parathyroid gland

G. Pineal gland

H. Posterior pituitary gland

Questions 4 and 5

Refer to the micrograph below to answer the following questions.

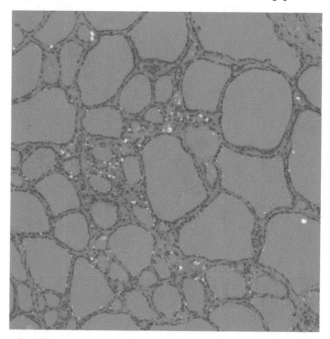

4. A 45-year-old woman visits her physician because she says she feels nervous, is always hot, and has lost about 11 kg in the past year, although she denies having changed her diet. The physician notices that the patient's eyes protrude and appear red and inflamed and that her pulse rate is elevated. A biopsy specimen of the organ shown in the micrograph shows an increase in lymphoid cells. The physician orders an array of tests. The tests are most likely to detect autoantibodies to what structure(s) in this organ?

A. Chief cells

B. Oxyphil cells

C. Parafollicular cells (C cells)

D. Thyroglobulin

E. Thyroid-stimulating hormone (TSH) receptors

F. Thyrotropin-releasing hormone (TRH) receptors

5. What is the most likely effect of insufficient dietary iodine on the organ shown in the micrograph?

   A. Decrease in production of calcitonin

   B. Decrease in production of parathyroid hormone (PTH)

   C. Decrease in production of thyroxine

   D. Increase in production of calcitonin

   E. Increase in production of PTH

   F. Increase in production of thyroxine

6. A patient complains that she is constantly fatigued, feels weak, is dizzy when standing, and has frequent bouts of diarrhea. The physician notices that the woman has a nice tan although it is the middle of winter. Physical examination indicates that she has lost about 7 kg since her last office visit and that her blood pressure is low. Laboratory studies show that her levels of cortisol and aldosterone are abnormally low. Insufficient secretion of hormones from which endocrine organ or region of organ is most likely responsible for the patient's symptoms?

   A. Adrenal cortex

   B. Adrenal medulla

   C. Anterior pituitary gland

   D. Follicular cells in the thyroid gland

   E. Parafollicular cells in the thyroid gland

   F. Parathyroid gland

   G. Pineal gland

   H. Posterior pituitary gland

7. A pheochromocytoma is a rare tumor of the adrenal medulla. Which of the following symptoms would most likely be observed in the presence of this tumor?

   A. Excretion of excessive amounts of dilute urine

   B. Galactorrhea, amenorrhea, and loss of libido

   C. Hypercalcemia, decalcification of bones, and kidney stones

   D. Weight gain, constipation, fatigue, and lethargy

   E. Weight loss, anxiety, elevated heart rate and blood pressure, and headaches

ANSWERS

1—D: Cells in the pars distalis (D) secrete somatotropins, prolactin, thyrotropin, follicle-stimulating hormone (FSH), luteinizing hormone (LH), and adrenocorticotropic hormone (ACTH). FSH and LH directly regulate gonad function. FSH promotes the growth of ovarian follicles and the secretion of estrogen in females, and spermatogenesis and secretion of androgen-binding protein by Sertoli cells in the testes in males. LH induces ovulation and formation of the corpora lutea in females and stimulates Leydig cells in the testes to synthesize and release testosterone in males.

2—B: The cells labeled B secrete oxytocin and antidiuretic hormone (ADH). ADH acts to concentrate urine by increasing the permeability of the collecting tubules of the kidney to water, causing water to be resorbed rather than excreted. Individuals with insufficient ADH are unable to concentrate urine, a condition known as central diabetes insipidus.

3—F: The parathyroid gland secretes parathyroid hormone (PTH), which functions to increase blood Ca^{2+} levels and promote phosphate excretion. PTH indirectly increases the number and activity of osteoclasts, which degrade bone matrix and release Ca^{2+} and phosphate into the blood. Excess PTH, which can result from a PTH-secreting adenoma, can cause hypercalcemia, decalcification of bone, and kidney stones.

4—E: The micrograph is an image of the thyroid gland. The symptoms this patient reports are characteristic of hyperthyroidism. The woman appears to have Graves' disease, an autoimmune disorder in which patients produce autoantibodies to TSH receptors that bind to the receptor and chronically stimulate thyroid follicular cells to synthesize and release thyroid hormones.

5—C: The micrograph is an image of the thyroid gland. Thyroid hormones (thyroxine and triiodothyronine) are iodinated derivatives of the amino acid tyrosine, and require iodine for their synthesis and secretion. Insufficient dietary iodine can result in hypothyroidism.

6—A: The patient is diagnosed with primary Addison's disease, a rare disorder in which the adrenal cortex does not secrete sufficient steroid hormones. Addison's disease most commonly results from autoimmune destruction of the adrenal cortex. The lack of cortisol secretion from the adrenal cortex results in high levels of CRH and ACTH production in the hypothalamus and pituitary respectively, due to the lack of negative feedback. Because melanocyte stimulating hormone (MSH) is derived from the same precursor as ACTH (pro-opiomelanocortin, POMC), higher levels of MSH are produced, which increases melanocyte activity. As a result, skin darkening and increased tanning are a common symptoms of this disease. A lack of ACTH production in the pituitary (choice C) produces many of the symptoms of primary Addison's disease, but not skin darkening, and this is called secondary Addison's disease. Similarly, a lack of CRH production in the hypothalamus produces tertiary Addison's disease.

7—E: Pheochromocytomas secrete excessive amounts of catecholamines, the hormones normally produced by the adrenal medulla. Catecholamines produce widespread effects throughout the body, including increased cardiac output and heart rate and release of glucose from the liver. The excessive catecholamines produced by pheochromocytomas also produce symptoms of weight loss, anxiety, and headaches.

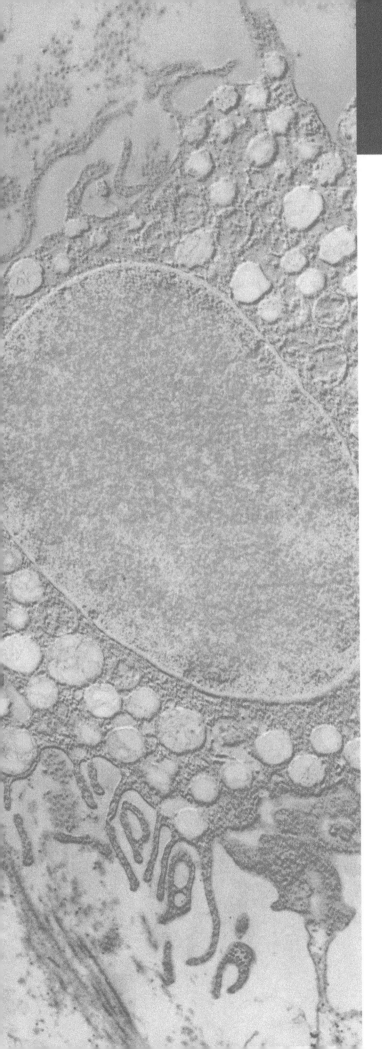

CHAPTER 15

FEMALE REPRODUCTIVE SYSTEM

OVERVIEW

The female reproductive system consists of a series of organs that function together to produce and nourish viable offspring. The organs that comprise the female reproductive system include the **ovaries**, the **uterine tubes**, the **uterus**, and the **vagina. Mammary glands** are also important for the nourishment of offspring, but are not part of the female reproductive system as such.

The **follicle** is the structural and functional unit of the ovary, and consists of a female gamete (an **oocyte**) surrounded by **follicular (granulosa) cells.** All oocytes and follicles originate prenatally and continue to develop and mature throughout a woman's reproductive life.

Female reproductive organs undergo monthly cyclic changes in their histologic organization and functional activity, known as the **ovarian** or **menstrual cycle**, which is regulated by the complex interplay of hypothalamic, pituitary, and ovarian hormones. As a result, usually a single oocyte is released each month, and the uterus is prepared to receive an embryo.

The ovary produces the female sex hormones **estrogen** and **progesterone.**

Estrogen promotes the development of female secondary sex characteristics, the maturation of ovarian follicles, and the restoration of the endometrium after menstruation.

Progesterone stimulates secretion by the endometrial glands to prepare the uterus for pregnancy.

OVARY

ORGANIZATION

The primary female reproductive organ is the ovary, which functions to produce female gametes (**oocytes, ova**) and the female sex hormones **estrogen** and **progesterone**. The paired ovaries are located in the pelvic cavity, one on each side of the uterus, and are held in place by the broad ligament (Figure 15-1A). The overall organization of the ovary is as follows (Figure 15-1B):

Each ovary is surrounded by a dense connective tissue capsule called the **tunica albuginea**, which is covered with a simple cuboidal epithelium called **germinal (surface) epithelium.**

The ovary is divided into two regions: the **cortex**, an outer cellular region containing thousands of **follicles** embedded in a highly cellular connective tissue stroma, and the **medulla**, a central region consisting of loose connective tissue, blood and lymphatic vessels, and nerves.

The **follicle** is the structural and functional unit of the ovary. Each follicle consists of a single **oocyte** surrounded by one or more layers of epithelial **follicular (granulosa) cells**, which produce sex hormones. During the reproductive years, follicles at various stages of maturation are present in the ovarian cortex.

OOGENESIS AND FOLLICLE MATURATION

Oocytes are formed by a process called **oogenesis**, which begins prenatally and is not completed until fertilization. In a female fetus, proliferating primordial germ cells arrest in the prophase of the first meiotic division midway through gestation and remain dormant until puberty. Each of these dormant **primary oocytes** is enclosed by a layer of flattened somatic follicle cells surrounded by a basal lamina. The follicles formed prenatally, the **primordial follicles**, are the only follicles present in the ovary until puberty (Figure 15-1C). Unlike spermatozoa, which are produced continuously in sexually mature males, all oocytes are generated before birth and their numbers decline continuously. Maturation of oocytes occurs cyclically, and usually a single mature oocyte is released each month. Throughout a woman's life, approximately only 0.1% of primordial follicles fully mature and are released; the remaining 99.9% die by a process termed **atresia**, and are removed by phagocytosis. Atresia begins prenatally and continues after menopause.

Primordial follicles begin to grow at puberty and undergo a stereotypical pattern of histologic and hormonal changes (Figure 15-1C–E). Primordial follicles initiate growth on a daily basis, so the ovary of a woman in her reproductive years contains follicles in all stages of maturation. During maturation, follicles undergo the following changes:

Growth. The oocyte enlarges from approximately 25 µm to 120 µm; mature graafian follicles reach 20–30 mm in diameter.

Proliferation of granulosa cells. Granulosa cells become cuboidal and proliferate and are interconnected by gap junctions. Oocytes surrounded by one or more layers of cuboidal granulosa cells are called **primary** or **preantral follicles.**

Formation of zona pellucida. The oocyte and granulosa cells secrete a mixture of glycoproteins that form a coat around the oocyte called the **zona pellucida**. The oocyte and granulosa cells communicate across the zona pellucida via gap junctions, which most likely provide nutrition for the oocyte. The zona pellucida remains with the oocyte after ovulation, and functions to bind sperm and initiate the **acrosomal reaction**. If fertilization occurs, the resulting zygote undergoes cleavage within the zona pellucida and hatches from the zona pellucida only after entering the uterus 4–5 days after ovulation.

Formation of the theca. Connective tissue cells adjacent to the basal lamina differentiate into a bilayered structure called the **theca folliculi**, which encapsulates the follicle. Cells in the inner layer (**theca interna**) secrete the androgen **androstenedione**, which granulosa cells transform to **estradiol**, the major circulating estrogen. The outer layer (**theca externa**) consists of fibroblasts and smooth muscle cells. Both layers of the theca are vascularized, whereas the granulosa is not.

Formation of the antrum. Fluid accumulates in the intercellular spaces in the granulosa, eventually forming a fluid-filled chamber called an **antrum**. Most growing follicles die by atresia before reaching the antral stage. However, at the start of each ovarian cycle, a cohort (5–12) of early antral follicles is rescued from atresia and selected to continue to grow and mature.

Reorganization of granulosa cells. As the antrum forms, the oocyte becomes suspended in the antral fluid and is supported by a small mass of granulosa cells, called the **cumulus oophorus**. The oocyte is surrounded by a layer of cells called the **corona radiata**, which accompanies the oocyte when it leaves the ovary at ovulation.

Dominance or death. Eventually, one member of the cohort of growing follicles becomes dominant and continues to grow and mature into a graafian follicle. The remaining cohort undergoes atresia.

Resumption of meiosis. The oocyte of the single dominant follicle resumes meiosis and divides asymmetrically to produce a **secondary oocyte**, which contains half the chromosomes, almost all of the cytoplasm, and a small **polar body**, which is discarded. The secondary oocyte soon initiates but arrests in the second meiotic division, which is not completed until fertilization.

Ovulation. The mature follicle ruptures and extrudes the oocyte and surrounding cumulus cells, which enter the uterine tube, where fertilization may occur. The empty follicle persists as the **corpus luteum**.

The entire process of growth from a primordial follicle to a fully mature graafian follicle requires approximately 3 months.

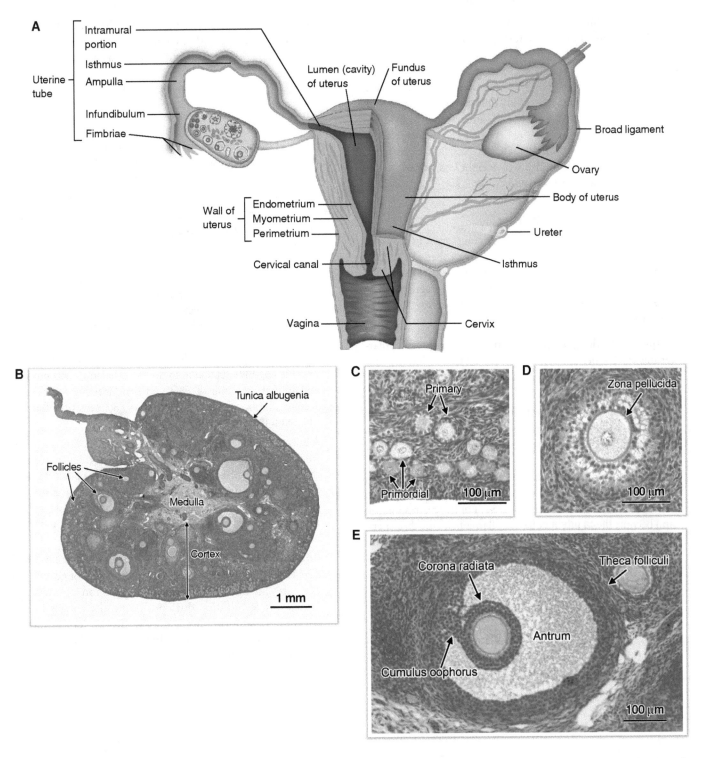

Figure 15-1: Arrangement of female reproductive organs, structure of the ovary and corpus luteum, and hormonal regulation of oogenesis. **A.** Female reproductive organs shown from the posterior. **B.** Section through a primate ovary. **C.** Primordial and primary follicles shown at higher magnification than in part B. **D.** Preantral follicle shown at higher magnification than in part B. **E.** Graafian follicle shown at higher magnification than in part B. (*continued on page 237*)

▽ Benign follicle cysts are common in the ovary and arise from graafian follicles that fail to rupture or that rupture and immediately seal. However, 3–6% of woman of reproductive age develop multiple follicle cysts as a result of the accumulation of immature follicles that have arrested at the antral stage, a syndrome termed **polycystic ovary disease (PCOD)**. Although the etiology of PCOD is not entirely understood, individuals with PCOD produce excessive androgen and may experience amenorrhea, infertility, obesity, insulin resistance, and hirsutism. PCOD usually is treated with weight loss, gonadotropin analogs, insulin mediators, or surgery. ▽

CORPUS LUTEUM

After ovulation, the large empty follicle transforms into a transient endocrine organ in the ovarian cortex called the **corpus luteum**. Secretions of the **corpus luteum** are essential to prepare the uterus for the initiation and maintenance of pregnancy. The appearance and fate of the corpus luteum depends on the length of time after ovulation (Figure 15-1F and G).

- **Corpus hemorrhagicum.** Blood from the disrupted capillaries fills the former antrum and forms a **central clot**, which is subsequently removed by phagocytes and replaced with connective tissue.

- **Corpus luteum of menstruation.** Cells of the granulosa and theca interna reorganize, alter their histologic appearance, and importantly, begin to secrete progesterone while continuing to produce estrogen and androgens, respectively. Granulosa cells enlarge to become **granulosa lutein cells**. Theca interna cells, which are smaller and stain more darkly than granulosa lutein cells, accumulate in the folds of the wall of the corpus luteum and become **theca lutein cells**. Capillaries from the theca penetrate the basal lamina and invade the granulosa. In the absence of fertilization, the corpus luteum functions for about 14 days and then degenerates and is replaced with a connective tissue scar called a **corpus albicans**.

- **Corpus luteum of pregnancy.** If fertilization and implantation occur, the trophoblast of the embryo (and later the developing placenta) produces **human chorionic gonadotropin (hCG)**, which supports continued function of the corpus luteum for the following 4–5 months, until the placenta has matured sufficiently to maintain pregnancy. The corpus luteum then degenerates and becomes a corpus albicans.

HORMONAL REGULATION OF FOLLICLE MATURATION

The mechanisms that trigger some primordial oocytes to begin to grow daily are not completely understood but appear to be independent of hormonal regulation. In contrast, maturation of the selected cohort of growing follicles beyond the preantral stage depends on the complex interplay of hypothalamic, pituitary, and ovarian hormones, as illustrated in Figures 15-1H and 15-3B.

- **Follicles grow in response to gonadotropins and estrogen** (pathway 1). Hypothalamic gonadotropin-releasing hormone (GnRH) stimulates the secretion of pituitary gonadotropins, follicle-stimulating hormone (FSH) and luteinizing hormone (LH). FSH promotes the cohort of antral follicles to grow and produce estrogen. LH promotes the theca interna to secrete androgens, which follicular granulosa cells convert to estrogens in a reaction catalyzed by the enzyme **aromatase**. Estrogen stimulates proliferation of granulosa cells, which in turn increases estrogen production. The increasing levels of estrogen provide **negative feedback** to the hypothalamus and pituitary and inhibit secretion of FSH and LH. Growing follicles also secrete inhibin, which further inhibits FSH secretion. **Only the dominant follicle survives this dip in FSH.**

- **A surge in LH triggers ovulation** (pathway 2). The dominant follicle continues to secrete estrogen, which midway through the cycle reaches a critical level that provides **positive feedback** to the hypothalamus and pituitary, causing a surge in LH (and to a lesser extent in FSH). This surge in LH triggers the meiosis of the dominant oocyte, followed by **ovulation** and the formation of a **corpus luteum**.

- **Progesterone and estrogen from the corpus luteum inhibit FSH and LH** (pathway 3). The decreased levels of gonadotropins prevent the development and ovulation of additional follicles. If fertilization does not occur, the corpus luteum degenerates and estrogen and progesterone levels decrease, freeing FSH and LH from inhibition, and the cycle commences once again.

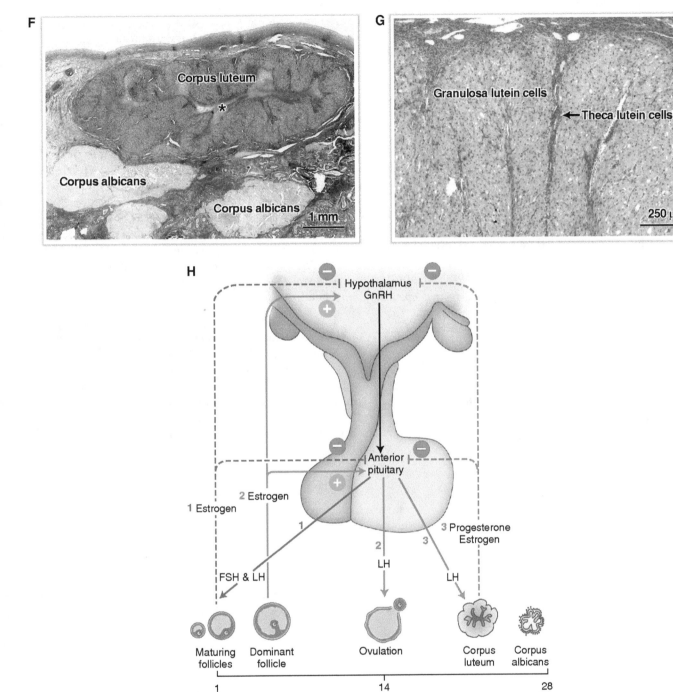

Figure 15-1: (*continued*) **F.** Section of a corpus hemorrhagicum. The asterisk indicates the central clot. Two corpora albicantia also are shown. **G.** Corpus luteum of menstruation shown at higher magnification than in part A. **H.** Interactions of hypothalamic, pituitary, and ovarian hormones that regulate follicle maturation during a typical 28-day ovarian cycle.

UTERINE TUBES

The uterine tubes (**fallopian tubes; oviducts**) consist of a pair of thin muscular tubes, each about 10–12 cm long, that stretch between the ovary and the uterus (Figure 15-1A). Uterine tubes function to provide an environment for fertilization of oocytes and transport of zygotes to the uterus.

STRUCTURE OF THE UTERINE TUBES

The wall of the uterine tube consists of an elaborately folded mucosa, a thick muscularis containing interwoven layers of circular and longitudinal smooth muscle, and a thin serosa. Each uterine tube is lined with simple ciliated columnar epithelium resting on a highly vascular lamina propria. There are two types of cells in the epithelium: **ciliated** and **secretory (peg)** (Figure 15-2A–C).

The uterine tube can be divided into four regions, as follows:

Infundibulum. The distal, funnel-shaped region that opens into the peritoneal cavity adjacent to the ovary. The infundibulum ends in a fringe of finger-like extensions, termed **fimbriae**.

Ampulla. An expanded central region where fertilization normally occurs.

Isthmus. The narrow proximal region near the uterine wall.

Intramural (uterine) portion. The segment that passes through the uterine musculature and opens into the interior of the uterus.

The height and complexity of the mucosal folds is greatest in the infundibulum and decreases toward the uterus. By contrast, the muscularis is thickest near the uterus. The epithelium undergoes cyclic changes with the menstrual cycle—epithelial cells hypertrophy and cilia elongate during the follicular phase, and atrophy and shed cilia in the late secretory phase.

FUNCTION OF THE UTERINE TUBES

At ovulation, fimbriae move closer to the ovary, engorge with blood, and surround the ovary like fingers of a glove. The beating cilia of the fimbria sweep the oocyte into the oviduct, where it can be fertilized. If the oocyte is not fertilized within 24 to 36 hours after ovulation, it degenerates. Secretions of the peg cells lubricate the lumen, provide nutrition and protection for spermatozoa, oocytes, and zygotes, and promote **capacitation** of sperm. Peristaltic contractions of the oviduct and beating of the cilia propel the oocyte or zygote toward the uterus, where implantation may occur.

▽ Occasionally, the zygote implants at sites other than the uterus, most commonly in the ampulla of the uterine tube, producing an **ectopic (tubal) pregnancy**. Tubal pregnancy is a painful, potentially life-threatening condition because the tube can rupture, resulting in hemorrhage and shock. Chronic inflammation and scarring of the uterine tube, which can result from a sexually transmitted disease such as gonorrhea or chlamydia, increase the chance that a tubal pregnancy will occur. ▼

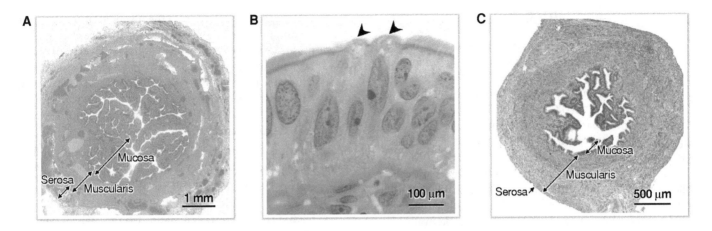

Figure 15-2: Uterine tube. **A.** Section through the ampulla of a uterine tube. Note the complex folds in the mucosa. **B.** Epithelium of the ampulla of a uterine tube shown at higher magnification than in part A. The arrowheads indicate secretory (peg) cells. **C.** Section through the isthmus region of a uterine tube. Note that the epithelium is less folded and the muscularis is thicker than in the ampulla (shown in part A).

UTERUS

The uterus is a pear-shaped organ that functions to support and nourish the growing fetus. The uterus can be divided into four regions: the upper dome-shaped **fundus**, the large **body** where implantation usually occurs, the narrow **isthmus**, and the cylinder-shaped **cervix** that inserts into the vagina (Figure 15-1A). In a nulliparous, nonpregnant female, the uterus is approximately $7 \times 6 \times 3$ cm and weighs about 50 gm. The uterus is larger after a pregnancy and smaller after menopause.

The wall of the uterus is composed of three layers: the mucosa called the **endometrium**, a thick muscularis called the **myometrium**, and an outer connective tissue layer called the **perimetrium** (either an adventitia or serosa). During each monthly cycle, ovarian hormones promote the proliferation of the endometrium and prime the uterus to receive an embryo. If implantation does not occur, most of the endometrium is sloughed and regenerates in the following 2 weeks. The fundus, body, and isthmus of the uterus share the same histologic organization and undergo similar changes during the ovarian cycle. The wall of the cervix differs from the rest of the uterus and changes less dramatically. Implantation results in the formation of a **placenta** derived from the endometrium and embryonic trophoblast.

MYOMETRIUM

The myometrium consists of several layers of smooth muscle (Figure 15-3A, C, and D). The central layers contain large arcuate arteries, which give rise to both the **straight** and **spiral (coiled) arteries** of the endometrium. The myometrium does not undergo dramatic changes during the ovarian cycle.

▽ The myometrium is the site of the most common benign tumors in females, **leiomyomas (uterine fibroids)**. These smooth muscle tumors usually are asymptomatic. In some cases, however, leiomyomas can cause heavy menstrual bleeding, painful menstruation, and impaired fertility. Like the smooth muscle cells from which they originate, leiomyomas are dependent on estrogen and progesterone and grow progressively until menopause, when they shrink. ▼

ENDOMETRIUM

The endometrium consists of a simple columnar epithelium composed of ciliated cells and secretory cells resting on a highly cellular lamina propria that contains simple tubular mucus-secreting **uterine (endometrial) glands** (Figure 15-3A). The structure, thickness, and functional state of the endometrium undergo marked changes during the ovarian cycle. At its thickest (5 mm), the endometrium has two distinct layers with separate blood supplies.

1. **Functional layer.** The superficial four-fifths of the endometrium, adjacent to the uterine lumen.

   • During the first half of each ovarian cycle the functional layer is shed and regenerated. In the second half, glands in the functional layer secrete products that nourish the embryo. Implantation and establishment of the placenta occur in the functional layer.

   • **Spiral arteries** derived from the myometrial arcuate arteries supply the functional layer. In the absence of pregnancy, progesterone levels decrease. Spiral arteries constrict in response to this decrease, which causes the functional layer to become ischemic, die, and slough off as **menses**. If pregnancy occurs, spiral arteries persist to supply the placenta.

2. **Basal layer.** The deepest region of the endometrium, adjacent to the myometrium.

   • The thin (0.5 mm) basal layer contains the base of the uterine glands, which are not shed at menstruation and proliferate to reconstitute the functional layer.

   • The blood supply is from the **straight arteries** derived from the myometrial arcuate arteries, which are not affected by the decrease in progesterone levels.

▽ **Endometriosis** is a condition in which endometrial tissue is growing in abnormal locations outside of the uterus, such as the ovaries or uterine tubes. Approximately 10% of women of reproductive age are affected by endometriosis, which can cause infertility, dysmenorrhea, and pelvic pain. Severe cases can result in hemorrhage and adhesions between the uterine tubes, ovaries, and other pelvic structures. ▼

THE OVARIAN CYCLE

The ovarian (menstrual) cycle refers to the changing pattern of hormone production in the hypothalamus, pituitary, and ovary that occurs each month during a woman's reproductive life. The changes in ovarian hormones drive the changes in the uterine endometrium (Figure 15-3B).

The ovarian cycle is **approximately 28 days** and typically is divided into three phases, with day 1 being the first day menstrual bleeding appears. Each phase has multiple names that reflect the functional state of the uterine endometrium, the functional state of the ovarian follicle, and the major hormone involved.

1. **Menstrual (resting) phase:** days 1–4. Menstrual discharge appears, consisting of the sloughed functional layer and blood from the ruptured spiral arteries. At the end of the menstrual phase, the endometrium consists of the thin basal layer without much overlying epithelium (Figure 15-3C).

2. **Proliferative (follicular, estrogenic) phase:** days 5–14. Cells in the lamina propria and glands in the basal layer proliferate in response to estrogen produced by the growing ovarian follicles and restore the functional layer. The proliferative phase ends with ovulation.

3. **Secretory (luteal, progestational) phase:** days 15–28. In response to progesterone produced by the corpus luteum, uterine glands begin to secrete glycogen and glycoproteins, which function to nourish the embryo before and during implantation. The glands become coiled and distended as they fill with the secretory product. The functional layer thickens due to the accumulation of secretory product, as well as to edema in the lamina propria (Figure 15-3D).

If fertilization does not occur, the corpus luteum degenerates, and the cycle returns to the first phase, the menstrual phase.

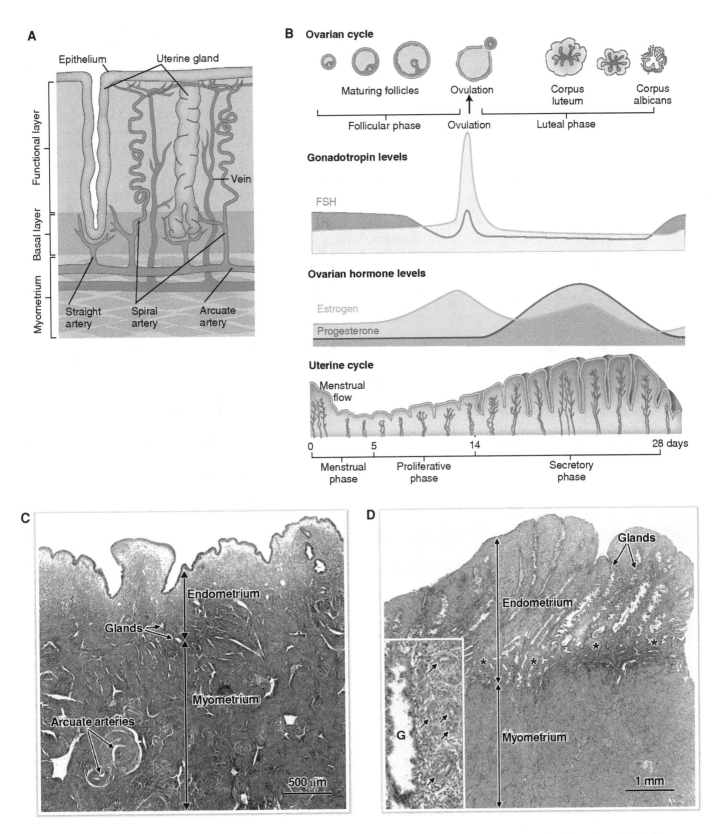

Figure 15-3: Structure of the uterus throughout the ovarian cycle and the cervix. **A.** Basic structure of the wall of the uterus during the secretory phase. **B.** Levels of pituitary and ovarian hormones and the resulting changes in ovarian follicles and uterine endometrium during a typical 28-day ovarian cycle. **C.** Section through the wall of the uterus during the menstrual or early proliferative phase of the ovarian cycle. The endometrium consists simply of the basal layer and base of the glands at this stage. Only part of the myometrium is shown. **D.** Section through the wall of the uterus during the secretory phase of the ovarian cycle. Asterisks indicate the border between the functional (upper) and basal (lower) layers of the endometrium. Only part of the myometrium is shown. Inset shows the spiral arteries (*arrows*) and a gland (G) at higher magnification. Note the difference in magnification between parts C and D. (*continued on page 243*)

CERVIX

The cervix is the cylindrically shaped lower portion of the uterus that bulges into the vagina. In contrast to other regions of the uterus, the walls of the cervix contain very little smooth muscle. Instead, they consist predominantly of dense connective tissue covered with epithelium, either nonkeratinized stratified squamous epithelium, similar to that found in the vagina, or simple columnar mucus-secreting epithelium, similar to that found in the uterus. The latter is characterized by branched tubular **mucous glands** (also called **crypts**), which may become dilated with mucus to form benign cysts (Figure 15-3E and F).

SQUAMOCOLUMNAR JUNCTION The squamocolumnar junction is the site where stratified and columnar epithelia meet; its location changes with age. At birth, the exposed ectocervix is covered with stratified epithelium, and the cervical canal (endocervix) is covered with columnar epithelium. Hormonal changes at puberty cause the columnar epithelium of the endocervix to spread and cover the ectocervix. However, in response to the harsh acidic vaginal environment that also develops at puberty, the epithelium of the ectocervix soon begins to transform to nonkeratinized stratified squamous epithelium. With increasing age, stratified epithelium covers progressively more of the ectocervix and can extend into the cervical canal after menopause. The transformation of columnar epithelium to stratified epithelium is an example of metaplasia, as described in Chapter 1, and occurs in a region called the **transformation zone**. This region is of clinical importance because most cases of cervical carcinoma originate in the transformation zone (Figure 15-3E and F).

SECRETORY CHANGES DURING THE OVARIAN CYCLE The cervical mucosa and glands do not undergo marked changes during the ovarian cycle and are not shed at menstruation. However, the composition of the mucus changes during the ovarian cycle, and plays an important role in fertility and early pregnancy. When ovulation occurs, high levels of estrogen and low levels of progesterone stimulate the glands to produce thin, watery mucus, which facilitates the entry of sperm into the uterus. Later in the ovarian cycle, high levels of progesterone stimulate the production of thicker mucus, which impedes both sperm and microorganisms from entering the uterus. During pregnancy, cervical glands secrete highly viscous mucus, which forms a plug in the cervical canal.

▽ **Cervical carcinoma** is estimated to cause the deaths of over 250,000 women worldwide each year, predominantly in developing countries. In developed countries, the incidence of cervical cancer has decreased dramatically as a result of the routine screening of women using the **Pap test**, a noninvasive diagnostic procedure developed by George Papanicolaou. Pap smears are performed by collecting epithelial cells scraped from the squamocolumnar junction and then stained on a microscope slide and examined for signs of abnormalities. Precancerous lesions can be surgically removed, preventing their progression to carcinoma. Infection with the **human papillomavirus (HPV)** plays a dominant role in the etiology of most cervical carcinomas. Immunization with a recently developed HPV vaccine (Gardasil) before the initiation of sexual activity reduces the risk of developing cervical cancer. ▽

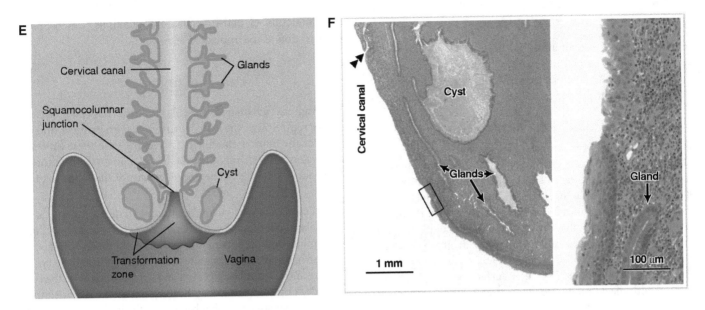

Figure 15-3: (*continued*) **E.** Longitudinal section through the cervix of an adult woman. Regions lined with simple mucus-secreting columnar epithelium are shown in dark pink; regions lined with stratified squamous epithelium are shown in blue. The transformation zone represents the regions in which the epithelium transitions between columnar and squamous during a woman's life. **F.** Longitudinal section through the lower portion of the cervix near the junction with the vagina. The double arrowhead indicates the opening of a gland into the cervical canal. The inset shows the boxed area containing the squamocolumnar junction at higher magnification.

PLACENTA

Fertilization and implantation of the embryo results in the development of the placenta within the endometrium (now called the **decidua**). The placenta is a complex, highly vascular "temporary" organ, derived from both maternal and fetal tissues.

STRUCTURE OF THE PLACENTA

The mature placenta is composed of thousands of branched **chorionic villi**, which are derived from the embryonic trophoblast and project into large blood-filled sinuses (**intervillous spaces**) in the decidua. Individual villi consist of fetal capillaries derived from umbilical vessels, covered by a thin syncytial membrane. Villi are bathed by maternal blood supplied by the spiral arteries. Placental septa originating from the decidua organize the villi into 15–25 incompletely separated compartments called **cotyledons**. The fetal surface of the placenta is covered with the amnion, into which the umbilical cord inserts (Figure 15-4A and B).

FUNCTION OF THE PLACENTA

The placenta joins the mother and fetus and serves the following functions:

- **Exchange of substances** between mother and fetus, including nutrients, maternal IgG antibodies, gases (O_2 and CO_2), and waste products. Fetal blood does not come in contact with maternal blood. Instead, exchange occurs across a barrier formed by the endothelium of the fetal capillaries in villi and the syncytial membrane.

- **Production of hormones**, including progesterone, estrogen, hCG, and human placental lactogen (also called human chorionic somatomammotropin).

- **Production of anti-inflammatory cytokines**, which prevent rejection of the implanted embryo.

VAGINA

The vagina is a flattened, muscular tube that has two main functions: reception of the penis during sexual intercourse, and passage of the fetus at childbirth. In keeping with these functions, the vagina has strong, distensible walls adapted to withstand abrasion. The walls of the vagina consist of the following three layers (Figure 15-4C and D):

- **Mucosa.** Consists of nonkeratinized stratified squamous epithelium continuous with that of the ectocervix and resting on lamina propria containing abundant elastic fibers, lymphocytes, and neutrophils. In response to estrogen, epithelial cells synthesize and accumulate glycogen and lipids, which gives the cells a pale, empty appearance in histologic sections. When epithelial cells desquamate, benign bacteria in the vagina metabolize the glycogen to lactic acid. The resulting low pH inhibits the growth of pathogenic microorganisms in the vaginal lumen.

- **Muscularis.** Consists of scattered bundles of circular smooth muscle below the mucosa and longitudinal bundles near the adventitia. Skeletal muscle forms a sphincter at the external opening.

- **Adventitia.** Consists of dense connective tissue with abundant elastic fibers, which gives the vagina its elasticity and distensibility.

Although the epithelium lacks glands, the vagina is kept moist and lubricated by exudate from thin-walled veins in the mucosa and muscularis, and by mucus produced in the cervix and in vestibular glands that line the vaginal opening.

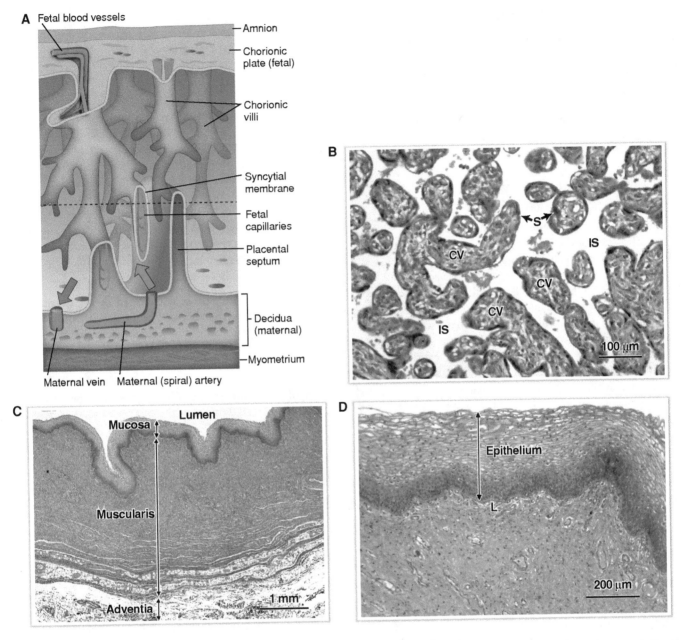

Figure 15-4: Structure of the placenta and vagina. **A.** Structure of the mature placenta. Chorionic villi derived from the embryo are bathed in maternal blood supplied by enlarged spiral arteries. Placental septa derived from the decidua organize the villi into partially separated compartments called cotyledons. The dashed line indicates the approximate plane of section in part B. **B.** Section of chorionic villi in a developing placenta. Intervillous spaces (IS) are filled with maternal blood. S, syncytial membrane; CV, chorionic villi. **C.** Section through the wall of the vagina. **D.** Wall of the vagina shown at higher magnification than in part C. L, lamina propria.

MAMMARY GLANDS

Mammary glands are highly modified apocrine sweat glands in the breast that function to produce milk to nourish infants. Mammary glands are present in both males and females, and are identical in the two sexes until puberty. Increased levels of estrogen at puberty trigger rapid growth of the breasts and glands in females; however, the glands differentiate fully only during pregnancy and lactation, regress when suckling has stopped, and involute after menopause.

STRUCTURE OF RESTING, INACTIVE MAMMARY GLANDS

In adult females, each of the paired mammary glands is composed of 15–25 lobes, separated by dense connective tissue and adipose tissue (essentially the reticular dermis). Each lobe represents a single branched, compound tubuloalveolar gland with its own opening in the nipple. Skin on the nipple is richly supplied with sensory nerve endings, and the surrounding pigmented region (the **areola**) contains abundant sweat and sebaceous glands. Smooth muscle fibers are present in the underlying connective tissue and are responsible for erection of the nipples. The histologic appearance of the mature mammary glands varies with age and the physiologic state.

In nonpregnant women, the glands contain only a few small rudimentary, secretory units and consist almost entirely of branched ducts (Figure 15-5A–C). The ducts in each lobe ultimately converge into the main **lactiferous duct**, which dilates and becomes the **lactiferous sinus** near its opening onto the nipple. The sinus is lined with stratified cuboidal epithelium, and the lactiferous duct and smaller ducts are lined with a simple cuboidal epithelium covered by myoepithelial cells. A few smooth muscle fibers also surround the larger ducts. Ducts are organized into lobules, and are surrounded by loose connective tissue.

CHANGES IN THE MAMMARY GLANDS DURING PREGNANCY

During pregnancy, increasing levels of hormones (e.g., estrogen, progesterone, human placental lactogen, and prolactin) stimulate mammary epithelial cells to proliferate and form secretory alveoli and **alveolar ducts**. A single layer of cuboidal secretory epithelium, surrounded by a network of myoepithelial cells, lines the alveoli and alveolar ducts. The stroma decreases, and the breasts become densely packed with glandular elements (Figure 15-5D–F). Lymphocytes and plasma cells infiltrate the loose connective tissue surrounding the ducts and alveoli. When suckling has stopped, most alveoli degenerate and the breasts return to a resting, prepregnancy state.

LACTATION

Milk production (lactation) begins late in pregnancy under the influence of prolactin from the anterior pituitary. For the first few days after the birth of an infant, alveoli and secretory ducts produce **colostrum**, a liquid rich in proteins, vitamin A, and sodium chloride. The production of mature milk, a complex emulsion of proteins (some found only in milk, such as casein and lactalbumin), lipids, and lactose begins a few days later. Both colostrum and mature milk contain immunoglobulins (IgA) produced by the plasma cells in the loose connective tissue surrounding the secretory elements, which provides the infant with antibodies and important immunologic protection for the first few months of life.

Let-down reflex Milk release is triggered by the hormone **oxytocin**, which is secreted from the posterior pituitary gland. Breast-feeding stimulates the sensory receptors in the nipple, which sends a neural signal to the hypothalamus, triggering the release of oxytocin from the axon terminals of the hypothalamic neurons, which produced it. Oxytocin stimulates the contraction of the myoepithelial cells of the secretory alveoli and smooth muscle cells of the lactiferous sinuses and ducts, causing the expulsion of milk. Interestingly, lipids in milk are released via apocrine secretion, whereas proteins are released via merocrine secretion (see Chapter 1).

▽ **Breast cancer** is the most common nonskin malignancy in women, and causes the deaths of over 40,000 women in the United States each year. Most breast cancers are classified as adenocarcinomas, and arise from the epithelial cells in the small secretory ducts. Major risk factors for developing breast cancer are genetic and hormonal. About 10% of breast cancers are due to hereditary defects, and about 25% of these cases arise from mutations in the BRCA1 and BRCA2 genes. Increased lifelong exposure to estrogen caused by the early onset of menstruation or by late menopause, or by never being pregnant or late first pregnancy, or by only a very short time breast-feeding or never breast-feeding increase a woman's risk of developing breast cancer. However, more than 70% of women who develop breast cancer have no known risk factors for the disease. Mortality from breast cancer usually is the result of spread of the carcinoma to vital organs, such as the brain or lung, via the circulatory or lymphatic system. ▽

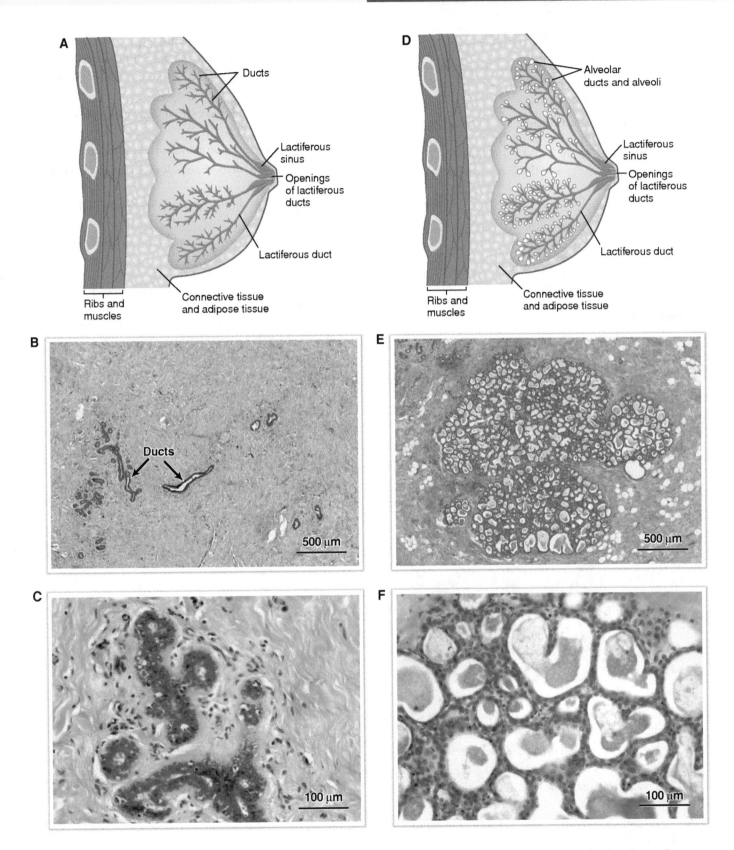

Figure 15-5: Structure of the mammary glands. **A.** Breast containing inactive resting mammary glands. **B.** Section of an inactive resting mammary gland. Note that the glands are composed almost entirely of ducts. **C.** Ducts in an inactive resting mammary gland shown at higher magnification than in part B. Note that the ducts are surrounded by highly cellular loose connective tissue in comparison to the dense connective tissue that comprises much of the remaining breast tissue. **D.** Breast containing active mammary glands in a lactating woman. Note the proliferation of secretory alveoli. **E.** Section of an active mammary gland. Note the abundant distended alveoli. **F.** Alveoli in an active mammary gland shown at higher magnification than in part E.

STUDY QUESTIONS

Directions: Each of the numbered items or incomplete statements is followed by lettered options. Select the **one** lettered option that is **best** in each case.

1. The pathologist has just received a biopsy specimen taken from the ovary of a normal healthy female. The ovarian cortex contains only atretic follicles and corpora albicantia. What is the age of the patient this biopsy specimen was most likely taken from?

   A. 1-month old neonate

   B. 9-year old prepubescent girl

   C. 25-year old woman

   D. 45-year old premenopausal woman

   E. 65-year old postmenopausal woman

Questions 2–4

Refer to the micrograph below when answering the following questions.

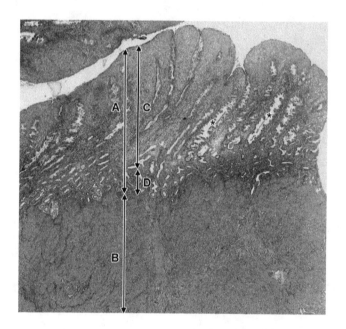

2. What is the effect on this organ of the failure to establish a pregnancy during an ovarian cycle?

   A. The region designated by letter A becomes ischemic and is sloughed off as menses.

   B. The region designated by letter B becomes ischemic and atrophies.

   C. The region designated by letter C becomes ischemic and is sloughed off as menses.

   D. The region designated by letter D becomes ischemic and is sloughed off as menses.

3. What stimulates the glands in this organ, noted with the *, to secrete during the secretory phase of the ovarian cycle?

   A. Androgens produced by granulosa cells of the graafian follicle

   B. Androgens produced by granulosa lutein cells of the corpus luteum

   C. Estrogen produced by granulosa cells of the graafian follicle

   D. Estrogen produced by granulosa lutein cells of the corpus luteum

   E. Progesterone produced by granulosa cells of the graafian follicle

   F. Progesterone produced by granulosa lutein cells of the corpus luteum

4. Identify the approximate day of the ovarian cycle from which this section was taken. Consider that menstrual bleeding begins on day 1 and the cycle is 28 days in length. Ignore the arrows and letters on the micrograph.

   A. Day 1

   B. Day 5

   C. Day 7

   D. Day 14

   E. Day 25

5. Glycogen is important in maintaining a low pH in the organ shown in the two micrographs below (two different magnifications). Which of the following mechanisms produces glycogen?

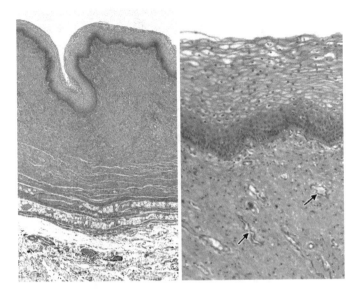

A. Desquamation of glycogen-containing epithelial cells

B. Exudation from the vessels indicated by the arrows (image on the right)

C. Secretion from cervical glands

D. Secretion from glands in the submucosa

E. Secretion from glands in the epithelium

6. Which hormone is most responsible for the regeneration of the functional layer of the endometrium after menstruation?

A. Androgen

B. Estrogen

C. Follicle-stimulating hormone

D. Luteinizing hormone

E. Progesterone

ANSWERS

1—E: Maturation of follicles ceases at menopause and the remaining follicles die by atresia. Thus, the ovary of a post-menopausal woman will contain corpora albicantia, the remnants of follicles that were ovulated during her reproductive years, and atretic follicles. Atresia of follicles occurs throughout a woman's life, but maturation of follicles does not begin until puberty. Thus, the ovary from a neonate or prepubescent girl will contain only primordial and atretic follicles, whereas the ovary of a 25-year old woman or a 45-year old premenopausal woman will contain follicles at all stages of maturation.

2—C: If pregnancy does not occur, the corpus luteum ceases to function and progesterone levels decrease. The decrease in progesterone causes the spiral arteries to constrict, which causes the functional layer of the endometrium (C) to become ischemic and slough off as menses. The basal layer of the endometrium (D) remains intact and proliferates to restore the functional layer. The myometrium (B) undergoes only minor changes during the ovarian cycle.

3—F: Granulosa lutein and theca lutein cells of the corpus luteum secrete progesterone during the secretory (luteal) phase of the ovarian cycle, which stimulates the endometrial glands to secrete glycogen and glycoproteins to nourish an embryo, should a pregnancy occur. Granulosa cells of graafian follicles produce estrogen but do not produce androgens or progesterone. Androgen is produced by cells of the theca interna and is converted to estrogen by granulosa cells. Estrogen promotes proliferation of granulosa cells and growth of follicles.

4—E: This is a section of uterus taken during the secretory (luteal) phase of the ovarian cycle (days 15–28). The functional layer is thick and the glands are coiled and distended in this section. On day 1 of the ovarian cycle, the functional layer would have been shed and only the basal layer would be present. The functional layer regenerates during the proliferative (follicular) phase (days 5–14); by day 14, the glands would be simple straight, narrow tubules.

5—A: The images show a section of vagina. There are no glands in the vagina. Epithelial cells synthesize glycogen, which is released into the lumen when the cells desquamate. Mucus from cervical glands and exudate from vessels in the mucosa and muscularis provide lubrication for the vagina.

6—B: Estrogen produced by the cohort of growing follicles stimulates proliferation of cells in the basal layer of the endometrium, which produces a new functional layer of endometrium.

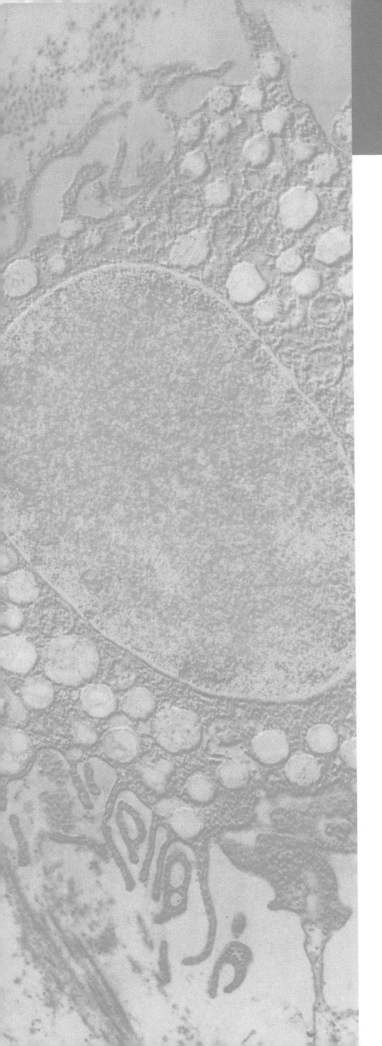

CHAPTER 16

MALE REPRODUCTIVE SYSTEM

OVERVIEW

The male reproductive system consists of a series of organs, tubules, and accessory glands that function in the production, nourishment, temporary storage, and intermittent discharge of **spermatozoa (sperm)**, the male germ cells. The organs and structures that comprise the male reproductive system include the following:

- **Testes.** Contain the **seminiferous tubules**, which produce the spermatozoa, and **Leydig cells**, which secrete the male sex hormone **testosterone**.

- **Intratesticular and extratesticular genital ducts.** Carry spermatozoa through the male reproductive system.

- **Accessory glands.** Produce secretions, which nourish spermatozoa and comprise the major portion of **semen**.

- **Penis.** Delivers semen to the female genital tract during intercourse; also conducts urine from the urinary bladder to outside the body.

Spermatozoa are produced in males continuously from adolescence throughout life through a complex process termed **spermatogenesis**, which involves a stereotypical pattern of mitotic and meiotic divisions of **spermatogenic cells** in the **seminiferous epithelium** of the testes.

Testosterone is the most important circulating **androgen** and is required for the development, maturation, and proper function of the male reproductive organs and for various secondary sex characteristics. The production of testosterone is under complex hormonal control involving the testes, pituitary gland, and hypothalamus.

TESTIS

ORGANIZATION OF THE TESTIS

The primary male sex organ is the testis, which functions as both an exocrine gland, with **spermatozoa** as the secretory product, and as an endocrine gland, which produces the male sex hormone **testosterone**. The paired testes, along with the epididymis and the initial portion of the ductus (vas) deferens, are located outside the body in the scrotum (Figure 16-1A). The overall organization of the testis is as follows:

- Each testis is surrounded by a dense, fibrous connective tissue capsule covered with mesothelium, called the **tunica albuginea**. The tunica albuginea thickens posteriorly to form the **mediastinum testis**, which bulges into the interior of the testis. Sperm ductules, blood and lymphatic vessels, and nerves enter and exit the testis through the mediastinum testis.

- The testis consists of numerous **seminiferous tubules** (the site of sperm production) supported by interstitial tissue containing **Leydig cells,** which produce testosterone.

- Connective tissue **septa** radiate from the mediastinum testis to divide the testis into approximately 250 **lobules**, each containing anywhere from one to four seminiferous tubules.

- A series of **genital ducts** within the testis (the tubuli recti, rete testis, and ductuli efferentes) convey sperm from the seminiferous tubules to the epididymis.

SEMINIFEROUS TUBULES

Seminiferous tubules are the structural and functional units of the testis (Figure 16-1B–D). Each unit is a highly coiled tubule about 80-cm long. Most tubules are loops that communicate at each end with the rete testis. Seminiferous tubules are lined with a complex stratified epithelium (**seminiferous epithelium**) composed of proliferating and differentiating **spermatogenic cells** and columnar, nonproliferating support cells called **Sertoli cells.** The seminiferous epithelium rests on a well-developed basal lamina and a lamina propria containing fibroblasts, collagen bundles, and **myoid cells** (Figure 16-1D). Contraction of the myoid cells produces rhythmic movements that help propel the contents of the seminiferous tubules toward the rete testis.

SEMINIFEROUS EPITHELIUM AND SPERMATOGENESIS

The seminiferous epithelium is composed largely of a proliferating population of **spermatogenic cells** in various stages of differentiation and development. In contrast to females, in whom germ cells become depleted as the female ages, germ cells are produced continuously in sexually mature males. Primordial germ cells undergo a stereotypical pattern of mitotic and meiotic divisions to produce mature spermatozoa through a process known as **spermatogenesis**. As developing gametes mature, they move from the basal lamina toward the lumen of the tubule. **Five stages in the production of sperm** can be identified as follows (Figure 16-1D):

1. **Spermatogonia.** Small diploid stem cells derived from primordial germ cells, which migrate into the testis during embryonic development and reside at the periphery of the seminiferous tubules in contact with the basal lamina. Spermatogonia divide by mitosis to produce both new stem cells and transit amplifying progenitor cells, which are also called spermatogonia. After several additional mitotic divisions, each progenitor spermatogonium divides a final time to produce two primary spermatocytes. The progeny of a progenitor spermatogonium apparently remain connected by intercellular bridges throughout spermatogenesis, which may synchronize their development.

2. **Primary spermatocytes.** Pale-staining cells, which are larger and closer to the lumen than are spermatogonia. Soon after their formation, primary spermatocytes enter the first meiotic prophase. Because this prophase lasts approximately 22 days, primary spermatocytes in various stages of chromosome condensation are frequently observed in histologic sections of the testis.

3. **Secondary spermatocytes.** Smaller than primary spermatocytes and have a pale-staining spherical nucleus. Secondary spermatocytes rapidly undergo the second meiotic division to produce haploid **spermatids**. Secondary spermatocytes are rarely observed because their life span is approximately only 8 hours.

4. **Spermatids.** Differentiating haploid cells that are embedded in invaginations of Sertoli cells near the lumen. As spermatids move toward the tubule lumen, they undergo an elaborate maturation process without mitosis, called **spermiogenesis,** to become mature **spermatozoa.** During spermiogenesis, spermatids elongate and discard most of their cytoplasm. In addition, they develop a flagellum and an **acrosome**, a cap-like organelle containing hydrolytic enzymes that facilitate the entry of sperm into an oocyte.

5. **Spermatozoa (mature sperm cells).** Highly specialized cells with a single flagellum and a small concentrated nucleus. Hundreds of millions of spermatozoa are produced and released into the lumen each day. Spermatozoa do not fully mature and become motile until they pass through the epididymis.

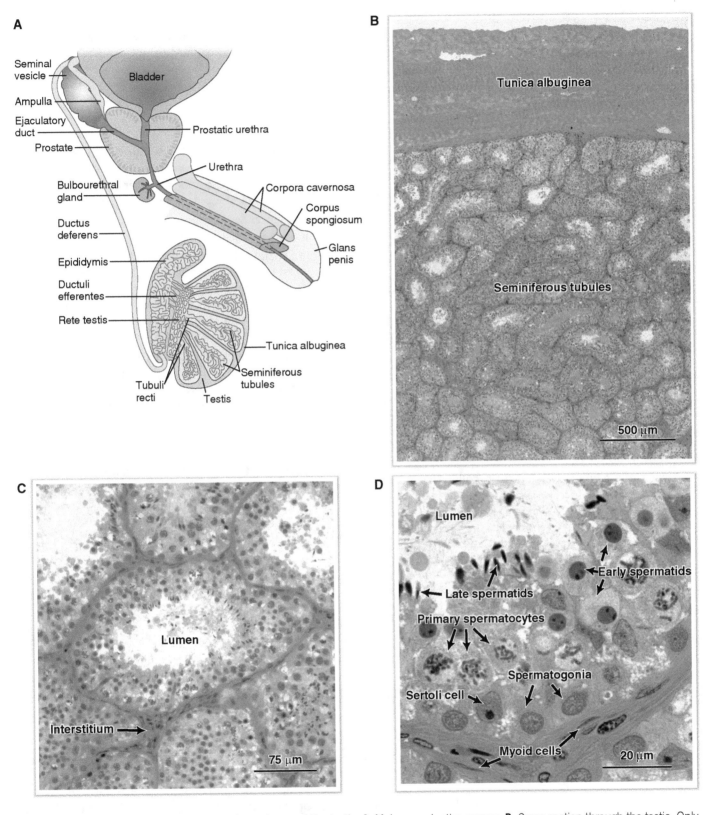

Figure 16-1: Overview of the male reproductive system and the testis. **A.** Male reproductive organs. **B.** Cross section through the testis. Only a small region of seminiferous tubules and the thick connective tissue capsule, called the tunica albuginea, are shown here. **C.** Cross section of seminiferous tubules in the testis shown at higher magnification than in part B. **D.** Seminiferous epithelium in the testis shown at higher magnification than in part C. A Sertoli cell and spermatogenic cells in different stages of maturation can be seen.

SERTOLI CELLS

In addition to spermatogenic cells, seminiferous epithelium contains a population of support cells called **Sertoli cells**, which, similar to spermatogonia, are in contact with the basal lamina. Sertoli cells can be distinguished from spermatogonia by the following **features** (Figure 16-2A and B):

- **Nucleus and nucleolus.** Sertoli cells have a large, basally located oval nucleus with deep infoldings and a prominent nucleolus.
- **Shape and size.** Sertoli cells extend from the basal lamina to the tubule lumen. The cells have complex shapes, with numerous invaginations that surround spermatocytes and spermatids.
- **Tight (occluding) junctions.** Sertoli cells form tight junctions with adjacent Sertoli cells that divide the seminiferous epithelium into a **basal compartment** containing spermatogonia, and an **adluminal compartment** containing spermatocytes and spermatids. During spermatogenesis, primary spermatocytes must pass through these tight junctions.

Sertoli cells have the following important **functions:**

- **Form the blood-testis barrier.** The barrier created by the tight junctions between Sertoli cells provides a special microenvironment for spermatogenesis in the interior of the tubule. The barrier protects sperm from immunologic attack by keeping sperm-specific antigens from entering the blood and becoming immunogens and by preventing circulating antibodies from entering the adluminal compartment.
- **Support and nourish spermatogenic cells.** The extensive infoldings of Sertoli cells physically support the interconnected spermatogenic cells and transport nutrients to developing germ cells and waste products away from these cells, which are isolated by the blood-testis barrier. Sertoli cells also secrete a fructose-rich fluid that nourishes and facilitates the transport of sperm to the genital ducts.
- **Produce cytoplasmic androgen-binding protein (ABP).** Sertoli cells secrete testicular fluid containing ABP into the lumen, which binds testosterone, thereby increasing the local concentration of testosterone at the site of spermatogenesis. Sertoli-cell secretion is stimulated by testosterone and by follicle-stimulating hormone (FSH).
- **Produce inhibin and activin.** These two polypeptide hormones suppress and activate FSH production, respectively.
- **Phagocytose degenerating germ cells and cytoplasm** shed by maturing spermatids during spermiogenesis.

LEYDIG CELLS

The interstitial spaces between seminiferous tubules are filled with a richly vascularized, loose connective tissue containing macrophages, mast cells, fenestrated capillaries, lymphatic vessels, nerves, and importantly, Leydig cells (Figure 16-2B).

- **Structure.** Leydig cells are polygonal cells with cytologic characteristics of steroid-secreting cells—abundant smooth endoplasmic reticulum and lipid droplets.
- **Function.** Leydig cells are endocrine cells that produce and secrete **testosterone**, the most important circulating **androgen**. Testosterone is required for the development, maturation and function of the male reproductive organs, male sexual function (libido), and regulation of gonadotropin secretion by the pituitary gland.

REGULATION OF SPERMATOGENESIS

Spermatogenesis is stimulated by testosterone. Testosterone levels in the seminiferous tubules are regulated by **luteinizing hormone (LH)** and **FSH** (also known as interstitial cell-stimulating hormone in males, or ICSH), as follows (see also Chapter 14):

- LH stimulates Leydig cells to produce testosterone, and FSH stimulates Sertoli cells to secrete ABP.
- Both LH and FSH are produced in the anterior pituitary in response to gonadotropin-releasing hormone (GnRH) from the hypothalamus. Testosterone exerts negative feedback and inhibits the production of LH, FSH, and GnRH. In addition, activin and inhibin from Sertoli cells stimulate and repress, respectively, the secretion of FSH.

Spermatogenesis is affected by temperature. The testes are housed outside the body within the scrotal sac, which helps maintain them at a temperature about 2° below core body temperature. Spermatogenesis requires this low temperature, although testosterone production by Leydig cells does not.

INTRATESTICULAR GENITAL DUCTS

Spermatozoa produced in the seminiferous tubules travel through the following genital ducts within the testis before entering the epididymis:

- **Tubuli recti.** Short, straight tubules that connect seminiferous tubules to the rete testis (Figure 16-1A). The initial portion of the tubuli recti is lined with epithelium composed entirely of Sertoli cells, whereas the more distal main segment is lined with simple cuboidal epithelium.
- **Rete testis.** Anastomosing channels in the mediastinum testis that empty into the ductuli efferentes (Figure 16-2C). The rete testis is lined with simple columnar or cuboidal epithelial cells that have short microvilli and a single flagellum.
- **Ductuli efferentes.** Approximately 10 to 20 short collecting ducts that connect proximally to the rete testis, pass through the tunica albuginea, and connect distally to the head of the epididymis. The epithelium of the ductuli efferentes is composed of patches of low cuboidal cells with microvilli alternating with patches of ciliated columnar cells, giving the epithelium a scalloped appearance, described poetically as **festooned epithelium** (Figure 16-2D). The microvilli-bearing cells resorb fluid secreted by the seminiferous epithelium, thus concentrating the sperm. Peristaltic contractions of the thin sheath of smooth muscle that surrounds each duct together with the cilia, which beat in the direction of the epididymis, help transport the sperm, which are not yet motile.

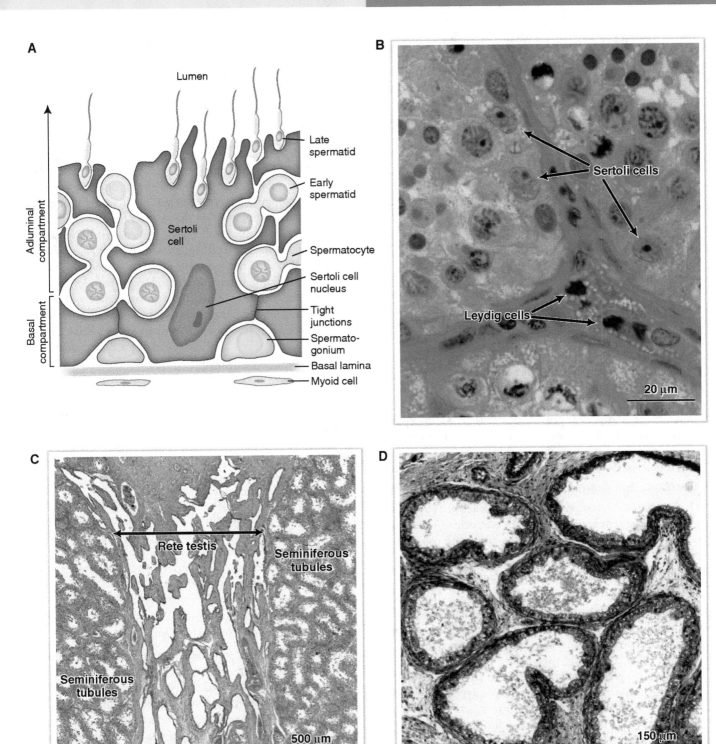

Figure 16-2: Cells and genital ducts in the testis. **A.** Sertoli cell surrounding and supporting spermatogenic cells in the seminiferous epithelium. Note that tight junctions between Sertoli cells divide the seminiferous tubule into a basal compartment and an adluminal compartment containing spermatogonia and the differentiating progeny, respectively. **B.** Seminiferous tubules and the interstitium of the testis. Sertoli cells are distinguished from spermatogonia by their large pale nucleus and prominent nucleolus. Leydig cells in the interstitium are characterized by vacuoles in their cytoplasm, an artifact of extraction of lipids during histologic processing. **C.** Section through the rete testis. Note the seminiferous tubules adjacent to the anastomosing channels of the rete testis. The microvilli and single flagellum of the cells that line the rete testis are not visible at this magnification. **D.** Section of ductuli efferentes. Note the characteristic scalloped appearance of the epithelium lining the ductuli efferentes. Collagen is stained blue in this section. (Azan stain)

EXTRATESTICULAR GENITAL DUCTS

Spermatozoa exit the ductuli efferentes and travel through a series of extratesticular genital ducts—the **epididymis**, the **ductus deferens** (also known as the **vas deferens**), the **ejaculatory duct**, and the **urethra**—before being released outside the body during ejaculation.

EPIDIDYMIS

Structure. Each epididymis is a single, highly convoluted tubule, approximately 4 m to 6 m long, and folded into a small, comma-shaped body located on the posterior aspect of each testis. Each epididymis is formed by the fusion of ductuli efferentes at its head and joins the ductus deferens at its tail (Figure 16-1A).

The lumen of the epididymis is lined with pseudostratified columnar epithelium, composed of **short basal cells** thought to function as stem cells, and **tall columnar cells** with long, branched, atypical microvilli (inaccurately referred to as **stereocilia**) (Figure 16-3A and B). The height of both the epithelial cells and stereocilia decreases from the region of the head to the tail. The epithelium rests on a well-defined basal lamina and a thin lamina propria, surrounded by a thin layer of circularly arranged smooth muscle. As the epididymis nears the ductus deferens, it acquires an outer and inner layer of longitudinally arranged smooth muscle. All of the muscle layers thicken markedly near the ductus deferens.

Function. The following functions occur in the epididymis:

1. **Resorption.** The remaining fluid secreted by the seminiferous epithelium is resorbed by the stereocilia in the head region of the epididymis, and the sperm are compacted into a dense mass.

2. **Phagocytosis.** Cell debris shed during spermiogenesis is taken up and digested.

3. **Sperm maturation and storage.** Sperm entering the epididymis are weakly motile and incapable of fertilization. They undergo their final maturation and become motile as they pass through the epididymis. This is an androgen-dependent process and presumably also depends on epididymal secretions. The muscular epididymal tail region serves as a sperm reservoir, and its contractions help conduct sperm to the ductus deferens during ejaculation.

SPERM TRANSPORT Sperm are passively transported from the seminiferous tubules into the epididymis by fluid that flows from an area of production (seminiferous tubules) to an area of resorption (ductuli efferentes and epididymis). This movement is assisted by the action of cilia and flagella in the rete testis and ductuli efferentes and by the spontaneous peristalsis of the smooth muscle coats of the ductuli efferentes and the epididymis.

DUCTUS DEFERENS

Structure. The ductus deferens is a straight, thick-walled muscular tube that passes out of the scrotum in the spermatic cord and connects the epididymis to the urethra via the ejaculatory duct within the prostate (Figure 16-1A). The lumen of each ductus deferens is lined with low pseudostratified columnar epithelium with stereocilia, similar to that in the tail of the epididymis (Figure 16-3D). The underlying fibroelastic lamina propria has numerous longitudinal folds, which give the lumen an irregular, stellate outline. The thick muscularis is composed of an inner and an outer longitudinal layer of smooth muscle separated by a thick circular layer (Figure 16-3C). The distal end of the ductus deferens is dilated to form the **ampulla**, which is characterized by highly folded, thickened epithelium. The ductus deferens is covered by a loose connective tissue adventitia.

Function. The ductus deferens transports sperm from the testis to the ejaculatory duct within the prostate gland. Autonomic nerve stimulation causes powerful contractions of the muscle wall of the ductus deferens, constituting one of the main propulsive forces for ejaculation.

▽ **Vasectomy** is a male form of birth control—a straightforward procedure in which each ductus deferens is surgically severed, and the cut ends are ligated or cauterized. A male who has had a vasectomy continues to produce sperm, but the procedure blocks the transportation of sperm from the testis to the ejaculatory duct, thereby resulting in sterility. Over time, sperm deteriorate and are phagocytosed. ▼

EJACULATORY DUCT

Each ductus deferens penetrates the prostate gland immediately distal to the ampulla. The portion of the ductus deferens within the prostate is called the ejaculatory duct. The paired ejaculatory ducts open into the prostatic urethra, thereby conveying sperm from the ductus deferens to the urethra (Figures 16-1A and 16-4B). Each duct is a short, straight tube lined with simple cuboidal epithelium without smooth muscle in its walls.

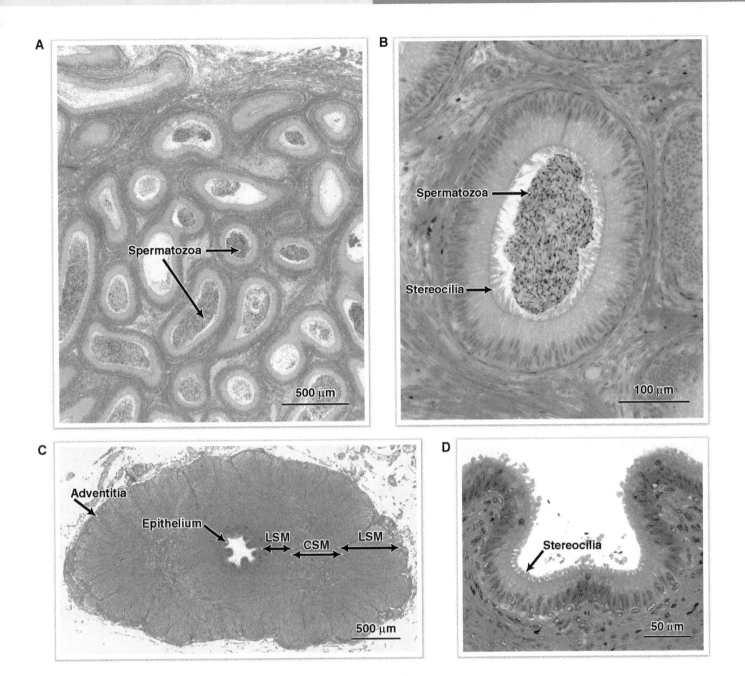

Figure 16-3: Extratesticular genital ducts. **A.** Section of the epididymis. All of the profiles shown here are cross sections of the one single tubule that comprises the epididymis. Note the presence of spermatozoa in the lumen of the epididymis. **B.** Cross section through the epididymis shown at high magnification. The long microvilli on the columnar cells in the epithelium are incorrectly referred to as stereocilia. **C.** Cross section through the ductus deferens. At low power, the ductus deferens can be confused with the ureter, which also has a narrow, irregular lumen and thick layers of smooth muscle (see Figure 13-4E). LSM, longitudinal smooth muscle; CSM, circular smooth muscle. **D.** Epithelium lining the ductus deferens shown at high magnification. The lumen is lined with pseudostratified columnar epithelium with stereocilia, allowing the ductus deferens to be readily distinguished from the ureter, which is lined with transitional epithelium (see Figure 13-4D).

ACCESSORY SEX GLANDS

The paired **seminal vesicles**, the single **prostate gland**, and the paired **bulbourethral glands** are accessory sex organs essential for reproductive function in males. The secretory products of these glands are released during ejaculation and contribute the major portion of the semen. The activity of each of the glands is dependent on the continued presence of testosterone.

SEMINAL VESICLES

- **Structure.** Seminal vesicles are two highly convoluted, unbranched tubules that develop as outpocketings of the ductus deferens at the distal end of the ampulla. The mucosa has tall complicated folds, which greatly increase the surface area for secretion but obscure the simple tubular organization of the vesicle when viewed in histologic sections. The lumen is lined with pseudostratified columnar secretory epithelium, resting on an elastin-rich lamina propria surrounded by a thin, inner circular layer and outer longitudinal layer of smooth muscle and a loose connective tissue adventitia (Figure 16-4A).

- **Function.** Seminal vesicles secrete a thick, creamy yellow fluid that comprises 70–80% of the ejaculate. The secretions contain a high concentration of **fructose**, a sugar that provides an energy source for sperm and nourishes sperm within the female reproductive tract. Contraction of smooth muscle in the seminal vesicle propels the secretions into the ejaculatory duct.

PROSTATE GLAND

- **Structure.** The prostate gland is a walnut-sized organ that surrounds the initial portion of the urethra just below the bladder, and is composed of 30 to 50 irregularly shaped, compound **tubuloalveolar glands** (Figure 16-4B and Figure 16-1A). The prostate is covered with a fibromuscular capsule, which extends septa into the prostate, dividing the gland into indistinct lobes. The glands are organized concentrically around the urethra in three zones, separated by thin layers of smooth muscle: an inner central (mucosal) zone; a transitional (submucosal) zone; and a peripheral (main) zone. Ducts of each zone empty directly into the urethra (Figure 16-4B). The glands are lined with elaborately folded, simple tall columnar to pseudostratified epithelium, and are embedded in a fibromuscular stroma (Figure 16-4C). In older men, calcified **prostatic concretions** may be present in the lumens of the glands.

- **Function.** The prostate produces and stores a thin, milky fluid that is rich in lipids, hydrolytic enzymes such as fibrinolysin, **prostatic acid phosphatase (PAP)**, and **prostate-specific antigen (PSA)**, as well as citrate and high levels of zinc. PAP and PSA are produced by both normal and malignant secretory epithelium and can be used as markers to follow the progression of prostatic cancer. Prostatic secretions, which are released during ejaculation, help liquefy the semen.

▽ **Benign prostatic hypertrophy (BPH) (nodular hyperplasia)** is an extremely common disorder in men older than age 50. BPH usually arises in the periurethral zones (the central and transitional zones). Hyperplasia of both secretory epithelium and stroma in these zones produces nodules that compress the urethra, making it difficult to void the bladder. In contrast, **prostatic adenocarcinoma**, one of the most common cancers in men, almost always originates in the peripheral main prostatic zone. Prostate cancer is characterized by increased serum levels of PAP and PSA. ▽

BULBOURETHRAL (COWPER'S) GLANDS

- **Structure.** The bulbourethral glands are paired, pea-sized compound tubuloalveolar glands that empty via long ducts into the urethra beyond the prostate (Fig. 16-1A). The glands are divided into lobes by septa containing skeletal and smooth muscle, and are lined with simple cuboidal to simple columnar mucus-secreting epithelium.

- **Function.** During sexual arousal, bulbourethral glands secrete a clear, viscous fluid containing abundant sugars prior to ejaculation. These secretions clean the urethra, facilitate the passage of sperm, and provide lubrication for sexual intercourse.

PENIS

- **Structure.** The penis is composed of three parallel, cylindrical masses of erectile tissue: the paired **corpora cavernosa of the penis**, located dorsally, and the single **corpus spongiosum**, located ventrally. Each erectile body is enclosed in a dense fibrous connective tissue capsule. The corpus spongiosum surrounds the urethra and is enlarged distally as the glans penis (Figure 16-4D and Figure 16-1A).

  The erectile tissue of the three corpora consists of large irregular, interconnecting venous spaces lined with endothelial cells and separated by trabeculae composed of connective tissue and smooth muscle (Figure 16-4D).

- **Function.** The penis has two important functions: as an **excretory** organ that voids urine from the urinary bladder to outside the body, and as a **copulatory** organ that transfers sperm to the female reproductive tract.

MECHANISM OF ERECTION

When the penis is flaccid, the erectile tissue is empty because blood in the dorsal penile arteries flows to the dorsal vein via an **arteriovenous anastomosis**, largely bypassing the corpora cavernosa and spongiosum. During sexual arousal, an erection is produced when parasympathetic innervation causes **relaxation** of the smooth muscle of the arteries that supply the erectile tissue, which allows blood to enter the corpora. As the corpora expand, they compress the penile veins, causing even more blood to be retained in the erectile tissues. After ejaculation, parasympathetic impulses to penile vasculature cease, allowing the arteriovenous shunts to reopen and blood to leave the corpora.

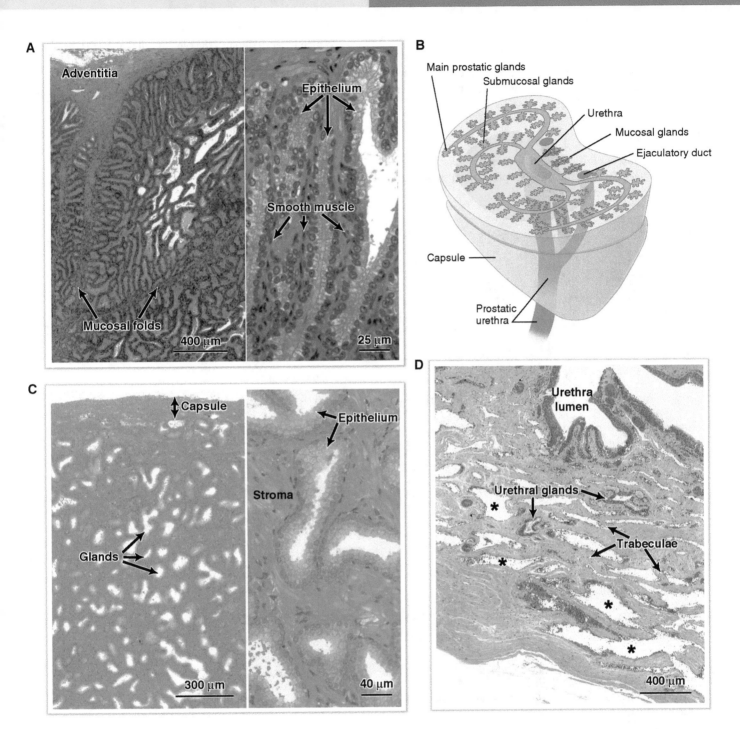

Figure 16-4: Male accessory sex glands and the penis. **A.** Section of the seminal vesicle shown at low-magnification (*left*) and at higher magnification (*right*). The mucosa of this single, unbranched tubule comprising the seminal vesicle is extensively folded in complicated patterns. **B.** Prostate gland. The cut surface shows the three zones of glands that surround the prostatic urethra and the location of the ejaculatory ducts. **C.** Peripheral (main) zone of the prostate gland. The prostate is composed of numerous compound tubuloalveolar glands; however, this organization is difficult to distinguish at either low magnification (*left*) or high magnification (*right*). **D.** Erectile tissue in the corpus spongiosum of the penis. Erectile tissue in the penis consists of anastomosing endothelium-lined venous spaces (*asterisks*) supported by trabeculae composed of connective tissue and smooth muscle. A portion of the urethra, which travels through the corpus spongiosum, is also shown in this image. Note the small urethral glands, which secrete mucus to protect the epithelium of the urethra from corrosive urine.

STUDY QUESTIONS

Directions: Each of the numbered items or incomplete statements is followed by lettered options. Select the **one** lettered option that is **best** in each case.

Questions 1 and 2

Refer to the micrograph below to answer the following questions. The lower panel of the micrograph shows a small region of the organ at high magnification.

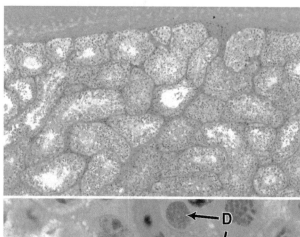

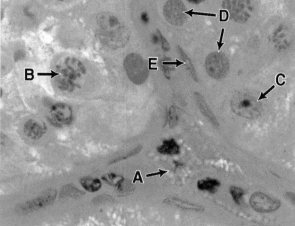

1. After reviewing a biopsy specimen, the pathologist diagnoses the tumor as originating from the cells labeled A. Which of the following substances would most likely be secreted by the cells in this tumor?

   A. Androgen-binding protein

   B. Cortisol

   C. Estradiol

   D. Fructose

   E. Testosterone

2. Which letter designates the type of cell that forms the blood-testis barrier?

3. Which organ listed below produces a fructose-rich fluid that comprises almost all of the ejaculate?

   A. Bulbourethral gland

   B. Epididymis

   C. Prostate gland

   D. Seminal vesicle

   E. Seminiferous tubules

4. Arrange in order the following structures and organs that a spermatozoon passes through en route to ejaculation.

   > 1 = seminiferous tubule
   >
   > 2 = epididymis
   >
   > 3 = ductuli efferentes
   >
   > 4 = ductus deferens
   >
   > 5 = rete testis
   >
   > 6 = urethra

   A. 1–2–3–4–5–6

   B. 1–3–5–2–4–6

   C. 1–5–2–3–4–6

   D. 1–5–3–2–4–6

   E. 1–5–4–2–3–6

5. The first clinically useful serum test for tumor detection was developed in the 1940s. The test was based on recognizing higher than normal levels of an acid phosphatase enzyme in blood samples obtained from men. This test identified tumors of which of the following organs?

   A. Bulbourethral glands

   B. Ductus deferens

   C. Epididymis

   D. Prostate gland

   E. Seminal vesicle

   F. Testis

ANSWERS

1—E: The cells labeled A in the micrograph are Leydig cells in the interstitium between the seminiferous tubules of the testis. Leydig cells normally synthesize and release the androgen testosterone.

2—C: The cell labeled C is a Sertoli cell, a nonproliferating support cell in the seminiferous epithelium of the testis. Tight junctions between Sertoli cells form the blood-testis barrier, which prevents sperm-specific antigens from entering the blood and becoming immunogens, and prevents circulating antibodies from entering the tubule lumen. Sertoli cells also secrete a fructose-rich fluid that nourishes developing sperm and facilitates transport of sperm to the genital ducts, androgen-binding, which binds testosterone and increases the concentration of testosterone in the tubule lumen, and inhibin and activin.

3—D: The seminal vesicles are two highly convoluted, unbranched tubules that secrete a thick, creamy fluid that comprises 70–80% of the ejaculate. The secretions contain a high concentration of fructose, which provides an energy source for sperm and nourishes sperm within the female reproductive tract.

4—D: Spermatogenesis occurs in the seminiferous tubules (1). The resulting spermatozoa pass into the rete testis (5), through the ductuli efferentes (3), and into the epididymis (2), where they become motile. The spermatozoa then pass along a long, straight tube, the ductus deferens (4), the short ejaculatory ducts (not listed), and into the urethra (6), from which they can be released via ejaculation.

5—D: The prostate gland is a walnut-shaped organ that surrounds the initial portion of the urethra just below the bladder. The prostate gland produces and stores a thick fluid that is rich in lipids and proteolytic enzymes, including prostatic acid phosphatase (PAP) and prostate-specific antigen (PSA), which help liquefy semen at ejaculation. Prostate cancer is characterized by increased serum levels of PAP and PSA in the blood.

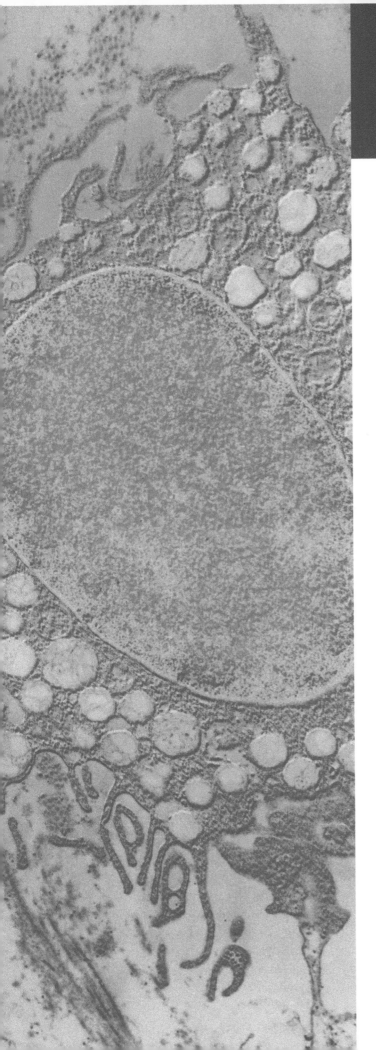

CHAPTER 17

FINAL EXAMINATION

STUDY QUESTIONS

Directions: Each of the numbered items or incomplete statements is followed by lettered options. Select the **one** lettered option that is **best** in each case.

1. A mother brought her 10-year-old son to the physician's office for his annual physical examination. While examining the boy, the physician discovered that the child had grown very little since his last visit. An MRI shows that the boy's epiphyseal growth plates are beginning to close prematurely. To slow the closure of the growth plate and to promote continued growth of his long bones, the physician prescribes somatotropin (growth hormone) as a treatment. What effect will this hormone have on the child's bone development?

   A. Stimulate the deposition of hydroxyapatite (calcium phosphate crystals) in the osteoid at the growth plate

   B. Stimulate the deposition of osteoid at the growth plate

   C. Stimulate the hypertrophy of chondrocytes in the growth plate

   D. Stimulate the maturation of osteoblasts to osteocytes

   E. Stimulate proliferation of chondrocytes in the growth plate

2. Asthma is a reactive airway disease resulting in intermittent reversible obstruction to airflow in the lungs due to hyperactivity and inflammation. A problem that occurs during an asthma attack is the constriction of the bronchioles inhibiting the flow of air into and out of the lungs. What histologic feature do the trachea and bronchi both possess that most likely prevents collapse during an asthma attack?

   A. Cilia

   B. Hyaline cartilage

   C. Mucus

   D. Simple columnar epithelial cells

   E. Surfactant

3. Nuclear extrusion is part of the normal process of producing human red blood cells. What is another process that normally involves the extrusion of the nucleus of a cell?

A. Apoptosis

B. Diapedesis

C. Holocrine secretion

D. Mast cell degranulation

E. Skeletal muscle cell fusion

4. A patient is brought to the emergency department after severing his median nerve. The neurosurgeon carefully realigns and sutures together the cut ends of different fascicles, and then predicts that the patient will regain use of his thumb. What is the mechanism by which the severed axons reconnect with their target muscles in the thumb?

A. The distal cut ends of the axons regenerate and reconnect with the motor neuron cell body

B. The proximal cut ends of the axons regenerate and reconnect with their target muscles

C. The proximal and distal cut ends of the severed axons fuse together

D. Schwann cells proliferate and "glue" the severed axons together

E. Stem cells in the spinal cord proliferate and generate new motor neurons, which grow into the median nerve and innervate the target muscles

5. A 45-year-old woman sees her physician because she has severe blistering over the buttocks. An analysis of sera with immunofluorescence shows autoantibodies localized as shown in the micrograph below. A biopsy indicates extensive inflammatory infiltrates with numerous eosinophils present. An abnormality in which of the structures listed below is mostly likely responsible for the patient's blisters?

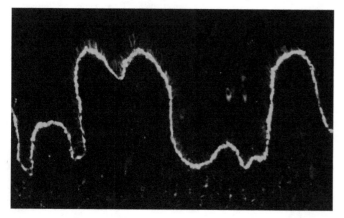

A. Desmosomes

B. Gap junctions

C. Hemidesmosomes

D. Zonula adherens

E. Zonula occludens

6. A 12-year-old boy is brought to the clinic because he has excessive urine production and low blood pressure. The physician requests laboratory studies for the analysis of antidiuretic hormone (ADH) and atrial natriuretic peptide (ANP) levels. What deviation from normal values would most likely be consistent with this patient's symptoms?

A. Either high ADH or high ANP

B. Either high ADH or low ANP

C. Either low ADH or high ANP

D. Either low ADH or low ANP

Questions 7 and 8

Refer to the micrographs below to answer the following questions. The lower image shows a small region of the upper image at higher magnification.

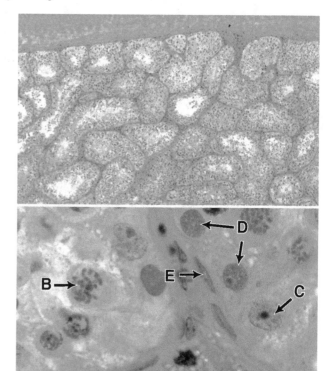

7. Which letter in the micrograph designates the type of cell that produces androgen-binding protein?

8. What is the function of cells labeled D in the micrograph?

A. Form the blood-testis barrier

B. Produce fructose

C. Produce testosterone

D. Undergo meiosis to produce spermatogonia and primary spermatocytes

E. Undergo mitosis to produce spermatogonia and primary spermatocytes

9. Arrange the following structures in proper sequence for the diffusion of CO_2 from an alveolar capillary to an alveolus (the following structures are in no particular order).

    1 = blood plasma

    2 = fused basal laminae of alveolar and endothelial cells

    3 = surfactant

    4 = cytoplasm of endothelial cell

    5 = cytoplasm of type I pneumocyte

    6 = air in alveolus

    A. 1–2–3–4–5–6

    B. 1–2–4–5–3–6

    C. 1–4–2–3–5–6

    D. 1–4–2–5–3–6

    E. 1–5–2–4–3–6

    F. 1–5–3–2–4–6

10. Focal adhesion complexes and hemidesmosomes are *both* associated with which of the following proteins?

    A. Actin

    B. Cadherins

    C. Integrins

    D. Keratin

    E. Tubulin

11. A researcher is studying eosinophils obtained from the blood of a patient and discovers a mutant gene. She determines that this mutant gene is not present in the patient's fibroblasts or epidermal cells, indicating that it was not present in the germ line DNA of the patient. The researcher then wants to test the hypothesis that the mutation occurred in a pluripotential hematopoietic stem cell. Verification of this hypothesis would most likely be obtained by discovering the mutation in which other cell type?

    A. Basophilic erythroblast

    B. Megakaryocyte

    C. Monoblast

    D. Polymorphonuclear neutrophil

    E. T cell

12. The left side of the diagram below shows longitudinal views of the three different kinds of large tubules in a kidney medullary ray (none is lined with low epithelium). Fluid is flowing in each tubule. Select the letter that best corresponds to the set of normal direction of flow for each of the three tubules.

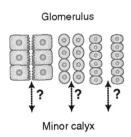

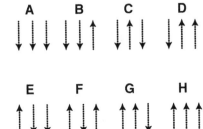

13. Medullary carcinoma is an endocrine cancer characterized by secretion of excessive amounts of calcitonin. In which of the following sites do the tumors in medullary carcinoma most likely arise?

    A. Adrenal cortex

    B. Adrenal medulla

    C. Anterior pituitary gland

    D. Follicular cells in the thyroid gland

    E. Parafollicular cells in the thyroid gland

    F. Parathyroid gland

    G. Pineal gland

    H. Posterior pituitary gland

14. Ehlers-Danlos syndrome is a group of inherited connective tissue disorders caused by the defective synthesis of collagen. Type 4 Ehlers-Danlos syndrome results from defective synthesis of type III collagen and is one of the more serious forms of the syndrome because blood vessels and other organs, such as the intestine, are prone to rupturing. Which connective tissue component is defective in patients with Ehler-Danlos type 4 syndrome?

    A. Collagen fibers

    B. Elastic fibers

    C. Ground substance

    D. Proteoglycans

    E. Reticular fibers

15. Bone participates in regulating serum Ca^{2+} levels by both rapid-acting and slower-acting mechanisms. What is the rapid response that occurs when Ca^{2+} levels in blood decrease?

    A. Release of Ca^{2+} by direct dissociation of calcium salts from bone surfaces, primarily in interstitial lamellae

    B. Release of Ca^{2+} by direct dissociation of calcium salts, primarily from newly formed bone surfaces

    C. Secretion of calcitonin, which reduces activity of osteoclasts

    D. Secretion of calcitonin, which stimulates activity of osteoclasts

    E. Secretion of parathyroid hormone, which reduces activity of osteoclasts

    F. Secretion of parathyroid hormone, which stimulates activity of osteoclasts

16. A 35-year-old woman visits her physician because she has noticed weakness and spasticity in the left lower extremity, visual impairment and throbbing in her left eye, and difficulties with balance, fatigue, and malaise. Laboratory studies show that there is an increase in cerebrospinal fluid protein, elevated gamma globulin, and moderate pleocytosis (increase in the number of mononuclear cells above normal levels). An MRI confirms areas of demyelination in the anterior corpus callosum. Which cells are specifically targeted in this patient's condition?

   A. Astrocytes

   B. Ependymal cells

   C. Microglia

   D. Oligodendrocytes

   E. Schwann cells

17. Blood flow in a patient with congestive heart failure becomes generally sluggish and as a result hepatocytes can die from lack of O_2. These dead cells are often replaced with connective tissue, ultimately resulting in a condition called congestive hepatopathy or cardiac cirrhosis. Refer to the diagram below illustrating the microanatomy of the liver to identify the letter that best demonstrates where you would expect to find the initial appearance of dead hepatocytes?

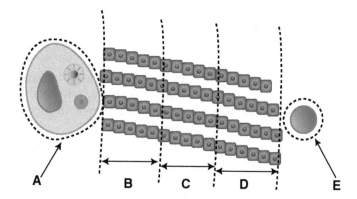

18. Benign prostatic hypertrophy (BPH) is a common disorder that occurs in men older than age 50. Hyperplasia of both secretory epithelium and stroma produce nodules that compress the genital duct (tube) that passes through the prostate, resulting in uncomfortable symptoms. Which genital tube is most directly impacted in BPH?

   A. Ductuli efferentes

   B. Ductus deferens

   C. Epididymis

   D. Tubuli recti

   E. Urethra

   F. Ureter

19. A patient whose home is in Kansas is planning a month-long visit to the Arctic Circle during the summer months. The trip will most likely directly affect the normal pattern of secretion of which hormone?

   A. Aldosterone

   B. Cortisol

   C. Epinephrine

   D. Melatonin

   E. Oxytocin

   F. Parathyroid hormone

   G. Thyroxine

20. A 55-year-old man arrives in your office complaining of intense pain in his lower back as well as pain radiating down his left leg. An MRI shows that he has two herniated discs. This patient's painful condition is most likely caused by tears or damage to which type of cartilage?

   A. Elastic cartilage

   B. Fibrocartilage

   C. Hyaline cartilage

21. In which of the following events is an abnormal growth of epithelial cells considered to be a malignant carcinoma?

   A. When it displays cell division at multiple sites

   B. When it expresses embryonic antigens

   C. When it loses differentiated characteristics

   D. When it penetrates the basal lamina

   E. When it recruits additional blood supply

22. Malignant tumors (cancers) are broadly classified as either leukemias or solid tumors. Leukemias involve the abnormal growth of white blood cells, arise in blood cell-producing sites, and the tumor cells are often found in the circulatory or lymphatic systems. Solid tumors, such as carcinomas, consist of localized, cohesive masses of abnormal cells growing in organs or body cavities. Given these distinctions, solid tumors would be expected to rely more heavily than leukemias on which of the following?

   A. Accumulation of mutations in cell cycle control genes

   B. Evasion of cell suicide programs

   C. Modification of chromosome ends to allow unlimited replication

   D. Production of angiogenesis factors

   E. Proliferation in the absence of specific growth factors

23. What is the function of the organ shown in the micrograph below?

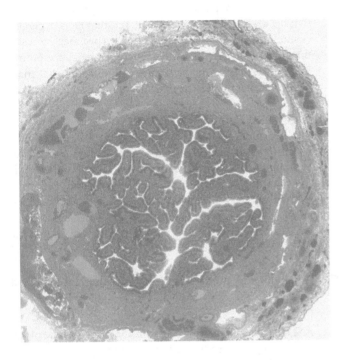

A. Exchange of nutrients between mother and fetus in females

B. Hormone secretion in males

C. Passage of urine in both males and females

D. Production of sperm in males

E. Transport of zygotes to the uterus in females

24. When skeletal muscle fibers are subjected to severe osmotic shock, the T tubules are disrupted and become disconnected and sealed off from the sarcolemma. In which of the following ways would disrupting the T tubules most likely affect the function of the muscle?

A. The muscle will disintegrate because sarcomeres are no longer held tightly together

B. The muscle will go into rigor mortis

C. The muscle will no longer contract in response to nerve stimulation, because excitation-contraction coupling is disrupted

D. Thick and thin filaments will separate, disrupting the normal striated appearance of the muscle cells

E. There would be no effect—the muscle will continue to be activated and contract normally

25. In some emergency situations, drugs (e.g., adrenaline) are injected directly into the chamber of a ventricle of the heart. In such a case, the needle could pass through the following typical ventricular components, listed in alphabetical order:

1. Adipose tissue
2. Cardiac muscle
3. Endothelial cells
4. Mesothelial cells
5. Purkinje fibers

Assuming that the needle passed through all the above components, from outside the heart to inside the chamber, which physical order below is most likely correct?

A. Outside –1 – 4 – 2 – 5 –3 – inside

B. Outside – 4 – 5 –1 – 2 – 3 – inside

C. Outside – 4 – 1 – 2 – 5 – 3 – inside

D. Outside – 3 – 5 – 2 – 1 – 4 – inside

E. Outside – 4 – 2 – 1 – 5 – 3 – inside

26. A patient comes to your office complaining of fever, muscle pain, and gastrointestinal discomfort. The patient lives on a farm and often butchers and cooks his own pork. He loves to hunt and often dines on the wild game he brings home. As part of your workup, you biopsy one of the man's painful muscles and discover that it is infected with the larvae of the parasitic roundworm *Trichinella*. Which connective tissue cell is most likely to be increased above normal at the site of the infection in this patient?

A. Adipocyte

B. Eosinophil

C. Fibroblast

D. Neutrophil

E. Plasma cell

27. Taste buds located near the tip of the tongue (anterior portion) are most commonly associated with which structures?

A. Circumvallate papillae

B. Filiform papillae

C. Fungiform papillae

D. Microfold cells

E. Von Ebner's glands

28. A 42-year old man complains of sudden onset of severe pain in the right upper quadrant, just under the ribs, of about 2 hours' duration, which then persists as a dull ache. The man states that these painful attacks occur whenever he eats fatty or fried foods. You order a series of tests, including a CT scan of the abdomen and an esophago-gastroduodenoscopy. Which abnormality do you most expect these tests to indicate?

A. Carcinoma of the head of the pancreas

B. Cirrhosis of the liver

C. Gallstones obstructing the common bile duct

D. Inflamed esophagus resulting from gastroesophageal reflux disease (heartburn)

E. Peptic ulcer

Questions 29–31

Refer to the illlustration below to answer the following questions.

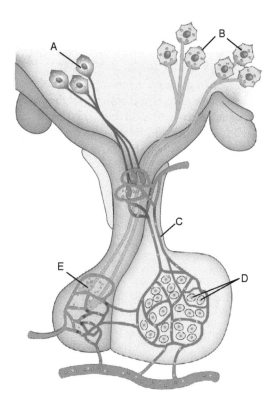

29. A lesion at site C in the above illustration would most likely *directly* affect secretion from which other endocrine tissue?

A. Adrenal cortex

B. Adrenal medulla

C. Anterior pituitary gland

D. Follicular cells in the thyroid gland

E. Parafollicular cells in the thyroid gland

F. Parathyroid gland

G. Pineal gland

H. Posterior pituitary gland

30. A patient visits his physician because he has noticed weight gain in his face, neck, and trunk. The man appears confused and suffers from hypertension. Laboratory studies indicate that his cortisol levels are elevated and that he is hyperglycemic. An MRI and subsequent biopsy specimen indicates that an adenoma is responsible for this patient's symptoms. Select the letter in the accompanying diagram that most likely indicates the cells from which the adenoma is derived?

31. After giving birth, a new mother is disappointed to find that her infant is unable to obtain milk from her swollen mammary glands, even after several days of painful attempts at suckling by the infant. A biopsy specimen indicates that the histology of the mammary glands is normal and presents as a fully developed mammary gland. Which letter on the accompanying diagram represents the site at which a defect in hormone release is most likely responsible for the inability to lactate?

32. A high level of antidiuretic hormone (ADH) in the blood would be expected to have which of the following effects in the kidney?

A. Decreased water recovery from the filtrate to the blood and closing of sodium channels in the distal convoluted tubule cell membranes

B. Decreased water recovery from the filtrate to the blood and decreased sodium pumping by the proximal convoluted tubule cells

C. Decreased water recovery from the filtrate to the blood and the removal of aquaporins from the collecting duct cell membranes

D. Increased water recovery from the filtrate to the blood and opening of sodium channels in the distal convoluted tubule cell membranes

E. Increased water recovery from the filtrate to the blood and increased sodium pumping by the proximal convoluted tubule cells

F. Increased water recovery from the filtrate to the blood and the addition of aquaporins into the collecting duct cell membranes

33. Which of the following proteins or structures enable the extracellular matrix and cytoskeleton to communicate across the cell membrane?

A Cadherins

B. Intermediate filaments

C. Integrins

D. Microtubules

E. Proteoglycans

34. Arrange in order the following events that occur during excitatory synaptic transmission, beginning with the synaptic cleft at a chemical synapse.

   1 = Neurotransmitter diffuses across the cleft

   2 = Presynaptic action potential arrives at the synapse

   3 = Presynaptic Ca^{2+} channels open and allow Ca^{2+} to enter

   4 = Postsynaptic receptors bind transmitter and open ion channels

   5 = Synaptic vesicles fuse with the membrane

   A. 2–1–3–5–4

   B. 2–3–5–1–4

   C. 2–3–1–5–4

   D. 2–5–1–3–4

   E. 2–5–3–1–4

35. An infant is born with a congenital defect in an integrin subunit that is expressed in neutrophils. The infant is brought to the clinic because she has a severe bacterial skin infection. What is the most likely situation regarding the neutrophil count in this infant's blood and at the site of the infection?

   A. Neutrophils will be high in the blood and low at the infected site

   B. Neutrophils will be high in the blood and high at the infected site

   C. Neutrophils will be low in the blood and low at the infected site

   D. Neutrophils will be low in the blood and high at the infected site

36. What is the major determinant of the extent and force of contraction of a skeletal muscle such as the gastrocnemius?

   A. The extent to which each myofiber contracts

   B. The extent to which each sarcomere contracts

   C. The number of myofibers that contract

   D. The number of myofibrils in each myofiber that contract

   E. The number of sarcomeres that contract in each myofiber

37. In which structure or organ does developing sperm undergo final maturation and become motile?

   A. Bulbourethral glands

   B. Ductus deferens

   C. Epididymis

   D. Prostate gland

   E. Seminal vesicle

   F. Testis

38. A researcher constructs a mouse colony in which newborn mice can be raised in a germ-free, antigen-free environment. Which immune system cell is most likely to be absent in these mice?

   A. Basophil

   B. B cell

   C. Macrophage

   D. Memory cell

   E. Polymorphonuclear neutrophil

   F. T cell

39. A 57-year-old woman who has smoked 2 packs of cigarettes a day for more than 35 years has trouble breathing and at times has violent coughing fits to remove the phlegm from her respiratory system. Her coughing episodes are most likely the result of destruction of which of the following structures?

   A. Cilia

   B. Clara cells

   C. Goblet cells

   D. Type I pneumocytes

   E. Type II pneumocytes

40. Refer to the image below of bone formation from the diaphysis of an embryonic canine limb bone to identify the letter that best indicates the cells that are responsible for the synthesis and secretion of osteoid and its initial mineralization.

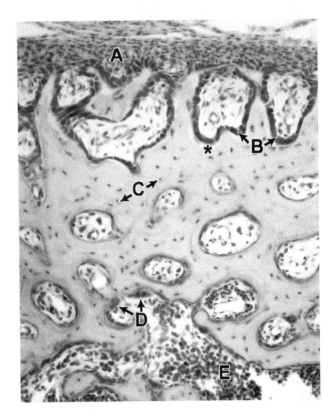

41. A researcher has developed a method for labeling all of the polymorphonuclear neutrophils (PMNs) present in the bone marrow of a mouse and is able to monitor the rate at which they exit the marrow and enter the bloodstream. When the researcher measures the concentration of PMNs in the blood, she finds the number is far lower than expected. She determines that few of these cells have entered connective tissue. What is the most likely explanation for the deficit in the expected numbers of freely circulating neutrophils?

A. Apoptosis

B. Diapedesis

C. Endomitosis

D. Margination

E. Respiratory burst

42. It is a relatively common practice to deliver medications to patients via a syringe and needle. In a subcutaneous injection, the medication would be deposited within which layer or sublayer of skin?

A. Hypodermis

B. Papillary dermis

C. Reticular dermis

D. Stratum basale

E. Stratum corneum

F. Stratum spinosum

43. Dust cells in the lungs are functionally equivalent to which cells in connective tissue proper?

A. Adipocytes

B. Eosinophils

C. Fibroblasts

D. Lymphocytes

E. Macrophages

F. Mast cells

G. Plasma cells

44. The filtrate in the thin descending limb (tDL) of the loop of Henle can be hypertonic, whereas the filtrate in the thick ascending limb (TAL) is hypotonic as the tubule enters the cortex. What are the states of active ion pumping and water impermeability in these two tubule elements that accounts for this situation?

	High Level of Ion Pumping		Water Impermeable Membranes	
	tDL	TAL	tDL	TAL
A.	+	−	+	−
B.	+	−	−	+
C.	−	+	+	−
D.	−	+	−	+

45. What is the function of the organ shown in the micrograph below? (The inset shows the lining of the lumen at higher magnification.)

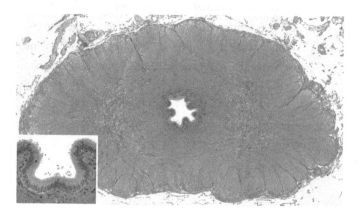

A. Passage of sperm from the epididymis to the urethra

B. Passage of sperm from the testis to the epididymis

C. Passage of urine and sperm

D. Passage of urine from the bladder to the urethra

E. Resorption of fluid and maturation and storage of sperm

46. The hypothalamus in the brain controls feeding behavior and regulates feelings of hunger and satiety. It responds to hormones produced by cells in which layer of the gastrointestinal tract?

A. Epithelium

B. Lamina propria

C. Muscularis mucosae

D. Submucosa

E. Muscularis externa

F. Adventitia

G. Serosa

47. At which site listed below does intramembranous ossification occur during development and growth of the diaphysis of a long bone?

A. Adjacent to the inner circumferential lamellae

B. Directly adjacent to the marrow cavity

C. Directly beneath the periosteum

D. In the epiphyseal growth plate

E. Inside an osteon

F. Within the Haversian canals

48. In a given muscle fiber at rest, the length of the I band is 1.0 μm and the A band is 1.5 μm. Contraction of that muscle fiber results in a 10% shortening of the length of the sarcomere. What is the length of the A band after the shortening produced by the muscle contraction?

   A. 1.50 μm

   B. 1.35 μm

   C. 1.00 μm

   D. 0.90 μm

   E. 0.45 μm

49. The parents of an infant who was born prematurely are devastated to learn that the infant's testes have failed to descend from the body cavity into the scrotum, a condition known as cryptorchidism. The pediatrician assures the parents that the testes may descend on their own within a few months; however, if they do not descend, she insists that the defect must be corrected surgically or the condition may eventually result in significant consequences. What symptom is this boy most likely to experience if his testes remain within the body cavity?

   A. Autoimmune response against sperm proteins

   B. Failure to develop secondary sexual characteristics

   C. Infertility

   D. Both infertility and failure to develop secondary sexual characteristics

   E. No physical symptoms but possible psychological problems

50. Cells undergoing viral infection or neoplastic transformation usually express "stress" proteins on their surface in response to the metabolic imbalances produced by these abnormal activities. Natural killer cells have receptors for these "stress" proteins that signal them to kill the stressed cells, but this killing may occur only if another condition is met. Which other condition of the stressed cells facilitates their killing?

   A. A foreign protein is expressed on the major histocompatability complex-I (MHC class I)

   B. Cytokines are expressed on MHC class I

   C. High levels of MHC class I are expressed

   D. Low levels of MHC class I are expressed

   E. The Fas ligand is expressed on MHC class I

51. The enzymes responsible for terminal digestion of carbohydrates and proteins in the small intestine are found linked to the surface of which cell type?

   A. Enterocyte

   B. Enteroendocrine cell

   C. Goblet cell

   D. Paneth cell

   E. Stem cell

52. You have received a biopsy specimen taken from the uterus of a 30-year-old woman. The pathology report states that the tissue is normal and describes the uterus as having a moderately thick endometrium with narrow, straight tubular glands. This biopsy specimen was most likely taken on which day of the woman's ovarian cycle. Assume that menstrual bleeding begins on day 1 of the cycle, and that the cycle is 28 days.

   A. Day 1

   B. Day 5

   C. Day 14

   D. Day 20

   E. Day 28

53. A 25-year-old woman is hiking in the desert when a sudden storm blows in—clouds cover the sun, she is soaked by rain, and the wind blows strongly. To conserve body heat, blood is diverted away from the surface of her skin via which of the following mechanisms?

   A. Arteriovenous anastomoses

   B. Glomerular systems

   C. Perivascular spaces

   D. Portal systems

   E. Sinusoids

   F. Shunts between small venules and medium veins

   G. Vasa vasorum

54. Which cartilaginous structure would be most severely affected by the defective synthesis of elastic fibers?

   A. Articular cartilage in joints

   B. Epiglottis

   C. Intervertebral discs

   D. Nasal septum

   E. Pubic symphysis

55. Activation of a naïve T cell requires two signals, known as costimulation. Binding of an antigen by the naïve cell's T-cell receptor (TCR) generates the first signal. What provides the second signal?

   A. Any cell undergoing an inflammatory reaction

   B. Cytokines released in response to an infection

   C. Epithelial reticular cells in the thymus

   D. Follicular dendritic cells

   E. Professional antigen-presenting cells

56. Which cells are most directly responsible for nourishing an oocyte or zygote during its journey to the uterus?

   A. Cells in the endometrial glands

   B. Ciliated cells in the uterine tube

   C. Granulosa lutein cells

   D. Peg cells in the uterine tube

   E. Theca lutein cells

57. Which cells in respiratory epithelium are most responsible for regulating the function of other cells in the epithelium?

A. Basal (short) cells

B. Brush cells

C. Ciliated columnar cells

D. Goblet cells

E. Small granule cells

58. Which component of the nervous system is primarily responsible for regulating the motility and other activities of the digestive tract?

A. Enteric neurons

B. Motor neurons

C. Projection neurons

D. Pyramidal neurons

E. Sensory neurons

59. You have received a biopsy specimen taken from the ovary of a normal healthy female. The ovarian cortex contains only primordial and atretic follicles. What is the age of the patient from whom this biopsy specimen was taken?

A. 1-month-old neonate

B. 13-year-old girl (just after puberty)

C. 25-year-old nonpregnant woman

D. 25-year-old pregnant woman

E. 50-year-old menopausal woman

60. What are the four basic types of tissue from which all organs are constructed?

A. Adipose tissue, connective tissue, epithelium, and muscle

B. Adipose tissue, epithelium, muscle, and nervous tissue

C. Adventitia, lamina propria, mucosa, and submucosa

D. Connective tissue, epithelium, muscle, and nervous tissue

E. Endothelium, epithelium, muscle, and nervous tissue

61. What is a major function of the connective tissue wrappings (endomysium, perimysium, epimysium) of skeletal muscle?

A. Conduct depolarization into the depths of the myofiber

B. Mediate excitation–contraction coupling

C. Sequester and release calcium

D. Synchronize contraction of all myofibers in a motor unit

E. Transfer the force of contraction to bone

62. Identify the predominant cell type that lines the lumen in the region of the organ associated with the gut tube shown in the micrograph below.

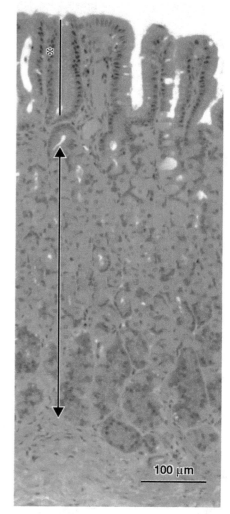

100 μm

A. Chief cells

B. Enteroendocrine cells

C. Mucous neck cells

D. Mucous surface cells

E. Parietal cells

F. Stem cells

63. The mushroom toxin phalloidin kills cells that are highly active in endocytosis. Cells of which zone in the liver and of which tubules in the kidney are most likely to be susceptible to phalloidin toxicity? (CD = collecting duct; DCT = distal convoluted tubule; PCT = proximal convoluted tubule)

	Hepatocytes		Kidney Tubule Cells		
	Zone 1	Zone 3	CD	DCT	PCT
A.	+	−	+	−	−
B.	+	−	−	+	−
C.	+	−	−	−	+
D.	−	+	+	−	−
E.	−	+	−	+	−
F.	−	+	−	−	+

64. Which of the following constituents of cartilage is most responsible for appositional growth?

    A. Chondroblasts

    B. Chondrocytes

    C. Interterritorial matrix

    D. Isogenous groups

    E. Proteoglycan aggregates

    F. Territorial matrix

65. Activated T_c cells with T-cell receptors (TCRs) that can recognize a specific viral peptide are continuously scanning cells in the body and will kill any cell that displays that viral peptide on its major histocompatibility complex-I (MHC class I) molecules. The first step for displaying the peptide on MHC class I is proteolysis of viral proteins in proteasomes. What is the next major step required for loading the peptide onto an MHC class I molecule?

    A. Diapedesis

    B. Endocytosis

    C. Glycosylation

    D. Membrane transport

    E. Opsonization

    F. Positive selection

66. Which hormone listed below is most important in triggering ovulation?

    A. Androgen

    B. Estrogen

    C. Follicle-stimulating hormone

    D. Luteinizing hormone

    E. Progesterone

67. Several substances imported from the gut lumen by absorptive columnar cells of the small intestine are carried directly to the liver via blood in the portal vein. Which imported substance does not follow this route?

    A. Amino acids

    B. Lipids

    C. Sugars

    D. Vitamins

    E. Water

68. Which connective tissue cells secrete hormones and cytokines involved in regulating metabolism and the immune system?

    A. Adipocytes

    B. Fibroblasts

    C. Macrophages

    D. Mast cells

    E. Plasma cells

69. Arrange in order the regions or zones that will be encountered as a pin passes into an epiphyseal plate from the diaphysis in a developing long bone.

    1 = zone of resting (reserve) cartilage
    2 = zone of maturation and hypertrophy
    3 = zone of calcification and death
    4 = zone of proliferation
    5 = zone of ossification

    A. 5–2–3–4–1

    B. 5–4–3–2–1

    C. 5–3–2–4–1

    D. 5–4–2–3–1

    E. 5–2–4–3–1

70. Which of the following three cells or structures comprise the blood-air barrier where gas exchange occurs in the lung?

    A. A type I pneumocyte, a Clara cell, and an endothelial cell

    B. A type I pneumocyte, an endothelial cell, and a basal lamina

    C. A type I pneumocyte, a type II pneumocyte, and a basal lamina

    D. A type I pneumocyte, a type II pneumocyte, and a dust cell

    E. A type II pneumocyte, an endothelial cell, and a basement membrane

71. A 5-year-old boy sustains a tear in his gastrocnemius muscle when he is involved in a bicycle accident. Which cellular mechanism allows regeneration and repair of the muscle?

    A. Dedifferentiation of myocytes into myoblasts

    B. Differentiation of fibroblasts to form myocytes

    C. Differentiation of satellite cells

    D. Fusion of damaged myofibers to form new myotubes

    E. Hyperplasia of existing myofibers

72. A 40-year-old man goes to the clinic because he has noticed large unpigmented patches of skin on various areas of his body. Which cell type is mostly likely defective or missing in this patient?

    A. Diffuse endocrine system (DNES) cells

    B. Keratinocytes

    C. Langerhans cells

    D. Melanocytes

    E. Merkel cells

73. An infant delivered prematurely from a 34-year-old woman suddenly develops severe respiratory difficulty. Premature infants with respiratory distress syndrome most likely have difficulty breathing due to immature formation of which of the following lung cells?

A. Alveolar phagocyte

B. Brush cell

C. Ciliated cell

D. Goblet cell

E. Type II pneumocyte

74. Which sublayer of epidermis produces substances that form an important part of the waterproof barrier that protects deeper tissues from dehydration?

A. Stratum basale

B. Stratum corneum

C. Stratum granulosum

D. Stratum lucidum

E. Stratum spinosum

75. What are the main organizational units of adult compact bone?

A. Inner circumferential lamellae

B. Interstitial lamellae

C. Osteons

D. Outer circumferential lamellae

E. Trabeculae

76. A researcher discovers a surface antigen that is unique to the hematopoietic stem cells (HSCs) in a mouse line. To prove that these are indeed HSCs, he isolates a single cell from the blood of a healthy mouse using an antibody reactive against the antigen. He then injects this cell into a mouse that has been irradiated with a normally lethal dose of x-rays sufficient to destroy all of its blood cell producing capacity. The mouse lives and is able to produce all blood cells. However, when the researcher places these HSCs in cell culture, no blood cells are produced. What is most likely missing from the culture condition that would account for this failure?

A. Basal lamina

B. Colony-stimulating factors

C. Endothelial cells

D. Hyaluronic acid

E. T-helper lymphocyte cells

77. Refer to the micrograph below to identify the clinical syndrome that results from autoimmune destruction of the predominant cells in the regions marked with asterisks.

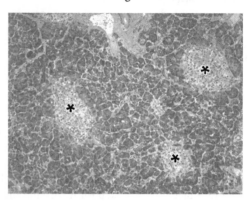

A. Acute pancreatitis

B. Crohn's disease

C. Hirschprung's disease

D. Type 1 diabetes mellitus

E. Type 2 diabetes mellitus

78. Which type of cell comprises an isogenous group (cell nest)?

A. Chondroblast

B. Chondrocyte

C. Osteoblast

D. Osteocyte

E. Osteoclast

79. An examination of fingernails is often part of a physical examination. The epithelium of the nail bed is thin and nail plates are nearly transparent, so it is possible to assess the amount of O_2 in blood by the color of blood in the underlying dermal vessels. The nail plates of fingernails and toenails are functionally equivalent to which sublayer of the epidermis?

A. Stratum basale

B. Stratum corneum

C. Stratum granulosum

D. Stratum lucidum

E. Stratum spinosum

80. Olympic athletes are tested for evidence of a variety of performance-enhancing substances with the use of sophisticated and sensitive chemical assays. Erythropoietin, or EPO, is one of the substances for which screening may be done. Prolonged treatment with abnormally high levels of EPO would be expected to result in what fairly obvious clinical condition?

A. Anemia

B. Eosinophilia

C. Jaundice

D. Polycythemia

E. Thalassaemia

81. Which of the following sensory structures in skin are responsible for detection of vibration and deep pressure?

A. Free nerve endings

B. Meissner's corpuscles

C. Merkel disks

D. Nissl bodies

E. Pacinian corpuscles

F. Ruffini endings

82. The presence of a germinal center in a lymph node is associated with which of the following events?

A. Early event in B-cell activation

B. Early event in T-cell activation

C. Late stage in B-cell activation and development

D. Late stage in T-cell activation and development

83. What is the most common mode of exocrine secretion?

A. Apocrine

B. Compound

C. Holocrine

D. Merocrine

E. Seromucous

84. In general, what is the thickest portion of large veins?

A. Tunica intima

B. Tunica media

C. Tunica adventitia

85. Cells in the theca interna of ovarian follicles secrete which of the following hormones?

A. Androgen

B. Estrogen

C. Follicle-stimulating hormone

D. Luteinizing hormone

E. Progesterone

86. Which component of the central nervous system listed below is most directly responsible for producing cerebrospinal fluid?

A. Astrocytes

B. Choroid plexus

C. Glia limitans

D. Neurons

E. Oligodendrocytes

F. Pia mater

87. In terms of its association with cytoplasmic elements, the zonula adherens of a columnar epithelial cell is most similar to which structure of a striated muscle cell?

A. A band

B. I band

C. Sarcoplasmic reticulum

D. T tubule

E. Z disk

ANSWERS

1—E: The growth and development of epiphyseal cartilage in the growth plate is influenced by various signaling molecules, including somatotropin. Somatotropin stimulates the liver to produce insulin-like growth factors (primarily IGF-1 or somatomedin C), which stimulate mitosis of chondrocytes in the epiphyseal growth plate, thereby promoting long-bone growth. Somatotropin levels (and hence IGF-1 levels) decrease at puberty, which contributes to closure of the epiphyseal plate.

2—B: The C-shaped hyaline cartilage rings (trachea) and cartilage plates (bronchi) physically keep the lumen of their airways open during an asthma attack and during exhalation. The bronchioles lack this cartilage support and, therefore, close during an asthma attack because there is no physical support to keep the lumen open.

3—C: Holocrine secretion occurs in sebaceous glands and involves the rupture of the entire cell and release of its contents, which then become the secretory product. In apoptosis, the cell nucleus disintegrates, and macrophages phagocytose pieces of the dying cell to avoid releasing cytoplasmic contents.

4—B: When a peripheral nerve is injured, the distal portions of the severed axons degenerate by a process called Wallerian degeneration. The proximal segments of the axons can grow new processes, and if the distances are fairly short, can reconnect with their original target muscles.

5—C: The antibodies are localized to the dermal–epidermal junctions. The patient is diagnosed with bullous pemphigoid in which bullous pemphigoid antibodies are produced to proteins specific to hemidesmosomes, the junctions that link cells in the stratum basale of the epidermis to the basement membrane. In bullous pemphigoid, the entire epidermis separates from the basement membrane. Other skin disorders result from autoantibodies to other types of junctions; for example, in pemphigus, antibodies attack desmosomes between epidermal cells.

6—C: Excessive urine production could be associated with low levels of antidiuretic hormone (ADH), which would result in failure to recover an adequate amount of water from urine in the medullary portions of the kidney. High levels of atrial natriuretic peptide (ANP) would reduce blood pressure via several mechanisms, including stimulating the excretion of Na^+ into the kidney tubules, which would result in an increasing water flow and greater urine production.

7—C: The cell designated C is a Sertoli cell. Sertoli cells produce testicular fluid containing androgen-binding protein, which binds testosterone and increases the concentration of testosterone in the tubule lumen. Sertoli cells also secrete a fructose-rich fluid that nourishes developing sperm and facilitates transport of sperm to the genital ducts, as well as inhibin and activin. Tight junctions between Sertoli cells form a blood-testis barrier that forms a protective environment for sperm production.

8—E: The cells labeled D are spermatogonia. Some spermatogonia divide by mitosis to produce additional spermatogonia, whereas others divide by mitosis to produce both spermatogonia and primary spermatocytes.

9—D: CO_2 in a pulmonary capillary must cross the blood-air barrier to reach the lumen of an alveolus. The blood-air barrier consists of the cytoplasm of an endothelial cell, the fused basal laminae of the endothelial cell and a type I pneumocyte, and the cytoplasm of the type I pneumocyte. Type II pneumocytes secrete surfactant that lines the internal surface of the alveolus. CO_2 must pass through the surfactant before entering the alveolus to be exhaled.

10—C: Focal adhesion complexes and hemidesmosomes contain integrins and serve to link cells to the basal lamina. Focal adhesions are associated with actin filaments, whereas hemidesmosomes are associated with keratin filaments.

11—E: The pluripotential hematopoietic stem cell (HSC) gives rise to two, more committed, oligopotent stem cells, the GEMM, which gives rise to other stem cells that produce the granulocytes, erythrocytes, monocytes, and megakaryocytes, and a lymphoid stem cell, which gives rise to stem cells that produce B and T lymphocytes. Choices A–D (basophilic erythroblast, megakaryocyte, monoblast, and polymorphonuclear neutrophil) all originate from the GEMM, as do eosinophils. Thus, finding the mutation in any of those cells would indicate that either a GEMM or an HSC cell could have undergone mutation, but that does not implicate the HSC as the sole target. If the mutation is found in T cells, it must have occurred in an HSC to be found in both sets of downstream cells; therefore, E is the best answer.

12—C: The tubule on the left has the tallest and widest cells; it represents a straight proximal tubule, and filtrate would be flowing down toward the minor calyx. The tubule in the middle has lower cells and the smallest diameter; it represents a thick ascending limb, and filtrate would be flowing up toward the medulla. The tubule on the right has the most cells per running length; it represents a collecting duct, and filtrate would be flowing down toward the medulla, on its final trip before entering a minor calyx.

13—E: Parafollicular cells in the thyroid gland secrete calcitonin and are the cells from which medullary carcinomas of the thyroid are derived.

14—E: Reticular fibers are constructed from type III collagen and are defective in Ehlers-Danlos type 4 syndrome.

15—B: When blood Ca^{2+} levels decrease, Ca^{2+} can be quickly released from bone by direct dissociation of calcium salts. This is thought to occur mainly from the surfaces of spongy bone and newly formed lamellar bone.

16—D: The patient has multiple sclerosis (MS), a demyelinating disease in which both CD4+ and CD8+ T cells, as well as autoantibodies, are targeted to oligodendrocytes. MS is almost twice as prevalent in women as in men. Demyelination is most commonly found in the anterior corpus callosum. Alterations in the cerebrospinal fluid show pleocytosis (an increase in the number of mononuclear cells above normal levels), increase in protein, and elevated gamma globulin. At autopsy, plaques are found that contain lymphocytes and monocytes in infiltrates around small veins in what is known as perivascular cuffing. Axons are generally preserved, but there is a paucity of oligodendrocytes, the cells that produce myelin in the central nervous system.

17—D: Cells of zone 3 are the last to receive blood. Zone 3 has the lowest pO_2, and it is in this region that the initial cell death is usually observed.

18—E: In benign prostatic hypertrophy, the enlarged periurethral portion of the prostate compresses the prostatic urethra, making it difficult to void the bladder.

19—D: The pineal gland secretes melatonin, which affects the rhythmic activity of other endocrine organs. Melatonin is secreted in a circadian pattern; secretion is promoted by darkness and inhibited by daylight. Therefore, exposure to continuous daylight will alter this normal circadian pattern of melatonin secretion.

20—B: Intervertebral discs are constructed from fibrocartilage, which can weaken over time and tear, allowing the nucleus pulposus to extrude. Inflammatory signals arising from the torn cartilage, as well as pressure of the extruded material on nearby spinal nerves, can be quite painful.

21—D: All of the events listed can be associated with the progression of a cancerous growth, but it is the ability to penetrate the basal lamina that most frequently signals that the growth has become malignant. At that point, the tumor cells can spread to other organs.

22—D: The ability to stimulate the formation of blood vessels, which requires the production of angiogenesis factors, is critical for the growth of solid tumors. Leukemia cells that lodge in lymph nodes or circulate in the blood will not require the induction of a blood supply. The other choices describe requirements common to both types of cancers.

23—E: The micrograph shows a section through the ampulla of the uterine tube. Oocytes are swept into the uterine tube after ovulation, where they can be fertilized. The resulting zygote is transported to the uterus via the uterine tube, where it may implant.

24—C: T tubules are thin projections of the sarcolemma that are in close contact with the sarcoplasmic reticulum. Membrane depolarization that occurs after the nerve action potential is carried inside the muscle by the T tubules and causes the sarcoplasmic reticulum to release the Ca^{2+} required for muscle contraction. If T tubules are separated from the sarcolemma, depolarization will not be carried inside the myofiber.

25—C: When drugs (e.g., adrenaline) are injected directly into the chamber of a ventricle of the heart in situations requiring emergency care, the needle will first pass through the mesothelial lining of the epicardium and encounter the large amount of adipose tissue normally present in that layer. Then, it will enter the myocardium and pass through cardiac muscle. Next, it will enter the endocardium; if it were to go through a Purkinje fiber, it will first encounter that in the subendocardial space before going through the endothelium to enter the ventricular chamber.

26—B: The patient has trichinosis, a parasitic disease caused by eating undercooked pork or wild game. Eosinophils rarely are found in normal connective tissue, but are recruited from the blood to sites of parasitic infections. Eosinophils can kill parasites by releasing cytotoxins from cytoplasmic granules.

27—C: Taste buds are usually associated with papillae. Both filiform and fungiform papillae are located on the anterior portion of the tongue, but filiform papillae lack taste buds. Circumvallate papillae, which also contain taste buds, are located farther back, toward the base of the tongue. Thus, the taste buds located near the tip of the tongue occur in fungiform papillae.

28—C: The gallbladder is located under the liver, the site of the patient's pain. The gallbladder stores bile, which is released in response to a fatty meal. Bile salts can precipitate and form gallstones, which can block the common bile duct and cause tremendous pain.

29—C: Hypothalamic neurons (site A) produce various releasing and inhibiting hormones that regulate secretion of endocrine cells in the anterior pituitary gland (site D). These releasing and inhibiting hormones are released into capillaries in the median eminence and transported to the anterior pituitary gland via the hypophyseal portal system. A lesion at site C would sever the hypophyseal veins and prevent the releasing and inhibiting hormones from reaching their targets in the anterior pituitary gland.

30—D: The patient has Cushing's syndrome, a disorder in which the adrenal cortex produces excess cortisol. Cortisol production is stimulated by adrenocorticotropic hormone (ACTH), which is secreted by cells in the anterior pituitary gland (cells labeled D). The most common form of Cushing's syndrome results from pituitary adenomas that produce excess ACTH.

31—E: Oxytocin causes contraction of myoepithelial cells in mammary glands during suckling, and is required for milk let-down and ejection. Oxytocin is synthesized primarily by neuro-secretory cells in the paraventricular nucleus (site B), but is stored and released from axon terminals in the pars nervosa (site E).

32—F: Antidiuretic hormone (ADH), also known as arginine vasopressin, increases the recovery of water from the filtrate to the blood. It acts on collecting duct cells to stimulate the fusion of cytoplasmic vesicles containing aquaporin water channels with the plasma membranes of these cells. The aquaporins facilitate the movement of water from the filtrate toward the interstitium of the medulla. The high NaCl concentration in the medulla creates an osmotic pressure that draws water out of the tubules.

33—C: Integrin receptors are transmembrane proteins that can bind to the extracellular matrix (ECM). When activated by contact with extracellular ligands, the cytoplasmic domain of integrin links to actin via a complex of cytoskeletal components. Integrins allow the regulated adhesion of cells to the ECM, which permits cell movements required in their development and wound repair.

34—B: When an action potential reaches a presynaptic terminal, it depolarizes the membrane, opening voltage-gated Ca^{2+} channels. The increased Ca^{2+} concentration causes synaptic vesicles to fuse with the membrane and release their contents of neurotransmitter into the synaptic cleft. The transmitter molecules diffuse across the synaptic cleft and bind to ligand-gated ion channels on the postsynaptic membrane. These channels open, allowing cations to enter. This depolarizes the postsynaptic membrane, opening voltage-gated Na^+ channels and spreading the depolarization away from the synapse.

35—A: Integrin activation and binding is required for cells to adhere tightly to endothelial cells at the site of an infection. Integrins are also required for the cells to migrate into tissues and move to the site of an infection. Therefore, neutrophil counts at an infection site in connective tissue will be low. Cytokines released from the site of infection will stimulate neutrophil production, so neutrophil counts in the blood will be very high. This is the situation found in the condition known as leukocyte adhesion deficiency.

36—C: Every skeletal muscle fiber receives a single synaptic contact from a single motor neuron. A motor neuron can contact several muscle fibers, called a motor unit. Activation of a motor neuron causes all the myofibers in the motor unit to contract maximally. That is, all of the sarcomeres in all of the myofibrils in all of the myofibers of the motor unit contract maximally. Thus, the extent to which a muscle contracts depends on the number of motor neurons that are activated and the size of their respective motor units; that is, on the number of myofibers that contract.

37—C: Sperm entering the epididymis are weakly motile and incapable of fertilization. They undergo their final maturation and become motile as they pass through the epididymis. This is an androgen-dependent process and presumably also depends on epididymal secretions.

38—D: The only cells that should be missing in the absence of antigen exposure are B- and T-memory cells. Memory cells arise during the process of clonal selection, which requires the presence of antigen.

39—A: Cilia line the apical surface of the respiratory epithelium and function to move debris (e.g., mucus; dust) away from the lungs. The chemicals in cigarette smoke damage the cilia. In addition, the airways produce more mucus as a result of the irritants from cigarette smoke. As a result, because of the damaged cilia and the increased mucus production, the only way to expel the mucus is through violent coughs.

40—B: The large cuboidal cells are osteoblasts, which are responsible for secreting and mineralizing osteoid (marked with an asterisk).

41—D: A large fraction of neutrophils is associated with the walls of blood vessels, or marginated. These cells roll along the walls of the vessels and constitute a reserve population of cells quickly available to enter the fluid phase.

42—A: The subcutaneous layer of loose connective that lies deep to the dermis is the hypodermis (in gross anatomy, it is referred to as superficial fascia). The loose nature of the hypodermis allows the skin to move over underlying tissues.

43—E: Dust cells (also known as heart failure cells) are alveolar macrophages.

44—D: The thick ascending limb (TAL) contains a high concentration of sodium pumps that actively remove Na^+ from the filtrate; Cl^- follows due to charge considerations. These tubules are also impermeable to water, which means that the loss of solute (Na^+ and Cl^-) without the efflux of water leaves the filtrate hypotonic. Water can exit and Na^+ can reenter the thin descending limb (tDL); Na^+ will be returned to the TAL along with fresh filtrate, allowing this segment to generate a high extracellular NaCl concentration—this is the basis of the countercurrent multiplier system of the kidney.

45—A: The micrograph is a cross-section through the ductus (vas) deferens. The ductus deferens connects the epididymis to the urethra via the ejaculatory duct within the prostate. The lumen is lined with low pseudostratified epithelium with stereocilia.

46—A: Enteroendocrine cells located in the epithelium of the mucosa release several hormones that stimulate cells in the hypothalamus and are involved in feelings of hunger and satiety. For example, gherlin is released from the stomach when it is empty and stimulates feelings of hunger.

47—C: The diaphysis of long bones grows in girth by intramembranous ossification (appositional growth) at the outer surface, just beneath the periosteum.

48—A: The average length of a sarcomere is 2.50 μm. This distance is measured from one Z line to the next Z line. If the resting length of the A band is 1.50 μm and the length of the I band is 1.0 μm, then the resting length of the sarcomere is determined by adding the length of the I band to the length of the A band. If there is a 20% contraction of the muscle (contraction to 80% of its length), then the sarcomere is reduced in length from 2.50 μm to 2.0 μm. The size of the A band remains unchanged (i.e., whether the contraction is 10% or 20%); therefore the length of the I band is reduced from 1.0 μm to 0.5 μm, making up for the 0.5 μm reduction in length during the muscle contraction.

49—C: Spermatogenesis occurs only below core body temperature. The testes normally are housed outside the body in the scrotal sac, which helps maintain a temperature about 2° below body temperature. Temperatures above that of the scrotum strongly inhibit spermatogenesis, but do not affect testosterone production by Leydig cells. Thus, a boy with cryptorchidism should develop normal secondary sexual characteristics, but is likely to be sterile.

50—D: Natural killer (NK) cells have receptors for the major histocompatability complex-I (MHC class I) molecule that inhibit their killing activity. The inhibition of MCH expression by a stressed cell permits the signal from stress receptors to activate killing responses and eliminate the stressed cell. This arrangement allows NK cells to kill infected or transformed cells that might evade T_c killing by down-regulating expression of MHC molecules.

51—A: Various digestive enzymes are associated with the brush border of enterocytes (absorptive columnar cell), and function to complete digestion of carbohydrates and proteins in the small intestine.

52—C: At the beginning of the ovarian cycle, the functional layer has been shed and only the basal layer remains, so the endometrium is very thin. The functional layer regenerates during the first half of the ovarian cycle; by day 14, the functional layer has regenerated and the glands are narrow, straight tubules. During the second half of the ovarian cycle (days 15–28), the endometrium becomes edematous and thickens, and the glands begin to secrete and become coiled and distended.

53—A: Arteriovenous anastomoses provide direct connections between arteries and veins that bypass capillary beds. In the skin, these keep blood from traveling near the surface, which prevents heat loss.

54—B: Elastic cartilage is an important component of the epiglottis, and also is present in the external ear and Eustachian tubes.

55—E: Professional antigen-presenting cells (APCs) express the B7 protein. This occurs when the APCs detect an inflammatory situation, such as when their toll-like receptors (TLRs) have bound molecules specific to pathogens. The interaction of B7 with CD28 on naïve T cells constitutes the second signal required to activate these cells—then, they express high affinity interleukin-2 (IL-2) receptors and secrete IL-2, which stimulates their mitosis. Only APCs can express B7. Think of B7 as providing the "second handshake" required to activate a naïve T cell. The requirement for two handshakes helps ensure that T cells only become activated when necessary and helps avoid the development of autoimmune reactions.

56—D: Secretions of the peg cells in the uterine tube lubricate the lumen of the uterine tube, provide nutrition and protection for spermatozoa, oocytes, and zygotes, and promote capacitation of sperm.

57—E: Small granule cells in respiratory epithelium are part of the diffuse neuroendocrine system, known as DNES. Their secretory granules contain diverse products, including serotonin, calcitonin, and bombesin, which are thought to control the function of other cells in respiratory epithelium.

58—A: Neurons in enteric ganglia (e.g., in Auerbach's plexus) are located in the wall of organs in the digestive tract and regulate motility and other activity of the digestive tract. Such ganglia receive input from the vagus nerve and are often classified as part of the parasympathetic branch of the autonomic nervous system. However, enteric neurons can function in the absence of outside input and are sometimes considered to be a separate system.

59—A: Maturation of follicles begins at puberty, but atresia begins prenatally. Thus, the ovary of a neonate will contain primordial and atretic follicles, but no maturing follicles. During the reproductive years, follicles at different stages of maturation will be present. After menopause, only atretic follicles and corpora albicantia will be present.

60—D: The four types of tissues from which organs are constructed are connective tissue, epithelium, muscle, and nervous tissue.

61—E: The endomysium, which surrounds individual myofibers, is continuous with the perimysium, which surrounds fascicles of fibers. The perimysium is continuous with the epimysium, which surrounds the entire muscle. The epimysium is continuous with the tendons. Thus, the force of contraction of muscle fibers is transmitted to bone via the connective tissue wrappings.

62—D: The micrograph shows a section through the body or fundus of the stomach. Note the long glands (double-headed arrow) and relatively short pits (indicated by the asterisk), and the abundance of parietal cells and chief cells in the glands. Mucous surface cells line the gastric pits (indicated by the asterisk) and cover the surface of the stomach. Chief cells, enteroendocrine cells, mucous neck cells, parietal cells, and stem cells are found in the glands.

63—C: Hepatocytes of zone 1 are the first to receive blood from the small intestine and are known to be the most active in endocytosis. The proximal convoluted tubule (PCT) cells are responsible for removing proteins that enter the initial filtrate and, therefore, are extensively involved in endocytosis.

64—A: Appositional growth occurs when chondroblasts in the perichondrium differentiate into chondrocytes, adding cells and matrix to the periphery of the existing cartilage body. This is the principal way that cartilage expands during development.

65—D: Transport of viral peptides by the TAP protein (transporter of antigen peptides) is the second step required for loading peptides onto the major histocompatability complex-I molecules (MHC class I molecules). This transporter moves peptides, produced by proteolysis of proteins in proteasomes, from the cytoplasm into the lumen of the rough endoplasmic reticulum, where the peptides can associate with newly synthesized MHC class I molecules.

66—D: A surge in luteinizing hormone in the middle of the ovarian cycle triggers ovulation.

67—B: Lipids are packaged as chylomicrons in enterocytes and released into the lamina propria of villi, where they enter lymphatic lacteals. The other substances listed will all predominately enter blood capillaries and be transported to the liver via the blood in the portal vein. The chylomicrons picked up by the lymphatic lacteals eventually enter the blood stream at the thoracic duct, which is near the heart.

68—A: Adipocytes secrete leptin, which inhibits food intake and increases energy expenditure, adiponectin, which increases insulin sensitivity and decreases blood glucose levels, and cytokines that affect the immune system.

69—C: During the process of endochondral ossification, chondrocytes in resting cartilage in the epiphysis proliferate, hypertrophy, become calcified, and die. The calcified cartilage is replaced with bone matrix. Thus, a pin passing into an epiphyseal growth plate from the diaphysis will pass through bone, calcified cartilage, hypertrophied chondrocytes, proliferating chondrocytes and resting cartilage.

70—B: The blood-air barrier consists of the cytoplasm of a type I pneumocyte (type I alveolar cell), an endothelial cell, and their fused basal laminae.

71—C: Satellite cells in skeletal muscle proliferate and reconstitute the damaged part of the myofibers. Satellite cells are supportive cells for maintenance of muscle and a source of new myofibers after injury or after increased load. There is no dedifferentiation of myocytes into myoblasts (choice A), or fusion of damaged myofibers to form new myotubes (choice D). Hypertrophy, not hyperplasia (choice E), occurs in existing myofibers in response to increased load. Proliferation of fibroblasts may occur in the damaged area but results in fibrosis, not repair of skeletal muscle. Fibroblasts do not differentiate into myocytes (choice B).

72—D: The patient has vitiligo, a common skin disorder characterized by the partial or complete loss of melanocytes, often by an autoimmune process that results in patches of unpigmented skin. In contrast, melanocytes are still present in individuals with albinism, but they are unable to synthesize melanin because of a lack or defect in the enzyme tyrosinase.

73—E: Some of the final cells to mature and develop in a fetus are the type II pneumocytes, which are responsible for the production of surfactant. Without surfactant, the alveoli collapse after each breath, resulting in labored breathing by premature infants.

74—C: Lamellar (membrane-coating) granules in cells in the stratum granulosum contain a mixture of hydrophobic glycolipids, which are discharged into the extracellular space between keratinocytes to form a waterproof barrier. This lipid barrier prevents loss of tissue fluids, but also prevents the diffusion of nutrient to more superficial cells, thereby hastening their death.

75—C: Most of the volume of adult compact bone is organized into osteons (Haversian systems), the main organizational unit of this bone. Osteons in the diaphysis of long bones are sandwiched between the inner and outer circumferential lamellae. Interstitial lamellae represent imprecise replacement of old with new osteons during normal bone remodeling and trabeculae are the thin struts of lamellar bone that support marrow cavities.

76—B: Cytokines, known as colony-stimulating factors (CSFs) in the case of hematopoietic cell proliferation, are required to drive the proliferation and differentiation of hematopoietic stem cells. The live mouse host synthesizes these factors, and the researcher must add them to the culture medium.

77—D: The asterisks mark the islets of Langerhans in the endocrine pancreas. β-cells, which produce insulin, are the predominant cells in the islets. Autoimmune destruction of cells results in type I diabetes mellitus. β-cells are still present in type 2 diabetes mellitus, but tissues fail to respond to insulin. Acute pancreatitis is a disorder of the exocrine pancreas. Crohn's disease is an autoimmune disease of the gastrointestinal tract. Hirschsprung's disease results from the failure of neural crest cells to migrate into the wall of the gut during embryonic development.

78—B: Isogenous group refers to a clone of chondrocytes in cartilage.

79—B: The nail plate is a highly modified stratum corneum.

80—D: Erythropoietin, also known as EPO, is a cytokine that stimulates the production of red blood cells; its prolonged administration will result in an unusually high hematocrit, or polycythemia.

81—E: Pacinian corpuscles are onion-shaped, encapsulated mechanoreceptors deep in the dermis and hypodermis that are highly sensitive to vibration and deep pressure.

82—C: Germinal centers contain centroblasts, large B cells that are undergoing isotype switching, which is a late event in B-cell development. Centroblasts also give rise to memory cells.

83—D: Merocrine secretion, the release of products primarily from secretory granules that does not involve the loss of general cell components, is the most common form of exocrine secretion.

84—C: The tunic adventitia is usually the thickest portion of large veins, and contains the bulk of the smooth muscle involved in controlling the diameter of these vessels.

85—A: Cells in the theca interna secrete androgen, which is converted to estrogen by granulosa cells.

86—B: The choroid plexus is formed by extensions of the pia mater that are covered with ependymal cells and extend into the ventricles. Ependymal cells move fluid from the capillaries into the ventricles to form cerebrospinal fluid (CSF), and also to remove some components (e.g., metabolic wastes) by transporting them back to the bloodstream. Thus, the choroid plexus both produces and regulates the composition of CSF.

87—E: The zonula adherens serves as an attachment for the actin fibers of the terminal web, which is analogous to the role of Z disks in organizing actin fibers in striated muscle cells.

INDEX

Note: Page numbers followed by "f" indicates figures.

CPSIA information can be obtained
at www.ICGtesting.com
Printed in the USA
FSHW021614220819
61294FS